AF735371

MEMORIAL VOLUME ON

Abdus Salam's

90th Birthday

MEMORIAL VOLUME ON

Abdus Salam's 90th Birthday

Editors

Lars Brink
Chalmers University of Technology, Sweden

Michael Duff
Imperial College London, UK

Kok Khoo Phua
Nanyang Technological University, Singapore

World Scientific

NEW JERSEY • LONDON • SINGAPORE • BEIJING • SHANGHAI • HONG KONG • TAIPEI • CHENNAI • TOKYO

Published by

World Scientific Publishing Co. Pte. Ltd.

5 Toh Tuck Link, Singapore 596224

USA office: 27 Warren Street, Suite 401-402, Hackensack, NJ 07601

UK office: 57 Shelton Street, Covent Garden, London WC2H 9HE

Library of Congress Cataloging-in-Publication Data

Names: Brink, Lars, 1945– editor. | Duff, M. J., editor. | Phua, K. K., editor. |
Salam, Abdus, 1926–1996, honouree.
Title: Memorial volume on Abdus Salam's 90th birthday / edited by Lars Brink (Chalmers University of Technology, Sweden), Michael Duff (Imperial College London, UK), Kok Khoo Phua (NTU, Singapore).
Description: Singapore ; Hackensack, NJ : World Scientific, [2017] | Includes bibliographical references.
Identifiers: LCCN 2016052024| ISBN 9789813144866 (hardcover) | ISBN 9813144866 (hardcover)
Subjects: LCSH: Salam, Abdus, 1926–1996. | Particles (Nuclear physics) | Condensed matter. |
Standard model (Nuclear physics) | Large Hadron Collider (France and Switzerland)
Classification: LCC QC16.S26 M46 2017 | DDC 539--dc23
LC record available at https://lccn.loc.gov/2016052024

British Library Cataloguing-in-Publication Data

A catalogue record for this book is available from the British Library.

Cover image: From Nationaal Archief. The image is licensed under the Creative Commons Attribution-Share Alike 3.0 Netherlands license.

Copyright © 2017 by World Scientific Publishing Co. Pte. Ltd.

All rights reserved. This book, or parts thereof, may not be reproduced in any form or by any means, electronic or mechanical, including photocopying, recording or any information storage and retrieval system now known or to be invented, without written permission from the publisher.

For photocopying of material in this volume, please pay a copying fee through the Copyright Clearance Center, Inc., 222 Rosewood Drive, Danvers, MA 01923, USA. In this case permission to photocopy is not required from the publisher.

Typeset by Stallion Press
Email: enquiries@stallionpress.com

Printed in Singapore by Mainland Press Pte Ltd.

Preface

Abdus Salam, born in 1926, was one of the most important, influential and exciting physicists of the last century. Growing up under modest conditions in what is now Pakistan, he broke all school records and went to Cambridge in England to finish his studies and to become a scientist. In 1957, he was appointed to a chair at Imperial College London, where he founded the Theoretical Physics Group and made it one of the world's centres of excellence in particle physics. At the age of 33, he was elected a Fellow of the Royal Society.

His contributions to physics are enormous. Already in his thesis in 1951, he finalised the full proof of the renormalisability of Quantum Electrodynamics. He was one of the first to understand the importance of parity violations in the weak interactions, and was one of the key persons to realise that the weak force should be formulated as a Yang–Mills theory. This led him eventually to formulate the consistent unification of the electromagnetic and weak nuclear forces, for which he shared the 1979 Nobel Prize with Sheldon Glashow and Steven Weinberg. After the proof of the renormalisabilty of the Yang–Mills theories, he was, together with Pati, one of the first to attack the problem of incorporating the strong interactions into the unification scheme. At the same time he initiated a program with Strathdee to study supersymmetric quantum field theories and introduced the concepts of superfields. Abdus Salam was always in the forefront of particle physics. He would not rest until "it was married with the fourth and most enigmatic force of gravity", a quest that is still at the forefront of current research.

He was very active not only in his homeland Pakistan, setting up all the government activities in science and technology in that new country, but with his boundless energy, was also an ambassador for science and education throughout the developing world. He firmly believed that every nation should be active in basic science in order to improve both its intellectual and economic standing. In 1964, he founded the International Centre for Theoretical Physics (ICTP) in Trieste, dedicated to that goal. The ICTP has served as the hub for this activity and now bears his name.

This book collected talks presented at the meeting organized to the memory of Abdus Salam's 90th birthday, held at the Institute of Advanced Studies, Nanyang Technological University (NTU) in Singapore from 25 to 28 January 2016. Many of his collaborators, students and friends came to give lectures. Four Nobel laureates, Gerard 't Hooft, David Gross, Carlo Rubbia and Tony Leggett, participated as well as many other famous physicists. The talks were partly historical. Michael Duff talked about Abdus Salam and his life and career and many of his old collaborators such as Jogesh Pati, Robert Delbourgo and Yu Lu interfoliated their scientific talks with reminiscences. Many of Salam's old students gave talks about their present works such as Peter West, Ali Chamseddine and Qaisar Shafi. One day was devoted to the present situation of particle physics with talks by Peter Jenni, Jim Virdee, Carlo Rubbia, David Gross and Hirotaka Sugawara. Also ICTP was well represented with talks by the present director Fernando Quevedo and former director Miguel Virasoro. Around 120 participants took part in the conference that was held in the Nanyang Executive Centre at NTU.

The Editors

Fig. 1. Gerard 't Hooft

Fig. 2. Carlo Rubbia

Fig. 3. David Gross

Fig. 4. Anthony Leggett

Salam was born in Jhang in what is now Pakistan in 1926 and came from what he himself described as humble beginnings. In fact, "I am a humble man" was something of a catchphrase for Salam and used whenever anyone tried to make physics explanations more complicated than necessary. He attended the Government College in Lahore and Panjab University before setting off for England and St. John's College, Cambridge, in 1946 where he gained a double first in Physics and Mathematics. He gained his PhD at the Cavendish Laboratory in 1952. He returned to Lahore for a couple of years but was appointed lecturer at Cambridge in 1954. Undoubtedly, the greatest influence on Salam at these early stages of his career was his mentor at St. John's, the great Paul Dirac, who remained Salam's hero throughout his life both as a great physicist and as a man who was largely disinterested in material wealth. (Salam himself never craved material riches, and was known to have paid for poor Third World students and postdocs out of his own pocket.)

Among Salam's earlier achievements was the role played by renormalization in quantum field theory when, in particular, he amazed his Cambridge contemporaries with the resolution of the notoriously thorny problem of overlapping divergences. His brilliance then burst on the scene once more when he proposed the famous hypothesis that *all neutrinos are left-handed*, a hypothesis which inevitably called for a violation of parity in the weak interactions. He was fond of recalling the occasion when he submitted (or should I say "humbly" submitted) his two-component neutrino idea to the formidable Wolfgang Pauli, whose verdict was: "Give my regards to my friend Salam and tell him to work on something better," adding that "this young man does not realize the sanctity of parity!" So Salam delayed publication until after Lee and Yang had conferred the mantle of respectability on parity violation. That taught Salam a valuable lesson and he would constantly advise his students never to listen to grand old men. (I hope this student, at least, has lived up to that advice!) It also taught him to adopt a policy of publish or perish, and his scientific output was prodigious with over 300 publications.

The Theoretical Physics group at Imperial College was founded in 1957 when the then head of physics, Lord Patrick Blackett, persuaded Abdus Salam to leave Cambridge and come to Imperial, notwithstanding attempts by his superiors at Cambridge to keep him there. Salam remained as Professor of Theoretical Physics until his death in 1996. He was elected to a Fellowship of the Royal Society in 1959 at the early age of 33. Salam's arrival at Imperial was celebrated in 2007 with the "Salam+50" conference.

Imperical College Physics.
Back row: Matthews, Blackman, McGee, Mason.
Front row: Salam, Butler, Blackett, Wright, Elliot.

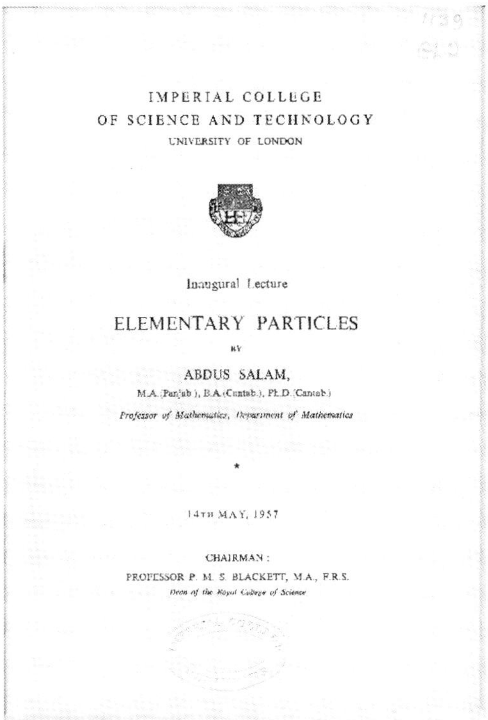

IMPERIAL COLLEGE
OF SCIENCE AND TECHNOLOGY
UNIVERSITY OF LONDON

Inaugural Lecture

ELEMENTARY PARTICLES

BY

ABDUS SALAM,

M.A. (Panjab), B.A. (Cantab.), Ph.D. (Cantab.)

Professor of Mathematics, Department of Mathematics

★

14TH MAY, 1957

CHAIRMAN:

PROFESSOR P. M. S. BLACKETT, M.A., F.R.S.

Dean of the Royal College of Science

reverse. People from all over the world would arrive and knock on his door to expound their latest theories, some of them quite bizarre. Yet Salam would treat them all with the same courtesy and respect. Perhaps it was because his own ideas always bordered on the outlandish that he was so tolerant of eccentricity in others; he could recognise pearls of wisdom where the rest of us saw only irritating grains of sand. Such an example was provided by the young military attache from the Israeli embassy in London who showed up one day with his ideas on particle physics. Salam was impressed enough to take him under his wing. The young man was Yuval Ne'eman and the result was flavour SU(3).

Let me recall just one example of a crazy Salam idea. In that period 1969–72, one of the hottest topics was the Veneziano Model and I distinctly remember Salam remarking on the apparent similarity between the mass and angular momentum relation of a Regge trajectory and that of an extreme black hole. Nowadays, of course, string theorists will juxtapose black holes and Regge slopes without batting an eyelid but to suggest that black holes could behave as elementary particles back in the late 1960's was considered preposterous by minds lesser than Salam's. As an interesting historical footnote let us recall that at the time Salam had to change the gravitational constant to match the hadronic scale, an idea which spawned his *strong gravity*; today the fashion is the reverse and we change the Regge slope to match the Planck scale!

Theoretical physicists are, by and large, an honest bunch: occasions when scientific facts are actually deliberately falsified are almost unheard of. Nevertheless, we are still human and consequently want to present our results in the best possible light when writing them up for publication. I recall a young student approaching Abdus Salam for advice on this ethical dilemma: "Professor Salam, these calculations confirm most of the arguments I have been making so far. Unfortunately, there are also these other calculations which do not quite seem to fit the picture. Should I also draw the reader's attention to these at the risk of spoiling the effect or should I wait? After all, they will probably turn out to be irrelevant". In a response which should be immortalized in *The Oxford Dictionary of Quotations*, Salam replied: "When all else fails, you can always tell the truth."

As Robert Walgate remarked in the New Scientist 1976; "Salam is a cultural amphibian. He has the heart of a poet and the mind of a scientist. He is an excellent physicist concerned with deep patterns; he is also a deeply compassionate man. These two threads intertwined throughout his life."

I think it was Hans Bethe who said that there are two kinds of genius. The first group (to which I would say Steven Weinberg, for example, belongs) produce results of such devastating logic and clarity that they leave you feeling that you could have done that too (if only you were smart enough!). The second kind are the "magicians" whose sources of inspiration are completely baffling. Salam, I believe, belonged to this magic circle and there was always an element of eastern mysticism in his ideas that left you wondering how to fathom his genius.

Slap on the back.

Acknowledgements

I am grateful to my conference co-organisers KK Phua and Lars Brink for affording me this opportunity to pay homage to Abdus Salam.

Life With Salam (1959–1976)

Robert Delbourgo
School of Physical Sciences, University of Tasmania,
Hobart, Tasmania 7001, Australia
bob.delbourgo@utas.edu.au

Having significantly interacted over 17 years with Abdus Salam, as an undergraduate, postgraduate, postdoc, and eventually as an academic colleague, I will try to paint a personal picture of Salam which may convey something about the man, his greatness and his humanity.

Keywords: Salam; standard model; achievements.

Personal Reminiscences

This conference commemorates the life and achievements of Abdus Salam. It is therefore incumbent upon me to begin by drawing a picture of the man with a few private reminiscences that may convey something about his greatness, genius and humanity. I know that *some* of you have worked with him in *some* capacity at *some* stage, but I suspect that only very few of you will have had the privilege of interacting 17 years with Abdus Salam as I have done: first as an undergraduate, then as a postgraduate, postdoc, and eventually as an academic colleague and scientific collaborator. I shall cover the period 1959–1976 when I was closely involved with him; there are a few others present here who can competently fill in the later years to leave you with a more complete portrait of Salam. If certain members of this audience have heard my vignettes of him before, I apologize in advance, but with these reminiscences you may at least enjoy reliving fond memories of him.

Salam was a man in a hurry; his reputation preceded him everywhere. As a lowly undergraduate student I first came across him in 1959 when we had to choose our third year specialty by making a selective tour of the various research departments at Imperial College. At that time Salam was housed in the Mathematics section (before moving to Physics in 1960) and I can vividly remember his verve and vivacity as he explained to us his latest pet project, which happened to be chiral symmetry and gamma-5 invariance for favouring massless left-handed neutrinos and leading to parity violation. That discourse went right over all our heads at the time but they coaxed me at least to turn to theoretical physics for my final year specialty. I am sure others have succumbed to Salam's persuasive abilities on whatever topic he expounded. That very year he taught us advanced quantum mechanics a-la-Dirac, whom he idolized; his lectures seemed pretty good to me, so in 1960 I embarked

Above all Salam impressed upon us the importance of tackling challenging problems: to prospect for new scientific fields and abandon raking over old coals, if one was to make one's mark. He lived up to that precept throughout his life, in spite of accusations that he had a scattergun approach to physics. It is a lesson that some young scientists today should heed. In his last years, despite his grave illness, he addressed the puzzle as to why certain life forms have particular handedness; what could be more fundamental or significant than that? I know that I miss his wise words, his friendship, his guidance, his generosity and his humanity. This conference is therefore a personal acknowledgement of how much he helped to shape my own career. More widely, it is a timely reminder of how much the scientific landscape and international developments owe to him. The ICTP is a permanent testament to that.

The Force and Gravity of Events

Robert Delbourgo
School of Physical Sciences, University of Tasmania,
Hobart, Tasmania 7001, Australia
bob.delbourgo@utas.edu.au

Local events are characterized by "where", "when" and "what". Just as (bosonic) spacetime forms the backdrop for location and time, (fermionic) property space can serve as the backdrop for the attributes of a system. With such a scenario I shall describe a scheme that is capable of unifying gravitation and the other forces of nature. The generalized metric contains the curvature of spacetime and property separately, with the gauge fields linking the bosonic and fermionic arenas. The super-Ricci scalar can then automatically yield the spacetime Lagrangian of gravitation and the Standard Model (plus a cosmological constant) upon integration over property coordinates.

Keywords: Properties; unification; gravity; forces.

1. An Algebraic Framework for Events

I now come to the scientific part of my talk. The material which I will present is sufficiently different from other attempts at unification of forces that I rather fancy, A. Salam might have given it a nod of approval. Two years ago, at the Dyson 90th anniversary conference, I outlined[1] how it is possible to unify gravity with the simplest of all forces, electromagnetism — Einstein's eldorado — simply by appending a single complex anti-commuting Lorentz *scalar* variable to spacetime, *not a spinor*; importantly no infinite KK modes arise. My partner in crime (Paul Stack) and I have made considerable progress since then and I will now try to summarize how to unify gravity with the other forces of nature through a relatively simple supermetric. Our attempts in this direction have been motivated by the present parlous state of particle physics and the snail's pace of progress in this area over the last 40 years. Here is a statement which may bring me some opprobium: namely, apart from the timely discovery of the Higgs boson, emergence of multiquark states and significant astrophysical advances, there is very little to celebrate in our attempts to unravel nature at the most basic level. This is in spite of determined, quasi-herculean efforts of theorists who have persistently espoused/promoted very clever ideas. So far, Nature stubbornly refuses to cooperate by providing us with unequivocal experimental signs of SUSY, strings/branes and other ingenious proposals. It seems that the simple Standard Model of particles and cosmology still rules. Nonetheless, its plethora of parameters have spurred theorists to search for generalizations of the Standard Model which may help to cut down the number of arbitrary constants the combination $\zeta^4\zeta^{\bar{1}}\zeta^{\bar{2}}\zeta^{\bar{3}}$ is recognizable as the Standard Model Higgs. However, the quark isomultiplets ψ which exist in Ψ are slightly different from the Standard Model! The up- and down-quarks come as two weak isodoublets/singlets and part of a weak *isotriplet/isodoublet/isosinglet* contained in SU(5) representations of dimension 45. Thus,

$$\begin{pmatrix} U^{[\bar{\mu}\bar{\nu}]} \sim \zeta^{\bar{\mu}}\zeta^{\bar{\nu}}\zeta^0 \\ D^{[\bar{\mu}\bar{\nu}]} \sim \zeta^{\bar{\mu}}\zeta^{\bar{\nu}}\zeta^4 \end{pmatrix}, \quad \begin{pmatrix} U'^{[\bar{\mu}\bar{\nu}]} \sim \zeta^{\bar{\mu}}\zeta^{\bar{\nu}}\zeta^0\zeta^{\bar{4}}\zeta^4 \\ D'^{[\bar{\mu}\bar{\nu}]} \sim \zeta^{\bar{\mu}}\zeta^{\bar{\nu}}\zeta^4\zeta^{\bar{0}}\zeta^0 \end{pmatrix}, \quad \begin{pmatrix} U''^{\lambda} \sim \zeta^{\lambda}\zeta^{\bar{4}}\zeta^0 \\ D''^{\lambda} \sim \zeta^{\lambda}(\zeta^{\bar{0}}\zeta^0, \zeta^{\bar{4}}\zeta^4) \\ X''^{\lambda} \sim \zeta^{\lambda}\zeta^{\bar{0}}\zeta^4 \end{pmatrix}$$

implies the existence of a brand new quark X'' (of charge $-4/3$) in a third generation. Though X'' may be more massive than even the top quark, the consequence at lower energy scales is that we do not expect the CKM matrix to be quite unitary.[c] Probably the best way to find X is via a high energy electron-positron collider? Other predictions of the scheme are that heavy leptons should be seen as well as unaccompanied (massive?) D-type quarks. If none of these signals eventuates then it is back to the drawing board and a re-examination to see if any of these ideas about property is salvageable or if the disease is terminal.

3. Force Fields

The most interesting feature of our scheme is the way that gauge fields enter and tie in with the quantum number assignments. We note that a flat metric in X is only invariant under global SU(N) unitary rotations of the N attributes. But as soon as we make them local or x-dependent, so that

$$\zeta^{\mu} \to \zeta'^{\mu} = [\exp(i\Theta(x))]^{\mu\bar{\nu}}\zeta^{\nu} \tag{4}$$

we find that there is an inconsistency in the transformation rules for the metric; we are forced to "curve" the space and introduce gauge fields to repair the fault. The way to do this is to write the generalized event (separation)^2 as

$$ds^2 = dX^M\, dX^N G_{NM}; \quad G_{NM} = \mathcal{E}_N{}^B \mathcal{E}_M{}^A \eta_{AB}(-1)^{[B][M]}, \tag{5}$$

where the metric arises through frame vectors $\mathcal{E}$ and the grading is defined in the usual way: $[m] = 0$, $[\mu] = 1$. Thus the transformations rules for G,

$$G'_{SR}(X') = \left(\frac{\partial X^M}{\partial X'^R}\right)\left(\frac{\partial X^N}{\partial X'^S}\right) G_{NM}(X)(-1)^{[S]([R]+[M])}, \tag{6}$$

under the local rotations of ζ demand that we introduce components $G_{n\mu}$, $G_{n\bar{\mu}}$ which have a vectorial character; they should be overall fermionic and must somehow involve the gauge field V as this is the communicator of property across spacetime.

---

[c]This meshes in with the observation that an isotriplet couples more strongly with the charged W-boson than an isodoublet and therefore the known decay width of the top quark requires a correspondingly smaller V_{tb} coupling to W.

A few moment's reflection (neglecting coupling constants for the present) leads one to the identification $\mathcal{E}_m{}^\alpha = -iV_m{}^{\alpha\bar{\nu}}\zeta^\nu$, which is very similar to the way that the em field makes an appearance in the original Klein–Kaluza model; there is really very little room for manoeuvre and the appearance is indeed entirely natural: gauge fields transmit property from one place and time to the next so they ought to arise in the spacetime-attribute sector. The only liberty permitted to us is to multiply by polynomials in property scalars $Z \equiv \zeta^{\bar{\mu}}\zeta^\mu$, since these are gauge invariant and carry no quantum numbers. We might say that inclusion of these polynomials corresponds to "curving" property space.

Using that freedom, the only metric which is fully consistent with local SU(N) gauge transformations is

$$\begin{pmatrix} G_{mn} & G_{m\nu} & G_{m\bar{\nu}} \\ G_{\mu n} & G_{\mu\nu} & G_{\mu\bar{\nu}} \\ G_{\bar{\mu}n} & G_{\bar{\mu}\nu} & G_{\bar{\mu}\bar{\nu}} \end{pmatrix}$$

$$= \begin{pmatrix} g_{mn}C + \ell^2\bar{\zeta}\{V_m, V_n\}\zeta C'/2 & -i\ell^2(\bar{\zeta}V_m)^{\bar{\nu}}C'/2 & i\ell^2(V_m\zeta)^\nu C'/2 \\ -i\ell^2(\bar{\zeta}V_n)^{\bar{\mu}}C'/2 & 0 & \ell^2\delta_\mu{}^\nu C'/2 \\ i\ell^2(V_n\zeta)^\mu C'/2 & -\ell^2\delta_\nu{}^\mu C'/2 & 0 \end{pmatrix}. \quad (7)$$

Here $C(Z) = 1 + \sum_{n=1}^N c_n Z^n$, $C'(Z) = 1 + \sum_{n=1}^N c'_n Z^n$ are independent polynomials of order N in Z which are allowed without destroying the gauge symmetry. One then readily checks that the rule (6) just corresponds to the usual gauge transformation: $iV'_m(x') = \exp[i\Theta(x)](iV_m(x) + \partial_m)\exp[-i\Theta(x)]$. If we just demand subgroup gauge symmetry, we can relax the conditions on the Z polynomials and have them invariant under local subgroup rotations, so more property curvature coefficients c_n can be entertained. We will come back to this when considering QCD plus QED and electroweak theory in such a framework.

The procedure from hereon is pretty straightforward,[5] paying very particular attention to orders of terms and signs that are due to grading. One first constructs the super-Ricci scalar $\mathcal{R}$ from the Christoffel symbols

$$2\Gamma_{MN}{}^K = [(-1)^{[M][N]}G_{ML,N} + G_{NL,M} \\ - (-1)^{[L]([M]+[N])}G_{MN,L}](-1)^{[L]}G^{LK} \quad (8)$$

via the Palatini form

$$\mathcal{R} = G^{MK}\mathcal{R}_{KM} = (-1)^{[L]}G^{MK}[(-1)^{[L][M]}\Gamma_{KL}{}^N\Gamma_{NM}{}^L - \Gamma_{KM}{}^N\Gamma_{NL}{}^L]. \quad (9)$$

Secondly one integrates $\mathcal{R}$ over property. This leads to the gravitational and gauge field Lagrangian plus a cosmological term.[6] (As a bonus, the stress tensor T_{mn} of the gauge fields is automatically incorporated in $\mathcal{R}_{mn}$ when we extract the resulting "equations of motion".) The coefficients in front of these terms depend on the number of properties[7] and on the property curvature coefficients but they all have the generic form

$$\left(\frac{\ell^2}{2}\right)^{2(N-1)} \int d^N\zeta\, d^N\bar{\zeta}\, \sqrt{G..}\, \mathcal{R} = \frac{\mathcal{A}R^{[g]}}{\ell^2} + \mathcal{B}\,\mathrm{Tr}(F.F) + \frac{\mathcal{C}}{\ell^4}, \tag{10}$$

where $F_{mn} \equiv V_{n,m} - V_{m,n} + i[V_m, V_n]$ and the N-dependent coefficients $\mathcal{A}$, $\mathcal{B}$, $\mathcal{C}$ are listed in Ref. 7.

The matter fields and their Lagrangians are then introduced,

$$\mathcal{L}_\phi = \int d^N\zeta\, d^N\bar{\zeta}\, \sqrt{-G..}\, G^{MN}\partial_N\Phi\partial_M\Phi, \tag{11}$$

$$\mathcal{L}_\psi = \int d^N\zeta\, d^N\bar{\zeta}\, \sqrt{-G..}\, \bar{\Psi}i\Gamma^A E_A{}^M\partial_M\Psi. \tag{12}$$

The gauge and gravitational interactions of the component fields (ϕ, ψ) then just fall out, but these sometimes require wave function renormalizations due to influence of the property curvature coefficients c_n — coefficients which are absent in flat space. The key point is that the gauge fields couple correctly to the matter fields through the vielbein term

$$E_A{}^M\partial_M \supset e_a{}^m[\partial_m + i(V_m\zeta)^\mu\,\partial_\mu - i(\bar{\zeta}V_m)^{\bar{\mu}}\,\partial_{\bar{\mu}}],$$

so the property derivative is compensated by a further property coordinate attached to the gauge field V; this is our version of covariant differentiation. Incidentally I ought to declare that such complicated calculations were originally carried using an algebraic computer package devised by Paul Stack and, after time-consuming computation, they always produced gauge- and coordinate-invariant results. Knowing this always happened, we have since been able to find a shorter analytic way of picking out the correct terms in (10)–(12) by a procedure which can be generalized to any number of attributes and dispense with Mathematica. Finally, to (11) and (12) we may add the renormalizable super-Yukawa self interactions $\Psi\Phi\Psi$ and $V(\Phi) \simeq \Phi^4$ in the usual manner, with the aim of generating a mass term through the expectation values held in the chargeless fields within $\langle\Phi\rangle$.

Before moving on, three comments about the fermion fields deserve particular mention. Firstly the adjoint field $\bar{\Psi}$ has to be carefully defined with appropriate signs[8] in property space to produce a series of terms $\bar{\psi}\psi$, after integrating over ζ. Secondly, $\bar{\zeta}\psi$ and their charge conjugates $\psi^{(c)}\zeta$ both appear in the full expansion of $\Psi(\zeta, \bar{\zeta})$ and they simply lead to a doubling of the eventual answers; thus we can simplify calculations by "halving" the expansion of Ψ to $\Psi \supset \bar{\zeta}\psi$ terms. Thirdly and intriguingly, we have to extend the concept of Dirac γ matrices to super Γ matrices, such that $(\Gamma^A P_A)^2 = \eta^{AB}P_B P_A$. In spacetime we get the standard $\Gamma^a = \gamma^a$ with

$\{\gamma^a, \gamma^b\} = 2\eta^{ab}$, but in the property sector one needs to ensure that the "square-rooted" Γ^α are *fermionic* and obey

$$[\Gamma^\alpha, \Gamma^\beta] = [\Gamma^{\bar\alpha}, \Gamma^{\bar\beta}] = 0, \quad [\Gamma^\alpha, \Gamma^{\bar\beta}] = 2\eta^{\alpha\bar\beta} = 2\delta_\beta{}^\alpha. \tag{13}$$

In the same way that Dirac introduced 4×4 matrices and made novel use of the Clifford algebra for spacetime, we must do something similar for property space. We can arrange for the commutators (13) to be satisfied by augmenting property space with auxiliary coordinates θ^α, setting $\Gamma^\alpha \equiv \sigma_+\theta^\alpha$, $\Gamma^{\bar\alpha} \equiv \sigma_-\partial/\partial\theta^\alpha$, and making sure that Ψ is multiplied by the projected singlet $\Theta \equiv (1+\sigma_3)\theta^1\theta^2\cdots\theta^N/2$, over which one eventually integrates.[d] There are probably less extravagant ways of doing this.

4. Electric and Chromic Relativity

To see how all this works out, consider QED and QCD which involve one attribute called electricity plus three "chromicity" properties (commonly termed red, green, blue). Thus, we confine ourselves to coordinates ζ^1 to ζ^4 and combine both chiralities in Dirac fields since those interactions are blind to parity. As we are confining ourselves to $U(1) \times SU(3)$ we are dealing with two sets of gauge fields within the fuller SU(4): the em field A and the gluon fields B, having coupling constants e and f respectively. One identifies the frame vectors $\mathcal{E}_m{}^\kappa = -i(fB_m - eA_m/3)^{\kappa\bar\iota}\zeta^\iota$, $\mathcal{E}_m{}^4 = ieA_m$, leading to the basic metric elements

$$G_{m4} = i\ell^2\zeta^{\bar 4}eA_m C'/2, \quad G_{m\iota} = i\ell^2[\zeta^{\bar\iota}eA_m/3 - \zeta^{\bar\kappa}fB_m{}^{\kappa\bar\iota}]C'/2, \tag{14}$$

which may be multiplied by polynomials in two *distinct* invariants $\zeta^{\bar\kappa}\zeta^\kappa$ and $\zeta^{\bar 4}\zeta^4$. I should point out that it is the interactions (14) which actually determine the charge and color assignments stated in (2) and (3). Also the coupling *must* accompany the gauge fields in order to produce the correct interactions with matter fields.

To simplify the subsequent argument about the resulting interactions, I will assume that the property curvature polynomials are common to spacetime and property space:

$$C = C' = 1 + \cdots + c_e(\zeta^{\bar 4}\zeta^4)(\zeta^{\bar\kappa}\zeta^\kappa)^2 + c_f(\zeta^{\bar\kappa}\zeta^\kappa)^3. \tag{15}$$

As we are dealing with four properties, we find that the Berezinian[9,10] is $\sqrt{-G..} = (2/\ell^2)^4\sqrt{-g..}C^{-2}$. A careful analytical calculation shows that the super-Ricci scalar contains the following gauge field combination:

$$\begin{aligned}\mathcal{R}\sqrt{-G..} \supset\ & [1 - 3c_e(\zeta^{\bar 4}\zeta^4)(\zeta^{\bar\kappa}\zeta^\kappa)^2 - 3c_f(\zeta^{\bar\kappa}\zeta^\kappa)^3 + \cdots] \\ & \sqrt{-g..}g^{km}g^{ln}[4e^2\zeta^{\bar 4}F_{kl}F_{mn}\zeta^4/3 + f^2\zeta^{\bar\kappa}(E_{kl}E_{mn})^{\kappa\bar\iota}\zeta^\iota],\end{aligned} \tag{16}$$

[d] Of course the adjoint $\bar\Psi$ contains the conjugate singlet $\bar\Theta = (1+\sigma_3)\theta^{\bar N}\cdots\theta^{\bar 2}\theta^{\bar 1}/2$.

where $F_{mn} = A_{n,m} - A_{m,n}$ and $E_{mn} = B_{n,m} - B_{m,n} + if[B_n, B_m]$ are the standard "curls" of the electromagnetic and gluon fields. The last step is to integrate over the four properties. Including appropriate scaling factors of ℓ^2 one gets

$$\int (d^4\zeta\, d^4\bar{\zeta})\mathcal{R}\sqrt{-G_{..}} \supset (-12\sqrt{-g_{..}}/\ell^2)[4c_f e^2 F.F + c_e f^2\,\mathrm{Tr}(E.E)]. \tag{17}$$

Last but not least, we must ensure gravitational universality; so we have to set $c_e f^2 = 4c_f e^2$, which is perfectly feasible without demanding equality of the color and electromagnetic couplings. If we relax the assumption that $C = C'$, it is even easier to ensure universality of Newton's constant G_{N}.

The color and electromagnetic interactions of the matter fields Ψ, Φ emerge from (11) and (12) exactly as expected. See Ref. 8. I shall not delve into that because the story is not quite complete and is therefore likely to be misleading: we have neglected neutrinicity (the fifth property ζ^0) so the ensuing generations are not the physical ones, as sketched in Sec. 3. To correct for this, we must turn to the leptons.

5. Electroweak Relativity

The application of our scheme to the original electroweak model[11–13] of leptons requires an interesting extension of previous work[14] and leads to an intriguing prediction about the weak mixing angle, not to mention the prediction of two leptonic generations. The fact that the weak isospin and hypercharge assignments of the leptons change with chirality obliges us to invoke distinct properties ζ_L and ζ_R for left- and right-handed leptons respectively to which the gauge fields latch on (through the frame vectors). The full SU(4) gauge field $V^{\mu\bar{\nu}}$, acting on the pair of doublets $(\zeta_L^0, \zeta_L^4, \zeta_R^0, \zeta_R^4)$ is not needed; only the restricted $\mathrm{SU}(2)_L \times \mathrm{U}(1)$ rotations demand attention. Thus, we re-interpret $V_m = L_m + R_m$, with

$$L_m = (g\mathbf{W}_m.\tau - g'B_m)/2, \quad R_m = g'B_m(\tau_3 - 1)/2, \tag{18}$$

possessing the standard weak hypercharge assignments

$$Y(\zeta_L^0, \zeta_L^4, \zeta_R^0, \zeta_R^4) = (-1, -1, 0, -2). \tag{19}$$

It must also be understood that L is to be associated with the left property derivative $\partial/\partial\zeta_L$ and R is to be associated with the right property derivative $\partial/\partial\zeta_R$; g and g' are the usual coupling constants tied to the weak triplet $\mathbf{W}$ and weak singlet hypercharge B respectively.

It is sufficiently general for our purposes to take the polynomial property curvatures C and C' to be equal and direct products of quadratic left- and right-handed polynomials:

$$C = C_R C_L = [1 + c_R Z_R + c_{RR} Z_R^2][1 + c_L + c_{LL} Z_L^2];$$
$$Z_R \equiv \bar{\zeta}_R \zeta_R,\ Z_L \equiv \bar{\zeta}_L \zeta_L. \tag{20}$$

These enter in the metric components:

$$G_{m\zeta_L} = -i\ell^2 \bar{\zeta}_L L_m C/2; \quad G_{m\zeta_R} = -i\ell^2 \bar{\zeta}_R R_m C/2, \tag{21}$$

$$G_{\zeta_L \bar{\zeta}_L} = G_{\zeta_R \bar{\zeta}_R} = \ell^2 C/2, \quad G_{\zeta_L \zeta_R} = G_{\bar{\zeta}_L \bar{\zeta}_R} = G_{\zeta_L \bar{\zeta}_R} = G_{\zeta_R \bar{\zeta}_L} = 0. \tag{22}$$

The remaining metric element reads

$$G_{mn} = C[g_{mn} + (\text{gauge field terms})]. \tag{23}$$

Factorizability of C simplifies the calculations enormously when we integrate over the whole eight properties: $\int d^2\zeta_R\, d^2\bar{\zeta}_R\, d^2\zeta_L\, d^2\bar{\zeta}_L$.

The various contributions to the super-Ricci scalar drop out as follows, bearing in mind that $\sqrt{-G..} = (2/\ell^2)^4 \sqrt{-g..}(C_R C_L)^{-3}$. There are three terms:

$$\int d^2\zeta_R..d^2\bar{\zeta}_L \sqrt{-G..}\mathcal{R} \supset 36\sqrt{-g..}(2/\ell^2)^4 R^{[g]}(2c_R^2 - c_{RR})(2c_L^2 - c_{LL}), \tag{24}$$

$$\int d^2\zeta_R..d^2\bar{\zeta}_L \sqrt{G..}\mathcal{R} \supset -\frac{3}{2}\sqrt{g..}\left(\frac{2}{\ell^2}\right)^3 [c_L(3c_R^2 - 2c_{RR})(g^2 \mathbf{W}_{mn}.\mathbf{W}^{mn}$$
$$+ g'^2 B_{mn}B^{mn}) + g'^2 2c_R(3c_L^2 - 2c_{LL})B_{mn}B^{mn}], \tag{25}$$

$$\int d^2\zeta_R..d^2\bar{\zeta}_L \sqrt{G..}\mathcal{R} \supset 12\sqrt{g..}(2/\ell^2)^5[(4c_L c_{LL} - 5c_L^2)(2c_R^2 - c_{RR})$$
$$+ (L \leftrightarrow R)], \tag{26}$$

where $W_{mn} \equiv W_{n,m} - W_{m,n} + ig[W_n, W_m]$ and $B_{mn} \equiv B_{n,m} - B_{m,n}$. The full answer is the sum of (24)–(26).

Universality of gravity at the semiclassical level anyway (and the correct normalization of the gauge fields) is guaranteed when we set

$$c_L(3c_R^2 - 2c_{RR})(g^2 - g'^2) = 2c_R(3c_L^2 - 2c_{LL})g'^2,$$

which is readily arranged. But much more intriguing is the fact that if the property curvature is parity conserving so that $c_R = c_L = c$, $c_{RR} = c_{LL} = c_2$ and implying that all parity violation comes from the gauge fields in the frame vectors, then $g^2 = 3g'^2$. Thus the weak angle reduces to 30°. It makes good sense because the property curvature C polynomial accompanies the gravitational field and, as far as we know, unquantized gravity does not know the left from the right. So this restriction seems very natural and the value of the weak angle is a consequence of gravitational universality in this framework; it is not a result of invoking a higher group or anomaly cancellation, as some other analyses[15,16] would have.

But when we come to the bosons we discover something new. Acting with the covariant derivative on the Bose superfield,

$$D_a\Phi = [E_a{}^m\partial_m + E_a{}^\mu\partial_\mu + E_a{}^{\bar{\mu}}\partial_{\bar{\mu}}]\Phi = [\partial_a + i(V_a\zeta)^\mu\partial_\mu - i(\bar{\zeta}V_a)^{\bar{\mu}}\partial_{\bar{\mu}}]\Phi.$$

Referring to Eq. (31) we obtain

$$2D\Phi.D\Phi = (1+2Z_L)(1+2Z_R)[\bar{\zeta}_R\{\partial\phi + i(\phi L - R\phi)\}\zeta_L\bar{\zeta}_L \times \{\partial\phi + i(\phi R - L\phi)\zeta_R\}] \qquad (35)$$

plus terms which disappear when integrated over property. If we concentrate on the uncharged fields held in the quartet ϕ, viz. $\phi^{0\bar{0}}$ and $\phi^{4\bar{4}}$, that occur on the diagonal (or equivalently ϕ_0 and ϕ_3), we find that

$$2\,\mathrm{Tr}[(\phi R - L\phi)(\phi L - R\phi)] \to \frac{1}{2}g^2W^+W^-(\phi_+^2 + \phi_-^2) + \frac{1}{4}\phi_+^2(gW_3 - g'B)^2 + \frac{1}{4}\phi_-^2(gW_3 + g'B)^2 - g'^2\phi_-^2\,; \quad \phi_\pm \equiv \phi_0 \pm \phi_3.$$

In order to recover the standard vector meson masses, we must therefore take

$$\langle\phi_-\rangle = 0 \quad \text{or} \quad \langle\phi_0\rangle = \langle\phi_3\rangle \quad \text{and} \quad \langle\phi_+\rangle = v,$$

for the expectation values, whereupon

$$\langle 2\,\mathrm{Tr}[(\phi R - L\phi)(\phi L - R\phi)]\rangle \to \frac{1}{2}v^2g^2W^+W^- + \frac{1}{4}v^2(g^2 + g'^2)Z^2 = \frac{e^2v^2}{2\sin^2\theta}W^+W^- + \frac{e^2v^2}{\sin^2 2\theta}Z^2. \qquad (36)$$

All is as it should be and the em field A remains massless.

Given these expectation values, we turn to the Yukawa interaction of the super-Bose field Φ with the super-fermion field Ψ. Before launching into this we need to remind ourselves that in order to get masses for leptons *as well as neutrinos*, we have to consider the Higgs doublet H as well as its doublet counterpart $i\tau_2H^*$. In our context it means that we have to consider ϕ as well as $\tau_2\phi^*\tau_2$. Since we will be integrating over the ζ and the fermion pieces involve $\bar{\zeta}_R\zeta_L$ or $\bar{\zeta}_L\zeta_R$, we need to pick out matching Bose pieces. Using the acceptable combination[f] $\hat{\phi} = c_l\tau_2\phi^*\tau_2 + s_l\phi$,

[f]With such a combination, $\mathrm{Tr}\,\hat{\phi}^2 = (c_l^2 + s_l^2)(\phi_0^2 + \phi^2) + 2c_ls_l(\phi_0^2 - \phi^2)$. So taking expectation values, $\mathrm{Tr}\,\langle\hat{\phi}\rangle^2 = 2(c_l^2 + s_l^2)v^2 \to 2v^2$ if we interpret $c_l \equiv \cos\theta_l$, $s_l \equiv \sin\theta_l$.

in place of ϕ, we then find that

$$-8\sqrt{2}\,\bar{\Psi}\Phi\Psi \supset (\bar{\zeta}_R\hat{\phi}\zeta_L + \bar{\zeta}_L\hat{\phi}^\dagger\zeta_R)(1+2Z_L)(1+2Z_R)$$
$$.\,[\bar{\zeta}_L\zeta_R(\overline{\psi_R} + Z_R\overline{\psi'_R})(\psi_L + Z_L\psi'_L) + (R \leftrightarrow L)].$$

Consequently, integrating over property produces a mixture of the two generations:

$$-16\int d^2\zeta_R..d^2\bar{\zeta}_L\;\bar{\Psi}\Phi\Psi = (2\bar{\psi}+\bar{\psi}')\hat{\phi}(2\psi+\psi') \equiv 5\bar{\hat{\psi}}\hat{\phi}\hat{\psi}.$$

Taking expectation values of Φ to generate a fermionic mass term, and recalling that $\langle\phi_-\rangle = 0$, the Yukawa term (including a coupling constant $\mathfrak{g}$) reduces to

$$5\mathfrak{g}(\overline{\nu_l},\bar{l})\begin{pmatrix} c_l\langle\phi_+\rangle & 0 \\ 0 & s_l\langle\phi_+\rangle \end{pmatrix}\begin{pmatrix} \nu_l \\ l \end{pmatrix} = 5v\mathfrak{g}(c_l\overline{\nu_l}\nu_l + s_l\bar{l}l). \tag{37}$$

The other mixture $\check{\psi} = (-\psi + 2\psi')/\sqrt{5}$ does not acquire a mass in this model. If we were to stretch credulity and pretend we have a decent model for leptons we would be inclined to associate $\hat{\psi}$ with the muonic doublet and $\check{\psi}$ with the electronic one; but all this is academic: we really need the three color properties to corall the known leptonic generations.

The last thing to consider is the effect of spacetime curvature (through ${e_m}^a$ or g_{mn}) and of property curvature $C(Z)$ on the above results. The effect of e is very simple: it just serves to make the interactions generally covariant and we have nothing more to add to that. The effect of C enters through the Berezinian

$$\sqrt{G..} = \sqrt{g..}(2/\ell^2)^4C^{-2}$$
$$\propto (1 - 2c_R - 2c_{RR}Z_{RR} + 3c_R^2Z_R^2)(1 - 2c_L - 2c_{LL}Z_{LL} + 3c_L^2Z_L^2).$$

It is subtler and causes mixing as well as wave function renormalization. To see what happens, consider the kinetic term of the fermions and simplify the argument by assuming the property curvature is blind to parity as we did before to recover a weak mixing angle of 30°. In that case, using the expanded

$$\begin{aligned}\sqrt{G..} = \sqrt{g..}(2/\ell^2)^4[1 &- 2c(Z_R+Z_L) + 4c^2Z_RZ_L \\ &+ (3c^2-2c_2)\{Z_R^2(1-2cZ_L) + Z_L^2(1-2cZ_R)\} \\ &+ \{(3c^2-2c_2)Z_LZ_R\}^2],\end{aligned} \tag{38}$$

we obtain, after ζ integration, the kinetic term $(D \equiv \partial - iV)$,

$$\begin{aligned}&\sqrt{g..}[(1-c)\{\bar{\psi}i\gamma.D\psi + \overline{\psi'}i\gamma.D\psi' - 2c(\bar{\psi}i\gamma.D\psi' + \overline{\psi'}i\gamma.D\psi)\} \\ &\quad + c(c^2+2c_2)\bar{\psi}i\gamma.D\psi]\end{aligned} \tag{39}$$

which reduces to (29) when $c \to 0$. Thus the curving of property engenders source field mixing and wave function renormalization, without affecting the coupling of the gauge field V configuration. Similar conclusions apply to the Bose sector.

Advantages of Unity With SU(4)-Color: Reflections Through Neutrino Oscillations, Baryogenesis and Proton Decay[a]

Jogesh C. Pati

SLAC, Stanford University, Menlo Park CA, 94025, USA

By way of paying tribute to Adbus Salam, I first recall the ideas of higher unification which the two of us introduced in 1972–73 to remove certain shortcomings in the status of particle physics prevailing then, and then present their current role in theory as well as experiments. These attempts initiated the *idea* of grand unification and provided the core symmetry-structure $G(2,2,4) = SU(2)_L \times SU(2)_R \times SU(4)$-color towards such a unification. Embodied with quark-lepton unification and left-right symmetry, the symmetry $G(2,2,4)$ is uniquely chosen as being the minimal one that permits members of a family to belong to a *single multiplet*. The minimal extension of $G(2,2,4)$ to a simple group is given by the attractive SO(10)-symmetry that was suggested a year later. The new concepts, and the many advantages introduced by this core symmetry (which are, of course, retained by SO(10) as well) are noted. These include explanations of the observed: (i) (rather weird) electroweak and color quantum numbers of the members of a family; (ii) quantization of electric charge; (iii) electron-proton charge-ratio being -1; (iv) the co-existence of quarks and leptons; (v) likewise that of the three basic forces — the weak, electromagnetic and strong; (vi) the non-trivial cancelation of the triangle anomalies within each family; *and* opening the door for (vii) the appealing concept of parity being an exact symmetry of nature at the fundamental level.

In addition, as a distinguishing feature, both because of SU(4)-color and independently because of $SU(2)_R$ as well, the symmetry $G(2,2,4)$ introduced, to my knowledge, for the first time in the literature: (viii) *a new kind of matter* — the right-handed (RH) neutrino (ν_R) — *as a compelling member of each family*, and together with it; (ix) (B-L) as a local symmetry. The RH neutrions — contrary to prejudices held in the 1970's against neutrinos being massive and thereby against the existence of ν_R's as well — have in fact turned out to be an asset. They are needed to (a) understand naturally the tiny mass-scales observed in neutrino oscillations by combining the seesaw mechanism *together with* the unification ideas based on the symmetry SU(4)-color, and also (b) to implement the attractive mechanism of baryogenesis via leptogenesis. The quantitative success of the attempts as regards understanding both (a) and (b) are discussed in Sec. 6. These provide a clear support simultaneously for the following three features: (i) the seesaw mechanism, (ii) the SU(4)-color route to higher unification based on a symmetry like SO(10) or a string-derived $G(2,2,4)$ symmetry in 4D, as opposed to alternative symmetries like SU(5) or even $[SU(3)]^3$, and (iii) the (B-L) - breaking scale being close to the unification scale $\sim 2 \times 10^{16}$ GeV.

The observed dramatic meeting of the three gauge couplings in the context of low-energy supersymmetry, at a scale $M_U \sim 2 \times 10^{16}$ GeV, providing strong evidence in favor

---

[a]Based on a talk delivered at the Adbus Salam 90th birthday Memorial Meeting at the IAS, Singapore, January 25–28, 2016.

of the ideas of both grand unification and supersymmetry, is discussed in Sec. 3. The implications of such a meeting in the context of string-unification are briefly mentioned. Weighing the possibility of a stringy origin of gauge coupling unification versus the familiar problem of doublet-triplet splitting in supersymmetric SO(10) (or SU(5)), I discuss the common advantages as well as relative merits and demerits of an effective SO(10) versus a string-derived $G(2,2,4)$ symmetry in 4D.

In Sec. 7, I discuss the hallmark prediction of grand unification, viz. proton decay, which is a generic feature of most models of grand unification. I present results of works carried out in collaboration with Babu and Wilczek and most recently with Babu and Tavartkiladze on expectations for decay modes and lifetimes for proton decay, including upper limits for such lifetimes, in the context of a well-motivated class of supersymmetric SO(10)-models. In view of such expectations, I stress the pressing need for having the next-generation large underground detectors — like DUNE and HyperKamiokande — coupled to long-baseline neutrino beams to search simultaneously with high sensitivity for (a) proton decay, (b) neutrino oscillations and (c) supernova neutrinos. It is remarked that the potential for major discoveries through these searches would be high.

Some concluding remarks on the invaluable roles of neutrinos and especially of proton decay in probing physics at the highest energy scales are made in the last section. The remarkable success of a class of supersymmetric grand unification models (discussed here) in explaining a large set of distinct phenomena is summarized. Noticing such a success and yet its limitations in addressing some fundamental issues within its premises, such as an understanding of the origin of the three families, and most importantly, the realization of a well-understood unified quantum theory of gravity describing reality, some wishes are expressed on the possible emergence and the desirable role of a string-derived grand-unified bridge between string/M-theory in higher dimensions and the world of phenomena at low energies.

I. Salam in Perspective

Abdus Salam was a rare phenomenon: a great scientist, a humanitarian and a strong promoter of his message that science is the common heritage of all mankind. He will surely be remembered for his many seminal contributions to physics, some of which have proven to be of lasting value. These include his pioneering work on electroweak unification for which he shared the Nobel Prize in physics in 1979 with Sheldon Glashow and Steven Weinberg.

But I believe his most valuable contribution to science and humanity, one that is perhaps unparalleled in the world, is the sacrifice he has made of his time, energy and personal comfort, including his family-life, in promoting the cause of science in the third world. His lifelong efforts in this direction have led to the creation of some outstanding research centres, including especially the International Centre for Theoretical Physics (ICTP) at Trieste, Italy,[b] an International Centre for Genetic Engineering and Biotechnology with components in Trieste and Delhi, and an International Centre for Science and High Technology in Trieste. All these centres focus on serving scientists from the third world. Salam dreamed of creating twenty international centres like the ICTP, spread throughout the world, emphasizing different areas of science and technology. Approaching developed as well as developing nations, for funding of such institutions, Salam often used the phrase: "science is not

[b]Now named the Abdus Salam International Centre for Theoretical Physics.

cheap, but expenditures on it will repay tenfold".[1] Salam's dreams in this regard could not come to fruition during his lifetime due to his illness in his later years. Fortunately, they have begun to be realized in part recently through the creation of ICTP partner institutions in Brazil, Mexico, Rwanda and China, thanks to the initiative of the present director of ICTP, Fernando Quevedo, and others.

Aspects of My Collaboration with Salam

My close collaboration with Salam started spontaneously through a tea conversation in the summer of 1972, during my short visit to ICTP, Trieste, and remained strong for over a decade.[c] Of this period, I treasure most the memory of many moments which were marked by the struggle and the joy of research that we both shared. While I have mentioned this in my previous writings,[2–4] befitting the present occasion, let me mention again one aspect of Salam's personality. During the ten year period of our collaboration, there have been many letters, faxes, arguments over the phone and in person and even heated exchanges, about tastes and judgements in physics, but always in a good natured spirit. In our discussions, Salam had some favorite phrases. For example, he would sometimes come up with an idea and get excited. If I expressed that I did not like it for such and such reason, he would get impatient and say to me: "My dear sir, what do you want: Blood?" I would sometimes reply by saying: "No Professor Salam, I would like something better". Whether I was right or wrong, he never took it ill. It is this attitude on his part that led to a healthy collaboration and a strong bond between us. Most important for me, by providing strong encouragement from the beginning, yet often arguing, he could bring out the best in a collaborator. For this I will remain grateful to him.

A Preview

As a preview of the topics to be covered in the following sections, I first recall in Sec. 2 the status of particle physics existing in mid-1972, together with its shortcomings, and then present the ideas of higher unification which Salam and I introduced in 1972–73[5–8] to remove some of these shortcomings. Current status of these ideas in the context of subsequent developments in theory and experiments are discussed in Secs. 2–6. To begin with, this would include a discussion in Sec. 2 of the chronological evolution of the unification idea during 1972–74 starting from: (1) those of the standard model symmetry $G(2,1,3) = SU(2)_L \times U(1)_Y \times SU(3)$-color; to (2) the minimal quark-lepton and left-right symmetric non-Abelian symmetry $\mathrm{G}(2,2,4) = \mathrm{SU}(2)_L \times \mathrm{SU}(2)_R \times \mathrm{SU}(4)$-color[7,8] that brought a host of attractive

[c]A brief account of how our collaboration started in May 1972 leading to the ideas of higher unification based on the symmetry SU(4)-color is given in my article in the Proceedings of the Salamfestschrift[2] which was held at ICTP in 1993 (that is probably the last scientific meeting that Salam attended), and a shorter version is given in the articles written in his honor after he passed away.[3,4] The first two sections of this talk are based in part on these three articles.

unique answer to the single question posed above, without being assumed. Aesthetically, this particular aspect appeared to be quite appealing to us and suggested, starting in 1972–73, that the SU(4)-color route to higher unification, embodying the symmetry G(2, 2, 4) as a desirable step, may well be used by nature at some level as being part of an ultimate picture. Fortunately, as we will see, such a route which of course includes extension of G(2, 2, 4) into a simple group, especially SO(10), can clearly be distinguished empirically from alternative ones such as those based on SU(5) devoid of SU(4)-color, through phenomena such as neutrino oscillations, leptogenesis, fermion masses and mixings, and even proton decay.

As an added desirable feature, since the symmetries $SU(2)_{L,R}$ and the vectorial SU(4)-color are individually anomaly-free, the gauge interactions of G(2, 2, 4), subject to L-R discrete symmetry, can be chiral (as desired), yet anomaly-free, and importantly parity conserving (see discussion below). As alluded to before, this in turn accounts on symmetry-grounds for the *non-trivial anomaly-cancelation* within each family for the SM symmetry.

Before discussing some of the additional major advantages of the symmetry G(2, 2, 4), let me mention one that is perhaps the most striking.

The Unwanted Right-Handed Neutrino

Either one of the symmetries $SU(4)$-color or $SU(2)_R$ implies, however, that *there must exist the right-handed counterpart (ν_R) of the left-handed neutrino (ν_L).* This is because the RH neutrino (ν_R) is the fourth color partner of the RH up quark; and it is also the $SU(2)_R$ doublet partner of the RH electron. Thus, given the symmetry $G(2, 2, 4)$, one necessarily had to postulate the existence of an *unobserved new member in each family* — the right-handed neutrino.

This in turn meant, especially within the SU(4)-color symmetry, that the neutrinos must be massive, like the quarks, posing the dilemma as to why they are so light. A natural resolution of this dilemma did emerge within the same G(2, 2, 4)-framework within four years through the realization of the seesaw mechanism[13] and I will discuss the same shortly. Meanwhile, there was a strong prejudice, however, in the 1970's, and even through the 1990's till neutrino oscillations were discovered, against neutrinos being massive and therefore against the existence of the RH neutrinos as well. Given that the upper limits on neutrino masses were known to be so small ($m(\nu_e)/m_e \lesssim 10^{-6}$, and after the "discovery" of ν_τ, $m(\nu_\tau)/m_{\rm top} < 10^{-9}$), many, perhaps most in the community, believed that they must be exactly massless.[i] This is infact what the two-component theory of the neutrino[53] as well as

[i]The extent to which this belief was ingrained among many leading physicists even in the 1990's may be assessed by an interesting remark by C.N. Yang at the 2002 Stony Brook conference on neutrinos.[51] See also Ref. 52.

the standard electroweak model of particle physics[28] and the SU(5) grand unification model,[9] possessing only LH neutrinos ($\nu_{\rm L}$'s), would naturally suggest.[j] In this sense, the RH neutrino was regarded perhaps by most as an unwanted child (the ugly duckling) in the 1970's, and I faced much resistance from the community in my seminars as regards the unavoidable need for such an unwanted object.

My only defense at that time was, what appeared in my view, the intrinsic beauty of the quark-lepton and left-right symmetric gauge structure $G(2,2,4)$ and of the associated neat organization of the members of a family into a (2×4)-structure of the LH and the RH fermions as noted below. As it turned out, with the realization of the seesaw mechanism,[13] and importantly the discovery of neutrino oscillations in 1998,[11] the RH neutrino has turned out to be an asset (a beautiful swan) to understand neutrino oscillations as well as the observed baryon asymmetry of the universe. I will discuss these in Secs. 5 and 6. Let me now return to presenting the other key features of the symmetry $G(2,2,4)$.

The introduction of the RH neutrino requires that there be sixteen two-component fermions in each family as opposed to fifteen for the standard model (SM) or the SU(5) symmetry. *Subject to left-right discrete symmetry ($L \leftrightarrow R$) which is natural to $G(2,2,4)$, all 16 members of the electron family now became parts of a whole — a single left-right self-conjugate multiplet $F = \{F_L \oplus F_R\}$, where*

$$F^e_{L,R} = \begin{bmatrix} u_r & u_y & u_b & \nu_e \\ d_r & d_y & d_b & e^- \end{bmatrix}_{\rm L,R}. \tag{3}$$

The multiplets F^e_L and F^e_R are left-right conjugates of each other transforming respectively as (2,1,4) and (1,2,4) of $G(2,2,4)$; likewise for the muon and the tau families. The symmetry $SU(2)_{L,R}$ treat each column of $F^e_{L,R}$ as a doublet; while the symmetry $SU(4)$-color unifies quarks and leptons by treating each row of F^e_L *and* F^e_R as a quartet; *thus lepton number is treated as the fourth color.* A very special feature of the symmetry $G(2,2,4)$ is now worth noting.

Left-Right Symmetry in the Fundamental Laws:

Because of the parallelism between the actions of $SU(2)_{\rm L}$ and $SU(2)_{\rm R}$ on the left and the right chiral fermions respectively, and because SU(4)-color is vectorial, the gauge symmetry $G(2,2,4)$, with the choice $g_{\rm L}^{(0)} = g_{\rm R}^{(0)}$, naturally opened the door for the *novel and attractive concept* that the laws of nature possess left $\leftrightarrow$ right discrete symmetry — i.e. parity invariance — at a fundamental level [56], that

[j] This is barring, of course, possible contributions[54] to the Majorana mass of ν_L from lepton-number violating quantum gravity effects $\sim (v^2_{\rm EW}/M_{Pl}) \sim (250\ {\rm GeV})^2/10^{19}\ {\rm GeV} \sim 10^{-5}$ eV, which are tiny, compared to the presently observed mass scales of atmospheric and even solar neutrino oscillations. As expressed elsewhere,[55] this smallness of (possible) quantum gravity effects prompts one to regard the atmospheric and solar neutrino oscillations as clear signals for physics beyond what one may expect within the standard model combined with quantum gravity.

interchanges $F^e_L \leftrightarrow F^e_R$ together with $W_L \leftrightarrow W_R$. Subject to suitable restrictions on the Higgs systems, the observed parity violation could then be interpreted as being a low-energy phenomenon arising entirely through a spontaneous breaking of the $L \leftrightarrow R$ discrete symmetry for a possible motivation for this sub-group (relevant to the pre-neutrino-oscillation era), see footnote 11 of Ref. [8], which should disappear at appropriately high energies. Thus within this picture, (using Pauli's words) God is no longer "weakly left-handed"; *left and right are treated on par at the fundamental level.*

Briefly, I may mention that if one did not insist on quark-lepton unification through SU(4)-color, parity invariance at a fundamental level can still be realized through the so-called "left-right symmetric (LRS)" model[8,56] based on the symmetry $G(2,2,1,3) = SU(2)_L \times SU(2)_R \times U(1)_{B-L} \times SU(3)^c$, a sub-group of $G(2,2,4)$. Much work,[57] especially in connection with having light W_R's which could possibly be observed at the LHC, has recently been carried out in the context of this minimal left-right symmetric (LRS) model.

I will return shortly to the relevance of having the RH neutrinos for an understanding of the neutrino masses. First, it is worth noting a few additional features of the symmetry $G(2,2,4)$ and its relationship to still higher symmetries.

1) *The Charge Formula*: The symmetry $G(2,2,4)$ introduces an elegant charge formula:

$$Q_{em} = I_{3L} + I_{3R} + (B-L)/2, \tag{4}$$

that applies to all forms of matter (including quarks and leptons of all six flavors, Higgs and gauge bosons).[k] Note that the quantum numbers of all members of a family, including the weak hypercharge $Y_W = I_{3R} + (B-L)/2$, are now completely determined by the symmetry group $G(2,2,4)$ and the tranformation-property of $(F_L \oplus F_R)$. This is in contrast to the case of the SM for which the 15 members of a family belong to five disconnected multiplets, with unrelated quantum numbers. Quite clearly the charges I_{3L}, I_{3R}, and $B-L$ being generators of $SU(2)_L$, $SU(2)_R$, and $SU(4)^c$ respectively are quantized; so also then is the electric charge Q_{em}.

2) *An Intimate Link Between SU(4)-color and L-R Symmetry*: At this point, an intimate link between $SU(4)^c$ and $SU(2)_L \times SU(2)_R$ is worth noting. As remarked before, assuming that $SU(4)^c$ is gauged and demanding an explanation of the quantization of electric charge as above leaves one with no other choice but to gauge minimally $SU(2)_L \times SU(2)_R$ (rather than $SU(2)_L \times U(1)_{I_{3R}}$). Likewise, assuming $SU(2)_L \times SU(2)_R$ and again demanding a compelling reason for the quantization of electric charge dictate that one must minimally gauge $SU(4)^c$

[k] The 15th diagonal generator of $SU(4)$-color entering into the electric charge formula of Ref. [8] is naturally proportional (as per Eq. (11)) to the charge $(B_q - 3L)$ where B_q = quark number. It was pointed out in Ref. 58 that this can be expressed in terms of the more familiar charge (B-L), since $B_q = 3B$, resulting in Eq. (4).

(rather than $SU(3)^c \times U(1)_{B-L}$). The resulting minimal gauge symmetry is then $G(224) = SU(2)_L \times SU(2)_R \times SU(4)^c$ that simultaneously achieves quantization of electric charge, quark-lepton unification and left-right symmetry. *In short, the concepts of SU(4)-color and left-right gauge symmetry in its minimal form (symbolized by $SU(2)_L \times SU(2)_R$) become inseparable from each other, if one demands that there be an underlying reason for the quantization of electric charge.* Assuming one automatically implies the other.

3) *Universality of Weak Interactions*: It is furthermore worth noting that the extension of the standard model symmetry to the level of the symmetry $G(2,2,4)$ provides a compelling reason for why weak interactions are universal with respect to quarks *and* leptons, though strong interactions are not. This is because $SU(2)_L$, generating weak interactions, commutes with SU(4)-color (as it must for the sake of renormalizability) and thus treats all four colors representing quarks and leptons universally. Non-universality of strong interactions can be attributed, as mentioned before, to spontaneous breaking SU(4)-color to SU(3)-color $\times$ $U(1)_{B-L}$ at high energies. These features, which were known from the 1940's and were essentially put in by hand to satisfy observations, thus found a rationale through quark-lepton unification as in $G(2,2,4)$.

4) *The Two Simple Mass Relations*: The symmetry SU(4)-color leads to two simple relations between the masses of quarks and leptons at the unification-scale M_U:

$$m_b(M_U) \approx m_\tau \tag{4a}$$

$$m(\nu^\tau_{Dirac}) \approx m_{top}(M_U) \tag{4b}$$

These two relations arise from SU(4)-color preserving leading entries in the fermion mass matrices which contribute to the masses of the third family [see Ref. 24 for a detailed discussion]. The sub-leading corrections that arise from SU(4)-color breaking in the (B-L) direction turn out to be important for the masses and mixings of only the first two families,[24] and that, of course, goes well with observations. Now, of the two relations given above, the first is successful empirically. As we will see in Sec. 4, the second is crucial to the success of the seesaw formula for $m(\nu^\tau_L)$ and thereby for the observed $\delta m^2(\nu)_{23}$.

5) *B-L as a Local Symmetry*: The symmetry SU(4)-color contains B-L as a generator, which provides some essential benefits. First, with B-L remaining intact at least upto the unification scale $M_U \approx 2 \times 10^{16}$ GeV, it serves to protect the RH neutrinos from acquiring a Planck or string-scale ($\sim 10^{18}$ Gev) Majorana mass through quantum gravity effects.[1] Following limits from the Eötvos-type experiments, however, one can argue that B-L must be violated at some scale M_{B-L} (considerations based on MSSM gauge coupling unification, that predicts the weak

[1]Such an ultraheavy Majorana mass ($\sim 10^{18}$ GeV) for the RH neutrino would be unacceptable because it would lead to too tiny a mass ($< 10^{-4}$ eV) for even the heaviest Dirac mass ~ 200 GeV for any neutrino species through the seeaw formula.

angle successfully, would in fact suggest that $M_{B-L} \approx M_U \approx 2 \times 10^{16}$ GeV, see Secs. 5 and 6). Now M_{B-L} sets the scale for the (superheavy) Majorana masses of the RH neutrinos, and thereby plays a crucial role in determining the masses of the light LH neutrinos via the seesaw formula (see Sec. 6). Furthermore, spontaneous breaking of B-L allows one to implement the mechanism of baryogenesis via leptogenesis in the presence of the elctroweak sphaleron effects which wipe out any (B-L)-conserving matter-antimatter asymmetry (see Sec. 6).

In brief, viewed against the background of particle physics in 1972–73, the symmetry $G(2,2,4)$ brought some attractive features for the first time. These include:

(i) Unification of all 16 members of a family within *one* left-right self-conjugate multiplet, with a neat explanation of all their quantum numbers;

(ii) Quantization of electric charge, with $Q_{e^-} + Q_p = 0$;

(iii) Quark-lepton unification through $SU(4)$-color;

(iv) A compelling reason for the universality of the weak interactions with respect to quarks *and* leptons;

(v) Automatic cancelation of the triangle anomalies on symmetry-grounds within each family;

(vi) Conservation of parity at a fundamental level;[56]

(vii) A rationale for the now-successful mass-relations 4a) and 4b);

(viii) B-L as a local symmetry;

(ix) The RH neutrino as a compelling member of each family; and

(x) The right set-up, due to (viii) and (ix), for implementing baryogenesis via leptogenesis.

These ten features constitute the hallmark of the symmetry $G(2,2,4)$, fulfilling major aesthetic ((i)–(vi)) as well as practical ((vii)–(x)) needs. As mentioned before, the three distinguishing features of $G(2,2,4)$ — i.e. the existence of the RH neutrinos and B-L as a local symmetry, together with the mass-relation 4b), now seem to be essential to understand naturally the tiny mass-scales observed in neutrino oscillations and to implement the mechanism of baryogenesis via leptogenesis. This will be discussed in more detail in Secs. 5 and 6. In addition to the features listed above, the symmetry $G(2,2,4)$ needing a spontaneous violation of B-L introduced a *new line of thinking* leading to a questioning of the conservations of baryon and lepton numbers,[7] which were otherwise held sacred owing to the extraordinary stability of the proton. Such a questioning in turn evolved in the context of unification-ideas going beyond that of the symmetry $G(2,2,4)$ leading to the hallmark prediction of grand unification — i.e. proton decay.[7,9] This is briefly noted below, together with other fundamental processes violating B and/or L. Proton decay will be discussed in more detail in Sec. 7.

Non-Conservations of B and L: Proton Decay as a Generic Feature

As mentioned before, B-L being a local symmetry in $G(2,2,4)$, the combination (B-L) and therefore B and/or L must be violated spontaneously. It was recognized that this feature is a reflection of a more general phenomenon involving non-conservations of baryon and lepton numbers which are most likely to occur in unified gauge theories,[7,9] including those going beyond the symmetry $G(2,2,4)$, that unify quarks and leptons as well as their three basic forces. Depending upon the nature of the gauge symmetry and the multiplet structure, the violations of B and/or L could be spontaneous,[m] as is the case for non-conservation of B-L in SU(4)-color, and those of B and L in the maximal one-family symmetry SU(16) which gauges both B and L Ref. [59]. Alternatively, the violations could be explicit; that is what happens for the subgroups of SU(16), like SU(5)[9] and SO(10),[10] for which the gauge boson interactions violate B and L. One way or another, baryon and lepton number conservations cannot be absolute in the context of such higher unification.

The simplest and most dramatic consequence of such non-conservation is proton decay ($\Delta B \neq 0, \Delta L \neq 0$), which as will see provides an indispensable tool to probe physics at truly high energies $\sim 10^{16}$ GeV, for proton decaying via the canonical dim $= 6(\Delta(B-L)=0)$ decay modes, such as: $p \to e^+\pi^\circ$ and $p \to \overline{\nu}K^+$. The other fundamental "processes" are the Majorana masses of the RH and LH neutrinos ($\Delta B = 0, |\Delta L| = 2$), which are relevant to an understanding of the neutrino masses via the seesaw mechanism, and in a related context the nutrinoless double beta-decay ($nn \to pp\ e^-e^-$),[62,63] both of which arise naturally within $G(2,2,4)$ and SO(10) symmetries. Based on our current understanding of masses and mixings of all fermions including neutrinos in the context of such unification symmetries, the latter process should occur at some level and could quite plausibly lie within the realm of observation in several ongoing or next-generation experiments.[63] The third equally fundamental process is $n - \bar{n}$ oscillation[64,65] ($|\Delta B| = 2, \Delta L = 0$) which probes into physics typically at energy-scales of $\sim 10^5$ GeV (or possibly lower). Several promising models have been constructed within symmetries like $G(2,2,4)$, SO(10) and $G(2,2,1,3)$ (see e.g. Ref. [65]) which suggest that $n - \bar{n}$ oscillation could well be discovered if the current sensitivity can be improved by one to two orders of magnitude, reaching free $n - \bar{n}$ oscillation time $\sim 10^9 - 10^{10}$ s. Now any of these three processes, if seen, would be a breakthrough in particle physics, shedding light on fundamental physics at a deeper level. Thus the need for an improved sensitive study of *each of these three processes* cannot be overemphasized.

In this talk, I will focus primarily on three aspects: (i) gauge coupling unification (Sec. 3), (ii) an understanding of the mass-scale (s) of neutrino oscillations (Secs. 5 and 6), and (iii) proton decay (Sec. 7). These turn out to be intimately linked in

[m] The case of spontaneous violation arises because the massless gauge particle coupled to any linear combination of B and L (which is gauged) must acquire a mass through SSB in order to conform with the limits from the Eötvos type experiments. The corresponding charge (B and/or L) must then be violated spontaneously.

symmetries, in particular those containing SU(4)-color, like SO(10) or E_6 or a string-derived G(2, 2, 4) symmetry (see below), in the context of the seesaw mechanism; and last but not least;

(iii) proton decay.

I will discuss the empirical verification of the first two features and the consequent implications in this and the next section. The third feature — proton decay — which now constitutes the missing piece of evidence for grand unification is discussed in Sec. 7.

Meeting of The Three Gauge Couplings: MSSM-GUT Versus String Unifications

It has been known for some time that the precision measurements of the standard model coupling constants (in particular $\sin^2\theta_W$) at LEP put severe constraints on the idea of grand unification. Owing to these constraints, the non-supersymmetric minimal $SU(5)$, and for similar reasons, the *one-step breaking* minimal non-supersymmetric $SO(10)$-model as well, are now excluded.[70] For example, minimal non-supersymmetric $SU(5)$ predicts: $\sin^2\theta_W(m_Z))\,|_{\bar{MS}} = 0.214 \pm 0.004$, where as current experimental data show: $\sin^2\theta_W(m_Z)_{expt}{}^{LEP} = 0.23153 \pm 0.00016$.[71] The disagreement with respect to $\sin^2\theta_W$ is reflected most clearly by the fact that the three gauge couplings (g_1, g_2 and g_3), extrapolated from below, fail to meet by a fairly wide margin in the context of minimal non-supersymmetric $SU(5)$ (see left panel of Fig. 1).

But the situation changes radically if one assumes that the standard model is replaced by the minimal supersymmetric standard model (MSSM), above a threshold of about 1 TeV, which is motivated independently by requiring naturalness in

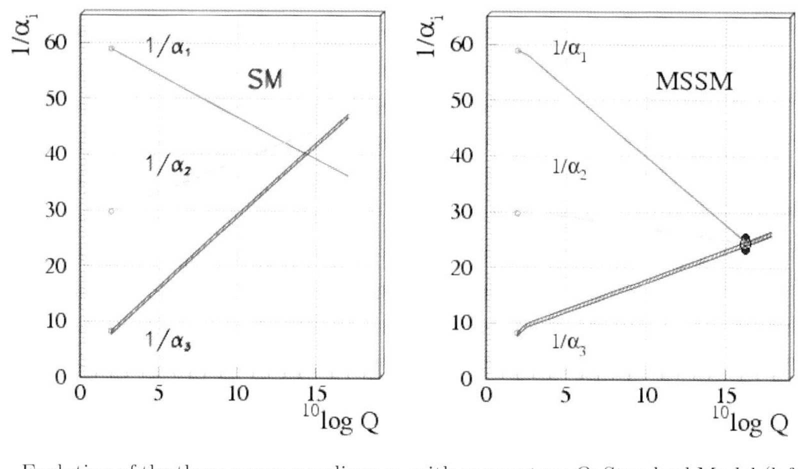

Fig. 1. Evolution of the three gauge couplings α_i with momentum Q: Standard Model (left panel) and Minimal Supersymmetric Standard Model (right panel)

the understanding of the Higgs mass.[20] In this case, the three gauge couplings are found to meet,[18] to a very good approximation (See Fig. 1, right panel) at a scale given by

$$M_U \approx 2 \times 10^{16}\,\text{GeV} \quad (\text{MSSM or SUSY SU(5)}) \tag{5}$$

This dramatic meeting of the three gauge couplings, or equivalently the excellent agreement of the MSSM-based prediction of $\sin^2\theta_W(m_Z)_{\text{Th}} = 0.2315 \pm 0.003$[18,70] with the observed value given above, provides a strong support for the ideas of both grand unification and low-energy supersymmetry, as being relevant to physics at short distances.

The simplest interpretation of the observed meeting of the three gauge couplings at the scale M_U, that preserves the successful prediction of the weakangle $\sin^2\theta_W$, is that a supersymmetric grand unification symmetry (often called GUT symmetry), like SU(5) or SO(10), breaks spontaneously *in one step*, (as opposed to possible multiple steps, relevant to SO(10)) at the scale M_U into the standard model symmetry $G(2,1,3)$.

Before discussing possible alternative interpretation of the observed meeting of the gauge couplings in the context of string/M theory, it is worthnoting that, unlike SU(5), a symmetry like SO(10) can in general break spontaneously in multiple steps (two or more) to the SM via a symmetry like $G(2,2,4)$ (see e.g. Refs. 72 and 73, and two recent works[74,75]) involving intermediate scale(s) of symmetry breaking. In these cases, even without supersymmetry, gauge coupling unification can be made to work by choosing appropriately the intermediate scale(s) of symmetry breaking. *But in such cases, the weak angle is no longer a prediction; it needs to be used as an input to fix the intermediate scale(s).*[p] Thus, in these cases, the success of the simplest interpretation of the observed meeting of the gauge couplings in predicting the weak angle in excellent agreement with experiments will have to be regarded only as an accident. Remarkably enough, an identical conclusion emerges in attempting to have a natural understanding of the mass-scale of the atmospheric neutrino oscillation (see Sec. 6). In short, *two independent phenomena* — gauge coupling unification and the mass-scale of the atmospheric neutrino oscillation — *seem to lead to one and the same conclusion favoring the simplest picture of the one-step breaking of a SUSY GUT Symmetry* (like SO(10)) or a string-unified $G(2,2,4)$ symmetry (see below)) to the SM symmetry at a high scale $M_U \approx 2 \times 10^{16}$ GeV.

Now, the simplest interpretation, mentioned above, would require a SUSY GUT-symmetry, like SO(10), to be effective at and above the unification-scale M_U in 4D.

[p]The case of Ref. 74 is somewhat of an exception in this regard because the single intermediate scale associated with the breaking of the $G(2,2,4)$ symmetry to the SM is identified with the Peccei-Quinn symmetry breaking scale. But, even in this case, intermediate-scale breaking of B-L does not go well with attempts to obtain a natural understanding of the mass-scale of the atmospheric neutrino oscillation.

An alternative interpretation that would permit a supersymmetric non-GUT symmetry like $G(2,2,4)$ breaking in one step into the SM symmetry at the unification-scale M_U is, however, possible in the context of string or M theory, which seems to be needed to unify all the forces of nature including gravity and also to obtain a good quantum theory of gravity. This is because, even if the effective symmetry in four dimensions emerging from a higher dimensional string theory is non-simple, like $G(2,2,4)$ or $G(2,1,4)$, or even the SM symmetry $G(2,1,3)$, string theory can still ensure familiar unification of the gauge couplings at the string scale. In this context, one could, therefore, first ask: can the effective symmetry below the string-scale in 4D be as small as the SM symmetry $G(2,1,3)$? I would argue (see Sec. 5 and 6) that attempts to understand naturally the neutrino mass-scales utilizing the seesaw mechanism, and to implement the mechanism of baryogenesis via leptogenesis successfully, provide an answer to the contrary. Both suggest clearly that the effective symmetry in 4D, below the string scale, should *minimally* contain either the $G(2,2,4)$ or its close relative $G(2,1,4) = SU(2)_L \times I_{3R} \times SU(4)^c$ symmetry. This is because, as mentioned before, B-L should survive at least up to the unification scale M_U so as to protect the RH neutrinos from acquiring Majorana masses of the string or Planck scale.[q] Furthermore, the full SU(4)-color symmetry, that contains B-L, is needed to ensure the two Dirac mass-relations given in Eqs. (4a) and (4b). The first is empirically successful, while the second is needed crucially for the success of the seesaw mechanism (see Sec. 6).

While keeping $G(2,1,4)$ in the picture, to simplify discussion, I will thus proceed to consider the two alternatives of a (hopefully) realistic string/M-theory solution possessing either an effective $G(2,2,4)$ symmetry or the GUT-symmetry SO(10), both with supersymmetry, emerging at the string-scale in 4D. Needless to say, among the vast landscape of string/M-theory solutions, which are essentially perturbative in nature, there is no criterion of selectivity at present, and identifying a completely realistic solution, even with only the SM symmetry in 4D, remains a challenge. But there do exist promising semi-realistic solutions containing an effective $G(2,2,4)$ symmetry in 4D,[76] with three generations, and some of them with the right set of Higgs-like multiplets to do the desired symmetry breaking in 4D. Semi-realistic string-derived SO(10) solutions also exist.[77] Thus at present, the existence of a completely realistic string-$G(2,2,4)$ or string-SO(10) solution in 4D must be regarded as an assumption. In view of the strong empirical motivations for either one of these two symmetries, and the theoretical motivations for the string/M theory being the mother theory, such an assumption seems worth pursuing, and I will do so in the following.

Let me now return to the question of a consistency that arises for a non-GUT $G(2,2,4)$-like string solution in 4D. While string theory will ensure gauge coupling unification (i.e. $g_{2\mathrm{L}} = g_{2\mathrm{R}} = g_4$) at the string-scale, one needs to account for

[q]Such an ultraheavy Majorana mass ($\sim 10^{18}$ GeV) for the RH neutrino would be unacceptable because it would lead to too tiny a mass ($< 10^{-4}$ eV) for even the heaviest LH neutrino through the seeaw formula.

the mismatch by about a factor of 20 between the MSSM unification scale M_U (given above), and the string-unification scale, given by $M_{st} \approx g_{st} \times 5.2 \times 10^{17}$ GeV $\approx 3.6 \times 10^{17}$ GeV (Here we have put $\alpha_{st} = \alpha_{GUT}(\text{MSSM}) \approx 0.04$).[78] Possible resolutions of this mismatch have been proposed. These include:

(i) utilizing the idea of *string-duality*[79] which allows a lowering of M_{st} compared to the value shown above, or alternatively

(ii) the idea of a *semi-perturbative* unification that assumes the existence of two vector-like families, transforming as $(16 + \overline{16})$, at the TeV-scale. The latter raises α_{GUT} to about 0.25–0.3 and simultaneously M_U, in two loop, to about $(1/2 - 2) \times 10^{17}$ GeV.[80]

(iii) **The case of a Few Extra Multiplets at GUT-scale**: A third possibility,[81] which has not appeared in a published form yet, but is quite feasible, would arise if a supersymmetric $G(2,2,4)$-solution emerges from string theory in 4D, possessing a few extra Higgs-like multiplets (as mentioned below) having GUT-scale masses. These extra multiplets are in addition, of course, to the familiar low-energy spectrum of MSSM possessing: (a) the three generations of quarks and leptons, (b) the multiplet $(2,2,1)_{\text{H}}$ of $G(2,2,4)$ which contains Hu and H_{d} of MSSM and serves to break the EW symmetry, and (c) their superpartners. The extra multiplets, that would serve the desired purpose of accounting for the mismatch (mentioned above), are given by the set:

$$\begin{aligned}\text{H}_{\text{extra}} = [2 \times (1,1,15)_{\text{H}} + \{(2,1,4)_{\text{H}} + (1,2,\overline{4})_{\text{H}}\}_{\text{A}} \\ + \{(2,1,\overline{4})_{\text{H}} + (1,2,4)_{\text{H}}\}_{\text{B}} + (2,2,1)'_{\text{H}}]\end{aligned}$$

Note that, of these extra multiplets, the $(1,1,15)_{\text{H}}$ is the analog of 45_{H} of SO(10) and the sub-sets A and B correspond precisely to 16_{H} and $\overline{16}_{\text{H}}$ of SO(10) respectively, while $(2,2,1')_{\text{H}} \subset 10'_{\text{H}}$ of SO(10). The VEVs of $\{(1,2,4)_{\text{H}} + (1,2,\overline{4})_{\text{H}}\}_{\text{A}}$ of GUT-scale are in fact needed to break the gauge symmetry $G(2,2,4)$ to the SM, while preserving supersymmetry. This is the exact analog of the roles of $\{16_{\text{H}} + \overline{16}_{\text{H}}\}$ for breaking SO(10) (see Sec. 4). Also one of the two $(1,1,15)_{\text{H}}$ ($\subset 45_{\text{H}}$ of SO(10)) having a GUT-scale VEV along the (B-L)-direction is needed to introduce the observed (B-L)-dependence in fermion masses and mixings, through the use of higher dimensional operators (see Sec. 5). The $(2,2,1)'_{\text{H}}$ is assumed to have zero VEV. In this sense, most of the multiplets exhibited in the set H_{extra} are in fact needed, minimally, to implement necessary symmetry breaking; only one $(1,1,15)_{\text{H}}$ and the $(2,2,1)'_{\text{H}}$ are really extra.

Now, the gauge couplings g_{2L}, g_{2R} and g_4 of the $G(2,2,4)$ symmetry, unified at the string scale $M_{\text{st}} \approx 4 \times 10^{17}$ GeV, would run down to lower energies with their respective β-functions. It may be verified that, in the presence of the multiplets given by H_{extra} and the MSSM-GUT spectrum, the one-loop beta functions for g_{2L}, g_{2R} and g_4 (including the gauge, Higgs and matter contributions) are equal, with $b_{2L} = b_{2R} = b_4 = 6$. Since, for MSSM, the gauge coupling α_U at the unification scale M_U is small $\approx$0.04, we expect that two-loop corrections will not alter the running between M_{st} to M_U significantly. Thus, the three gauge couplings of $G(2,2,4)$, unified at the string-scale, will still remain unified while running down upto the conventional MSSM GUT-scale $M_U \approx 2 \times 10^{16}$ GeV, where the symmetry $G(2,2,4)$ will break to the SM symmetry by utilizing the VEVs of the sub-sets A and B and of $(1,1,15)$ listed in $\mathrm{H}_{\mathrm{extra}}$. This resolves the mismatch between the MSSM-unification at M_U and the string-unification of $G(2,2,4)$ at $M_{\mathrm{st}} \approx 4 \times 10^{17}$ GeV.

There is an additional benefit of the extra multiplets listed above. It is well known that with a "typical" SUSY spectrum at the electroweak scale, and without any GUT-scale threshold corrections, the MSSM gauge couplings unify, with 2-loop running from the weak to the GUT-scale, for $\alpha_3(m_Z) \approx 0.127$. This is somewhat higher than the observed PDG average value of $\alpha_3(m_Z)_{\mathrm{expt}} = 0.1176 \pm 0.002$ (see second reference in Ref. 71). Now, the one-loop GUT-scale threshold corrections, *including contributions from the extra multiplets*, can be calculated in terms of two parameters: (i) the ratio of the two GUT-scale VEVs $\langle(1,2,\bar{4})_{\mathrm{H}}\rangle/\langle(1,1,15)_{\mathrm{H}}\rangle$ and (ii) the ratio M/M_U, where M denotes an "average" mass of the superheavy GUT-scale particles, assumed to have a common mass, for simplicity. One finds[81] that, allowing for a reasonably wide range of variation of these two ratios, the GUT-scale threshold correction to $\alpha_3(m_Z)$ ranges between $-(0.006 - 0.012)$, which gives $\alpha_3(m_Z)$ within one standard deviation of the observed value. This demonstrates that with the addition of just two extra multiplets (one $(1,1,15)_{\mathrm{H}}$ and one $(2,2,1)'_{\mathrm{H}}$) of an assumed string-origin, together with one $(1,1,15)_{\mathrm{H}}$ and the sub-sets A and B listed in $\mathrm{H}_{\mathrm{extra}}$ that are needed to implement symmetry breaking, a string-unified $G(2,2,4)$ symmetry can be fully consistent with the observed GUT-based MSSM gauge coupling unification at $M_U \approx 2 \times 10^{16}$ GeV, with a clear benefit as regards the predicted value of $\alpha_3(m_Z)$.

I should add that of the three possible resolutions of the mismatch between M_{st} and M_U for a string-unified $G(2,2,4)$ symmetry, the third one needing two extra multiplets has a practical advantage. For the first two, one way or another, the ratio (M_U/M_{st}) is near unity, *where as for the third the conventional values of* $M_U \approx 2 \times 10^{16}$ *GeV and* $M_{st} \approx 4 \times 10^{17}$ *GeV are preserved*, and thereby their ratio of $M_U/M_{st} \approx 1/20$ as well. This latter value of the ratio of M_U/M_{st} clearly goes well with the hierarchical entries that are needed for an understanding of the masses and mixings of the charged fermions (Sec. 5) and of the neutrinos (Sec. 6) through higher dimensional operators.

While the mismatch can thus quite plausibly be removed for a non-GUT string-derived symmetry like $G(2,2,4)$, a GUT symmetry like SU(5) or SO(10) would have an advantage in this regard because it is guaranteed to keep the gauge couplings together between M_{st} and M_U (even if $M_U \sim M_{st}/20$), and thus not even encounter the problem of a mismatch between the two scales. A supersymmetric GUT-solution (like SU(5) or SO(10)), however, has a possible disadvantage as well, because it needs certain color triplets to become superheavy by the so-called double-triplet splitting mechanism (see Ref. 25 and references therein and discussion in Sec. 7), in order to avoid the problem of rapid proton decay. However, no such mechanism has emerged yet, in string theory, for the GUT-like solutions.[77]

Non-GUT string solutions, based on symmetries like $G(2,2,4)$ or $G(2,1,4)$ for example, have a distinct advantage in this regard, in that the dangerous color triplets, which would induce rapid proton decay, are often naturally projected out for such solution through string compactification.[82,83] Furthermore, the non-GUT solutions invariably possess new "flavor" gauge symmetries, which distinguish between families. These symmetries are immensely helpful in explaining qualitatively the observed fermion mass-hierarchy (see e.g. Ref. 83) and resolving the so-called naturalness problems of supersymmetry such as those pertaining to the issues of squark-degeneracy,[84] and quantum gravity-induced rapid proton decay.[85]

Weighing the advantages and possible disadvantages of both, it seems hard at present to make a priori a clear choice between a presumed GUT (like SO(10)) versus a non-GUT (like $G(2,2,4)$) string-solution. As expressed elsewhere,[4] it therefore seems prudent to keep both options open and pursue their phenomenological consequences. Now, the advantages of an effective $G(2,2,4)$ or SO(10) symmetry in understanding masses and mixings of all fermions including neutrinos turn out to be essentially identical, especially if one uses low-dimensional Higgs multiplets (as opposed to large-dimensional ones) to break SO(10) or $G(2,2,4)$ to the SM in 4D (See discussion in Secs. 4–6 and Ref. 24). I will thus proceed by assuming that either a suitable $G(2,2,4)$-solution with a mechanism to resolve the mismatch between M_U and $M_{\rm string}$ of the sort mentioned above (most preferably the third one), or a realistic SO(10)-solution with the needed doublet-triplet mechanism, will emerge from string theory. As we will see in Sec. 7, a study of proton decay can help distinguish between these two alternatives. I next discuss the spontaneous breaking of SO(10) or $G(2,2,4)$ to the SM, which will be relevant to our understanding of the masses and mixings of all fermions including neutrinos (Secs. 5 and 6).

IV. Breaking SO(10) or $G(2,2,4)$ to the SM

For compactness of notation, I will present the Higgs system that would be responsible for breaking SO(10) to the SM. A completely analogous set of Higgs multiplets can be used to break $G(2,2,4)$ to the SM. Now two distinct sets of multiplets have been used to break SO(10) to the SM, each having some advantages over the other.

One of these, which I will follow in the rest of my talk, utilizes a minimal set of low-dimensional Higgs multiplets (LOH)[86–92] to break a supersymmetric SO(10) symmetry to $\mathrm{SU}(3)^c \times \mathrm{U}(1)_{\rm em}$. The same set is also used to generate a predictive framework for understanding masses and mixings of all fermions including neutrinos.[24] Minimally, it consists of the multiplets:

$$\mathrm{H_{LOH}} = \{10_{\rm H}, 16_{\rm H}, \overline{16}_{\rm H}, 45_{\rm H}, S_{\rm H}\} \tag{6}$$

The $16_{\rm H}$ is paired with $\overline{16}_{\rm H}$ to preserve supersymmetry at the GUT-scale. Of these, the VEV of $\langle 45_{\rm H}\rangle \sim \mathrm{M_U}$ in the (B-L)-direction breaks SO(10) to $G(2,2,1,3)$ and those of $\langle 16_{\rm H}\rangle = \langle\overline{16}_{\rm H}\rangle \sim \mathrm{M_U}$ along the $\langle\tilde{\overline{\nu}}_{\rm RH}\rangle$ and $\langle\tilde{\nu}_{\rm RH}\rangle$-components break masses and mixings of the quarks and some of the light neutrinos:[24]

$$
\begin{aligned}
& m_b^\circ(M_U) \approx m_\tau^\circ(1-8\epsilon^2) \Rightarrow m_b(m_b) \approx (4.7\text{-}4.9)\ \text{GeV}, \\
& \sqrt{\Delta m^2_{\text{atm}}} \equiv \sqrt{\Delta m^2_{31}} \approx m(\nu_3) \approx (1/24\ \text{eV})(1/2-2),\ (\text{see Sec. 6}) \\
& V_{cb} \approx \left| \sqrt{\frac{m_s}{m_b}} \left|\frac{\eta+\epsilon}{\eta-\epsilon}\right|^{1/2} - \sqrt{\frac{m_c}{m_t}} \left|\frac{\sigma+\epsilon}{\sigma-\epsilon}\right|^{1/2} \right| \approx 0.044, \\
& \begin{cases} \theta^\nu_{23} \approx \left| \sqrt{\frac{m_\mu}{m_\tau}} \left|\frac{\eta-3\epsilon}{\eta+3\epsilon}\right|^{1/2} + \sqrt{\frac{m_{\nu_2}}{m_{\nu_3}}} \right| \approx |0.437+0.408|, \\ \text{Thus,}^{105} \sin^2 2\theta^\nu_{23} \approx 0.994,\quad (\text{for } m(\nu_2)/m(\nu_3) \approx 1/6), \end{cases} \\
& V_{us} \approx \left| \sqrt{\frac{m_d}{m_s}} - \sqrt{\frac{m_u}{m_c}} \right| \approx 0.20, \\
& \left| \frac{V_{ub}}{V_{cb}} \right| \approx \sqrt{\frac{m_u}{m_c}} \approx 0.07, \\
& m_d(1\ \text{GeV}) \approx 8\ \text{MeV}.
\end{aligned}
\tag{12}
$$

Before discussing the empirical successes of the predictions listed above, let me first comment on some aspects of the neutrino-sector pertaining to the first family. As regards $\nu_e - \nu_\mu$ and $\nu_e - \nu_\tau$ oscillations, the standard seesaw-contributions, based on the hierarchical mass-matrices M_ν^D (Eq. (8)) and M_R^ν (Eq. (10)), typically lead to rather small oscillation angles ($\lesssim$0.1).[24] It has however, been noted[93] that non-seesaw contribution, e.g. to $\nu_L^e \nu_L^\mu$ mixing mass-term $\sim (2-6)\times 10^{-3}$ eV, can quite plausibly arise through higher dimensional operators in accord with flavor symmetry (see Ref. 93 and 106). Such a contribution, combined with the standard seesaw contributions based an M_ν^D and M_{R}^ν, that yields a diagonal $(\nu_L^\mu \nu_L^\mu)$ mass $\sim (3-10)\times^{-3}$ eV, can plausibly lead to θ^ν_{12} as large as $\sim$(30°–35°), in accord with the LMA MSW solution of the solar neutrino problem (For a review of current experimental status, see Refs. 107 and 108). Similarly, one can obtain through non-seesaw contribution $\theta^\nu_{13} \approx$ (2°–10°), to be compared with the observed value of $\approx$9°.[107,108] Thus, including the seesaw (from M_ν^D and M_{R}^ν) and non-seesaw contributions, together with a natural value of $y \approx -1/18$ (which is expected to be of order 1/10 by flavor symmetry[102]), one gets:

$$
\begin{aligned}
m(\nu_2) &\approx (8.5)\times 10^{-3}\,\text{eV (seesaw)} \\
m(\nu_1) &\approx (1-\text{few})\times 10^{-3}\,\text{eV; thus } \Delta m^2_{21} \approx 7\times 10^{-5}\,\text{eV}^2 \\
\theta^\nu_{12} &\approx (20^\circ\text{–}35^\circ)\ (\text{non-seesaw}) \\
\theta^\nu_{13} &\approx (2^\circ\text{–}10^\circ)\ (\text{non-seesaw})
\end{aligned}
\tag{13}
$$

Here, $m(\nu_2)$ should be regarded as a reasonable expectation within the model, within a factor of 2 or so either way, corresponding to a natural choice of $|y| \approx 1/10 - 1/25$ (say). Unlike the results listed in Eq. (12), which are compelling predictions (postdictions) of the model, however, those on θ^ν_{12} and θ^ν_{13} noted above should be considered only as plausible and consistent possibilities within the model.

The Majorana masses of the RH neutrinos ($N_i \equiv \nu^i_{\rm R}$) are given by (see discussion in Sec. 6).

$$\begin{aligned} M_3 &\approx M_R \approx 10^{15}\ \text{GeV}\ (1/2\text{-}2), \\ M_2 &\approx |y^2| M_3 \approx (3 \times 10^{12}\,\text{GeV})(1/2\text{-}1), \\ M_1 &\approx |x - z^2| M_3 \sim (1/4\text{-}2) 10^{-5} M_3 \sim 10^{10}\ \text{GeV}(1/10\text{-}2). \end{aligned} \tag{14}$$

where $y \approx -1/18$ and $x \sim z^2 \sim 10^{-4}(1/2 - 2)$ have been used in accord with flavor symmetry.[102] *Note that we necessarily have a hierarchical spectrum for the light as well as the heavy neutrinos with normal hierarchy* $m_1 < m_2 \ll m_3$ *and* $M_1 \ll M_2 \ll M_3$. Leaving out the oscillation angles θ^ν_{12} and θ^ν_{13}, it seems remarkable that the seven predictions in Eq. (12) agree with observations, pretty well. Particularly intriguing is the (B-L)-dependent *group-theoretic correlation* between the contribution from the first term in V_{cb} and that in θ^ν_{23}, which explains simultaneously why one is small (V_{cb}) and the other is large (θ^ν_{23}).

Another interesting point of the hierarchical BPW model[24] is that with $|y|$ being hierarchical (of order 1/10 as opposed to being of order 1) and $m(\nu_2)/m(\nu_3)$ being of order 1/5–1/10, *the mixing angle from the neutrino sector* $\sqrt{m(\nu_2)/m(\nu_3)}$ *necessarily adds (rather than subtracts) to the contribution from the charged lepton sector (see Eq. (12))*, as shown in Ref. 24. As a result, in the BPW model, both charged lepton and neutrino-sectors give medium-large contribution (≈ 0.4) which *add* to naturally yield a maximal θ^ν_{23}. At the same time, analogous entries for $V_{\rm cb}$ subtract, leading to its smallness (see Eq. (12)). This thus becomes a simple and compelling prediction of the model, based essentially on the group theory of the minimal Higgs system in the context of SO(10) or G(224) and the hierarchical nature of the mass-matrices.[u]

The success of the model as regards especially the first six predictions listed above provides some confidence in the *gross pattern* of the Dirac and Majorana mass matrices presented above. I now proceed to discuss primarily the implications of the mass-scale of the atmospheric neutrino oscillation in the next section.

[u]The explanation of the largeness of θ^ν_{23} together with the smallness of $V_{\rm cb}$ outlined above, based on medium-large contributions from the charged lepton and neutrino sectors, is quite distinct from alternative explanations. In paricular, in the lop-sided Albright-Barr model,[95] the largeness of θ^ν_{23} arises almost entirely from the lop-sidedness of the charged lepton mass matrix. This distinction between the BPW and the AB models leads to markedly different predictions for the rate of $\mu \to e\gamma$ decay in the two models.[109]

with an inverse rate $\lesssim 10^{35}$ yrs, it would first of all mean that an intact GUT-symmetry like SO(10) is very likely not operative in 4D; instead an effective symmetry like $G(2,2,4)$, or its close relative $G(2,1,4)$, very likely having a string-origin, is operative in 4D with low-energy supersymmetry; *and it is the neutrino-mass related* $d = 5$ *operator (see Sec. 7.1) that induces such a decay.*

In short, proton decay, if seen, will bring a wealth of knowledge of a fundamental nature that can not be gained by any other means.

As we have seen, the potential for discovery of proton decay, within a well-motivated class of grand unification models, is high. This is why an improved search for proton decay is now most pressing. This can only be done with a large detector built deep underground. Most desirably we would need both Water Cherenkov (as in HyperKamiokande) and Liquid Argon (as in LBNF-DUNE) detectors, because the former is specially sensitive to the $(e^+\pi^\circ)$-mode, and the latter to the $(\overline{\nu}K^+)$-mode. Such a detector, coupled to a long-baseline intense neutrino beam, can simultaneously sensitively study neutrino oscillations so as to shed light on neutrino mixing parameters, mass-ordering, and most importantly CP violation in the neutrino system. And it can help efficiently study supernova neutrinos. In short such a detector (or rather two such detectors of the type mentioned above) would have a unique multi-purpose value with high discovery potential in all three areas.

VIII. Concluding Remarks

Neutrinos seem to be as elusive as revealing. Simply by virtue of their tiny masses, they provide some crucial information on the nature of the unification symmetry and on the unification scale, more precisely on the (B-L)-breaking scale. In particular, as argued in Sec. 6, combined with the b/τ mass-ratio, the mass-scale $(\sqrt{\Delta m^2_{\rm Atm}} \approx \sqrt{\Delta m^2_{31}} \approx 1/20\,{\rm eV})$ of the atmospheric neutrino oscillation provides clear support for the following three features:

(i) the existence of the SU(4)-color symmetry in 4D above the GUT-scale which provides not only the RH neutrinos but also B-L as a local symmetry and a value for $m(\nu^\tau_{\rm Dirac})$ (Eq. (4b));
(ii) the (B-L)-breaking scale being close to the familiar SUSY unification-scale $M_U \approx 2 \times 10^{16}\,{\rm GeV}$, which provides the mass-scale M_R for the Majorana mass of the heaviest RH neutrino (see Eqs. (14) and (15)); and importantly
(iii) the seesaw mechanism that successfully explains the atmospheric neutrino oscillation mass-scale by utilizing both (i) and (ii) (see Eqs. (16)–(18)).

In turn this chain of argument selects out the effective symmetry in 4D being either a string-derived $G(2,2,4)$ or $SO(10)$-symmetry, as opposed to other alternatives like $SU(5)$, $[SU(3)]^3$ or even flipped $SU(5) \times U(1)$. As a corollary, this supports the idea that Nature intrinsically is left-right symmetric (parity-conserving).[56]

Furthermore, the success of the $G(2,2,4)/SO(10)$-based seesaw mechanism in accounting for the neutrino oscillation mass-scales (both atmospheric and solar, see Sec. 6) implies that the masses of both the heavy RH and the light LH neutrinos are Majorana, not Dirac, which violate lepton number by two units. That in turn provides, by utilizing the out-of-equilibrium decays of the RH neutrinos and the electroweak sphaleron process, the promising mechanism of baryogenesis via leptogenesis, which can naturally yield $Y_B \approx 10^{-10}$ (see Sec. 6). In short, the neutrinos may well be at the root of our own origin.

Including the insight gained from the neutrinos as above, we are now in possession of a set of facts which may be viewed as the *matching pieces of a puzzle* in that all of them can be understood simply by just one idea — that of supersymmetric grand unification. These include: (i) the quantum numbers of all the members in a family including the RH neutrino; (ii) the quantization of electric charge, with $Q_{e^-} = -Q_p$; (iii) the dramatic meeting of the three gauge couplings (Fig. 1, right panel) or equivalently the success of the associated prediction of the weak angle; (iv) $m_b^\circ \approx m_\tau^\circ$; (v) $\sqrt{\Delta m^2_{\rm atm}} \approx 1/20\,\text{eV}$; (vi) a nearly maximal $\theta^\nu_{23} \approx \pi/4$ with a minimal $V_{\rm cb} \approx 0.04$; and (vii) baryogenesis via leptogenesis leading to $Y_B \approx 10^{-10}$ (see Sec. 6).

All of these features and more hang together within a *single unified framework* based on a presumed string-derived $G(2,2,4)$ or SO(10) symmetry, with low-energy supersymmetry. Unless this neat fitting of all the pieces within a single simple picture is just a mere coincidence, it is hard to think that that can be the case, it is pressing that dedicated searches be made to find the two missing pieces of this picture — that is: proton decay and supersymmetry.

The search for supersymmetry, which is now in progress at the LHC, needs to be continued as intensely as possible at the LHC and beyond to cover the multi-TeV region, if need be, at future accelerators and linear colliders. That for proton decay, as noted in the previous section, needs megaton-size deep-underground detectors, like HyperKamiokande and LBNF-DUNE and their successors (if need be), not only to search for this process as sensitively as possible, but also to study the branching ratios of different decay modes, should proton decay be discovered. As discussed in the previous section, the prospects for discovery of proton decay with improvements of current SuperKamiokande limits by factors of 5 to 10 are high.

The discovery of proton decay will no doubt constitute a landmark in the history of physics. That of supersymmetry will do the same. The discovery of these two features — supersymmetry and proton decay — will fill the two missing pieces of a pretty picture — a gauge unification of matter and of its forces.

On the theoretical side, it is but natural to dream that a deeper fundamental theory should provide a unity of all the forces of Nature including gravity, together with a good quantum theory of gravity, while providing a predictive and realistic description of Nature. That should necessarily include an understanding of the origin of the three families, and of their hierarchical masses and mixings. The string/M-theory, with its majestic beauty (in the words of Edward Witten), is undoubtedly the best candidate we now have for such a deeper fundamental theory.

Notwithstanding the limited understanding we presently have of this theory (especially in its non-perturbative aspects), because of the hanging-together of several pieces within one and the same picture as mentioned above, it stands to reason to ask: can a "preferred solution" of string/M-theory (if it could exist) lead to something like the picture as above based on an effective $G(2,2,4)$ or SO(10)-like symmetry in 4D, with the necessary ingredients to go with reality? If such a solution does emerge, it would no doubt provide a *very useful and desirable bridge* between string theory and the low-energy world described by the standard model. This is because, as noted above, such a bridge is found to work in describing a set of phenomena in a non-trivial manner, though some very fundamental issues — such as the origin of the three families and an understanding of the cosmological constant — are outside of its reach. May be a string-derived grand-unified bridge could fill these gaps. Could some clear glimpses of such an utopian picture, based on developments in experiments and theory, with both proton decay and supersymmetry in hand, and with a better understanding of string theory describing reality in our possession, emerge by the 120th birthday of Abdus Salam, or even by the 150th?

Acknowledgement

I would like to thank especially Kaladi Babu, Alon Faraggi, Rabindra Mohapatra, Zurab Tavartkiladze and Frank Wilczek for collaborative discussions over the years which have shaped the content of this article. I have greatly benefitted from discussions with Pasquale Di Bari, Pran Nath, Stuart Raby, Goran Senjanovic, Qaisar Shafi and Edward Witten on various aspects of the physics in the paper. Communications from Edward Kearn and M. Shiozawa on the latest SuperKamiokande results on searches for proton decay have been helpful.

I am grateful to late Abdus Salam for a most fruitful collaboration and the joy of a warm friendship between us that lasted till he parted from this world. With gratitude and respect I dedicate this talk to his memory.

I am thankful to Lars Brink, Mike Duff and Kok Khoo Phua for their hospitality, and to Chee-Hok Lim and the members of the Stallion Press, Chennai, India for their help in the processing of this paper. The research is supported in part by a Department of Energy Grant: Contract No. DE-AC02-76SF00515.

References

1. Source: "Ideals and Realities" — Selected Essays by Abdus Salam, Ed. by C.H. Lai, Publ. by World Scientific.
2. J. C. Pati in Salam Festschrift, Ed. by A. Ali, J. Ellis and S. Randjbar-Daemi, Publ. by World Scientific, pp. 368–391, 1993.
3. J. C. Pati, "*Recollections of Abdus Salam: Scientist and Humanitarian*" — Physics Today (abridged version), August (1997); News Letter of the American Chapter of the Indian Physics Association (1997).

4. J. C. Pati, "*With Neutrino Masses Revealed, Proton Decay is the Missing Link*", hep-ph/9811442; Proc. Salam Memorial Meeting (1998), World Scientific; *Int'l Journal of Modern Physics A*, vol. 14, 2949 (1999); J. C. Pati.
5. J. C. Pati and Abdus Salam, Proc. 15th High Energy Conference, Batavia, reported by J. D. Bjorken, Vol. 2, p. 301 (1972).
6. J. C. Pati and Abdus Salam, *Phys. Rev. D* **8**, 1240 (1973).
7. J. C. Pati and Abdus Salam, *Phys. Rev. Lett.* **31**, 661 (1973).
8. J. C. Pati and Abdus Salam, *Phys. Rev. D* **10**, 275 (1974).
9. H. Georgi and S. L. Glashow, *Phys. Rev. Lett.* **32**, 438 (1974).
10. H. Georgi, in Particles and Fields, ed. by C. Carlson (AIP, NY, 1975), p. 575; H. Fritzsch and P. Minkowski, *Ann. Phys.* **93**, 193 (1975).
11. The SuperKamiokande Collaboration: "Evidence For Oscillation of Atmospheric Neutrinos", ICRR-Report-422-98-18 (July, 1998); Y. Fukuda *et al.* (Super-Kamiokande), *Phys. Rev. Lett.* **81**, 1562 (1998), hep-ex/9807003; K. Nishikawa (K2K) Talk at Neutrino 2002, Munich, Germany.
12. Q. R. Ahmad *et al.* (SNO), *Phys. Rev. Lett.* **81**, 011301 (2002); B. T. Cleveland *et al.* (Homestake), *Astrophys. J.* **496**, 505 (1998); W. Hampel *et al.* (GALLEX), *Phys. Lett. B* **447**, 127, (1999); J. N. Abdurashitov *et al.* (SAGE) (2000), astro-ph/0204245; M. Altmann *et al.* (GNO), *Phys. Lett. B* **490**, 16 (2000); S. Fukuda *et al.* (SuperKamiokande), *Phys. Lett. B* **539**, 179 (2002). Disappearance of $\bar{\nu}_e$'s produced in earth-based reactors is established by the KamLAND data: K. Eguchi *et al.*, hep-ex/0212021.
13. P. Minkowski, *Phys. Lett. B* **67**, 421 (1977); M. Gell-Mann, P. Ramond and R. Slansky, in: *Supergravity*, eds. F. van Nieuwenhuizen and D. Freedman (Amsterdam, North Holland, 1979) p. 315; T. Yanagida, in: *Workshop on the Unified Theory and Baryon Number in the Universe*, eds. O. Sawada and A. Sugamoto (KEK, Tsukuba) 95 (1979); S. L. Glashow, in *Quarks and Leptons*, Cargése 1979, eds. M. Levy *et al.* (Plenum 1980) p. 707, R. N. Mohapatra and G. Senjanovic, *Phys. Rev. Lett.* **44**, 912 (1980).
14. M. Fukugita and T. Yanagida, *Phys. Lett. B* **174**, 45 (1986); M. A. Luty, *Phys. Rev. D* **45**, 455 (1992); W. Buchmuller and M. Plumacher, hep-ph/9608308.
15. V. Kuzmin, Va. Rubakov and M. Shaposhnikov, *Phys. Lett.* **BM155**, 36 (1985).
16. F. Gursey, P. Ramond and R. Slansky, *Phys. Lett. B* **60**, 177 (1976); Y. Achiman and B. Stech, *Phys. Lett. B* **77**, 389 (1978); Q. Shafi, *Phys. Lett. B* **79**, 301 (1978); A. deRujula, H. Georgi and S. L. Glashow, 5th Workshop on Grand Unification, edited by K. Kang *et al.*, World Scientific, 1984, p. 88.
17. H. Georgi, H. Quinn and S. Weinberg, *Phys. Rev. Lett.* **33**, 451 (1974).
18. S. Dimopoulos, S. Raby and F. Wilczek, *Phys. Rev. D* **24**, 1681 (1981); W. Marciano and G. Senjanovic, *Phys. Rev. D* **25**, 3092 (1982) and M. Einhorn and D. R. T. Jones, *Nucl. Phys. B* **196**, 475 (1982). For work in recent years, see P. Langacker and M. Luo, *Phys. Rev. D* **44**, 817 (1991); U. Amaldi, W. de Boer and H. Furtenau, *Phys. Rev. Lett. B* **260**, 131 (1991); F. Anselmo, L. Cifarelli, A. Peterman and A. Zichichi, *Nuov. Cim. A* **104** 1817 (1991).
19. Y. A. Golfand and E. S. Likhtman, *JETP Lett.* **13**, 323 (1971). J. Wess and B. Zumino, *Nucl. Phys. B* **70**, 139 (1974); D. Volkov and V. P. Akulov, *JETP Lett.* **16**, 438 (1972).
20. Independently of gauge coupling unification, low-energy supersymmetry with super-partners having masses at the TeV-scale is motivated as being perhaps the best idea known to naturally protect the gauge-hierarchy of an input ratio of $(m_{\rm W}/{\rm M_U}) \sim 10^{-14}$ in the presence of all radiative corrections: E. Witten, *Nucl. Phys. B* **185**, 513 (1981); R. K. Kaul, *Phys. Lett. B* **109**, 19 (1982). M. Dine, W. Fiscler and

M. Srednicki, *Nucl. Phys. B* **189**, 575 (1981); S. Dimopoulos and S. Raby, *ibid. B* **192**, 353 (1981).

21. For a concise overview on GUT, see e.g. S. Raby in Particle Data Group Review (2016); earlier reviews include: P. Langacker, *Phys. Rep.* **72**, 16 (1981); P. Nath and Pavel Fileviez Perez, *Phys. Rep.* **441**, 191 (2007). For a current review of theory and experiments, see Ref. 65.
22. M. Green and J. H. Schwarz, *Phys. Lett. B* **149**, 117 (1984); D. J. Gross, J. A. Harvey, E. Martinec and R. Rohm, *Phys. Rev. Lett.* **54**, 502 (1985); P. Candelas, G.T. Horowitz, A. Strominger and E. Witten, *Nucl. Phys. B* **258**, 46 (1985). For introductions and reviews, see: M. B. Green, J. H. Schwarz and E. Witten, "Superstring Theory" Vols. 1 and 2 (Cambridge University Press); J. Polchinski, "String Theory", Vols 1 and 2 (Cambridge University Press).
23. K. S. Babu, J. C. Pati and F. Wilczek, "*Suggested New Modes in Supersymmetric Proton Decay*", *Phys. Lett. B* **423**, 337 (1998).
24. K. S. Babu, J. C. Pati and F. Wilczek, "*Fermion Masses, Neutrino Oscillations and Proton Decay in the Light of the SuperKamiokande*" hep-ph/981538V3; *Nucl. Phys. B* (to appear).
25. K. S. Babu, J. C. Pati and Z. Tavartkiladze, *JHEP* **1006**, 084 (2010) [arXiv: 1003.2625 [hep-ph]].
26. Update of Ref. 25 in light of LHC: K. S. Babu, J. C. Pati and Z. Tavartkiladze, To appear.
27. S. L. Glashow, *Nucl. Phys.* **22** 57a (1961); S. Weinberg, *Phys. Rev. Lett.* **19**, 1269 (1967); Abdus Salam, in Elementary Particle Theory, Nobel Symposium, ed. by N. Svartholm (Almqvist, Stockholm, 1968), p. 367.
28. P. W. Higgs, *Phys. Lett.* **12**, 132 (1964); **13**, 508 (1964); F. Englert and R. Brout, *Phys. Rev. Lett.* **13**, 321 (1964); G. S. Guralnik, C. R. Hagen and T. W. B. Kibble, *Phys. Rev. Lett.* **13**, 585 (1965); T. W. B. Kibble, *Phys. Rev.* **155**, 1554 (1967).
29. G. 't Hooft, *Nucl. Phys. B* **35**, 167 (1971); G. 't Hooft and M. Veltman, *Nucl. Phys. B* **44**, 189 (1972).
30. J. J. Sakurai *Ann. Phys.* **1**, 11 (1960).
31. C. N. Yang and R. L. Mills, *Phys. Rev.* **96**, 191 (1954).
32. For attempts to derive the current algebra framework using the singlet U(1) vector gluon see e.g. H. Fritzsch and M. Gell-Mann, Proc. 15th High Energy Conf., Batavia (1972), pages 135–165, and reference there in. The difficulty of SU(9) or SU(12) global symmetry associated with the U(1) vector gluon coupling was one reason for abandoning this possibility: See J.C. Pati and Abdus Salam, ICTP preprint IC/73/81 (unpublished), and L.B Okun and V.I Zacharov, *Phys. Lett. B* **47**, 258 (1973).
33. O. W. Greenberg, *Phys. Rev. Lett.* **13**, 598 (1964).
34. M. Han and Y. Nambu, *Phys. Rev.* **139**, B1006 (1965).
35. H. Georgi and S.L. Glashow, *Phys. Lett.* **28**, 1494 (1972).
36. J. C. Pati and A. Salam (Ref. 5 and 6); H. Fritzsch and M. Gell-Mann, (Ref. 32); H. Fritzsch, M. Gell-Mann, and H. Leutwyler, *Phys. Lett. B* **47**, 365 (1973).
37. To clarify the background of this suggestion, as noted in the text, the idea of generating a "superstrong" force by gauging SU(3)-color symmetry was first initiated by Han and Nambu in Ref. 34. In their work, however, SU(3)-color was broken explicitly, rather than spontaneously, owing to the fundamental coupling of the photon to integer-charge quarks (icq); this spoiled the renormalizability of the theory. Furthermore, they introduced an additional fundamental strong interaction by utilizing the SU(3)-flavor gauge symmetry. Such a possibility is, however, exclude[d] if one uses

part of the flavor symmetry to generate electroweak interactions, as in $SU(2)_L \times U(1)_Y$. These two shortcomings were removed and the advantages of the combined gauge structure $SU(2)_L \times U(1)_Y \times SU(3)^c$ were noted by Salam and me starting with a contribution to the 1972 Batavia conference,[5] followed by a paper in Ref. 6. See comments on the question of quark-charges in Ref. 38.

38. At the time of our suggestion of the combined gauge symmetry at the 1972 Batavia Conference (Ref. 5), Fritzsch and Gell-Mann also noted the possibility of $SU(3)^c$ gauging at the same conference.[36] They, and many others, favored the alternative of fractionally charged quarks (fcq) and unbroken SU(3)-color, with permanent confinement of quarks and gluons. With little understanding of confinement in those days, we thought that the case of fcq and permanent confinement should be regarded only as an alternative to the case of spontaneously broken $SU(3)^c$, which would lead to gauge integer charge quarks (icq), with possibly semi-confined but ultimately liberated quarks. Despite the presence of fundamental scalars, that are needed to break $SU(3)^c$ spontaneously in the case of icq, the loss of asymptotic freedom and thus departure from scaling are rather mild at the energy scales ($<1-2$ GeV) probed at that time. The two alternatives arise within the same theory depending upon the nature of SSB. Salam and I, (*Phys. Rev. Lett.* **36**, 11 (1976)), as well as others, most notably G. Rajasekaran and P. Roy (Pramana **6**, 303 (1975)), therefore proposed key experiments which could distinguish between the two alternatives. Eventually, by the 1980's on the basis of these tests, especially the two-photon experiments (see e.g. R. Godbole, J. Pati, S. Rindani, T. Jayaramam and G. Rajasekaran, *Phys. Lett.* **142**, 91 (1984)), fcq was clearly favored over gauge icq. The basic ideas of higher unification, including the advantages of the symmetry-structure $G(2,2,4)$ listed in the text, as also the success of the (combined) standard model symmetry, are of course independent of the nature of quark-charges.
39. By the time SU(5) was proposed, about a year later,[9] the situation had changed. Asymptotic freedom had been discovered,[40] as also the weak neutral current interaction,[41] both of which gave strong support to the standard model.
40. D. Gross and F. Wilczek, *Phys. Lett.* **30**, 1343 (1973); D. Politzer, *ibid*, **30**, 1346 (1973).
41. F. J. Hasert *et al.*, *Phys. Lett. B* **46**, 121, 138 (1973); P. Muset *et al.*, *J. Phys. (Paris)* **11/12**, T34 (1973).
42. J. C. Pati and Abdus Salam, Refs. 5 and 6; C. Bouchiat, J. Iliopoulos and P. Meyer, *Phys. Lett. B* **38**, 519 (1972); S. Weinberg, in *Fundamental Interactions in Physics and Astrophysics*, eds. G. Iverson *et al.* (Plenum Press, New York, 1973), p. 157.
43. S. Glashow, J. Iliopoulos and L. Maiani, *Phys. Rev. D* **2**, 1285 (1970).
44. N. Cabibbo, *Phys. Rev. Lett.* **10**, 531 (1963).
45. M. Kobayashi and K. Masakawa, *Prog. Theor. Phys.* **49**, 282 (1972).
46. S. Weinberg, *Phys. Rev. Lett.* **31**, 494 (1973); R. N. Mohaptra, J. C. Pati and P. Vinciarelli, *Phys. Rev. D* **8**, 3652 (1973).
47. J. D. Bjorken, *Phys. Rev. Lett.* **16**, 408 (1966); J. D. Bjorken, *Phys. Rev.* **179**, 1547 (1969).
48. R. P. Feynman, *Phys. Rev. Lett.* **23**, 1415 (1969); J. D. Bjorken and E. A. Paschos, *Phys. Rev.* **185**, 1975 (1969).
49. S. Weinberg, *Phys. Rev. Lett.* **31**, 494 (1973); D. J. Gross and F. Wilczek, *Phys. Rev. D* **8**, 3633 (1973); H. Fritzsch, M. Gell-Mann and H. Leutwyler, *Phys. Lett. B* **47**, 365 (1973). The unconventional idea that quarks are somehow confined was first conceived of by M. Gell-Mann from the very early days, prior to the ideas of

color and QCD (private communications; this may be inferred from his statement on "mathematical" quarks).
50. See e.g. M. Creutz, *Phys. Rev. Lett.* **45**, 313 (1980); *Phys. Rev. D* **21**, 2308 (1980); M. Lüscher and P. Weisz, *JHEP* 0207, 049 (2002); for a review, see e.g. T. DeGrand, *Int. J. Mod. Phys. A* **19**, 1337 (2004).
51. C. N. Yang, *"Necessary Subtlety and Unnecessary Subtlety"*, Proceedings of the Int'l Conference on "Neutrinos and Implications for Physics Beyond the Standard Model", Ed. by R. Shrock, Publ. By World Scientific, pp5 (October 11, 2002). In this talk, Yang states: "Personally I tend to disbelieve unnecessary subtleties. That is perhaps why Dyson called me conservative. I did not believe in neutrino oscillations, even after Davis' painstaking work and Bahcall's careful analysis: the oscillations were, I believed, uncalled for. Now, after the beautiful experiments which we shall hear about in the next few days, I have to surrender and accept neutrino oscillations, as reality, but still as unnecessary subtleties". Relative to Yang's remark it seems appropriate to mention that neutrino oscillations (and thus neutrino masses) now appear to be a very *necessary subtlety* of nature in the sense that, (a) as discussed in Sec. 6, the tiny neutrino masses seem to be the nature's way to provide a clue to physics at the unification scale including the nature of the unification symmetry, and (b) they also seem to be at the root of the origin of matter-antimatter asymmetry, and thus at the root of our own origin.
52. Initially, owing to his fondness for the two-component theory of the neutrino, of which he was a co-father,[53] even Salam, despite his liking for SU(4)-color, resisted against the introduction of the RH neutrino. Happily, his reservations in this regard softened over a few months. Sadly he did not live to see, however, the empirical support of these ideas that came through the discovery of the atmospheric neutrino oscillation in 1998.
53. A. Salam, Nuovo cim. **5**, 299 (1957); L. D. Landau, *Nucl. Phys.* **3**, 127 (1957); T. D. Lee and C. N. Yang, Phys. Rev. **105**, 1671 (1957).
54. S. Weinberg, *Phys. Rev. Lett.* **43**, 1566 (1979); E. Akhmedov, Z. Berezhiani and G. Senjanovic, *Phys. Rev. D* **D 47**, 3245 (1993).
55. J. C. Pati, *"Implications of the SuperKamiokande Result on the Nature of New Physics"*, Proceedings of the Neutrino-98 Conference, held at Takayama, Japan (June 4–9, 1998), *Published in Nucl. Phys. Proc. Suppl.* **77**, 299 (1999).
56. J. C. Pati and Abdus Salam, *Phys. Rev. D* **10**, 275 (1974); R. N. Mohapatra and J. C. Pati, *Phys. Rev. D* **11**, 566 and 2558 (1975); G. Senjanovic and R. N. Mohapatra, *Phys. Rev. D* **12**, 1502 (1975).
57. For an incomplete list of references on consideration of TeV-scale W_R, see e.g., Y. Zhang, H. An, X. Ji and R. N. Mohapatra, *Nucl. Phys. B* **802**, 247 (2008); A. Maiezza, M. Nemevsek, F. Nesti and G. Senjavnovic, *Phys. Rev. D* **82**, 055022 (2010); C. H. Lee, P. S. Bhupal DeV and R. N. Mohapatra, *Phys. Rev. D* **88**, no. 9, 093010 (2013); M. Lindner, F. S. Queiroz, W. Rodejohann and C. E. Yaguna, *JHEP* **1606**, 140 (2016); D. Borah, S. Patra and U. Sarkar, *Phys. Rev. D* **83**, 035007 (2011); J. Chakrabortty, H.Z. Devi, S. Goswami and S. Patra, *JHEP* **1208**, 008 (2012); R. L. Awasthi, M. K. Parida and S. Patra, *JHEP* **1308**, 122 (2013); J. Brehmer, J. Hewett, J. Kopp, T. Rizzo and J. Tattersal, *JHEP* **1510**, 182 (2015); B. A. Dobrescu and Z. Liu, *Phys. Rev. Lett.* **115**, no. 21, 211802 (2015).
58. A. Davidson, *Phys. Rev. D* **20**, 776 (1979); R. E. Marshack and R. N. Mohapatra, *Phys. Lett. B* **91**, 222 (1980).
59. J. C. Pati, A. Salam and J. A. Strathdee, *Nuovo Cim. A* **26**, 72 (1975); *Phys. Lett. B* **108**, 121 (1982). The maximal local symmetry SU(16) of sixteen two-component

fermions in a family was considered as a possible parent symmetry for grand unification in the early days, despite its being much bigger than its subgroups of SU(5) and SO(10), because it treats B and L as local rather than global symmetries and thereby provides a fundamental reason for their origins. These symmetries are thus exact in the lagrangian, but must be violated spontaneously for reasons explained in footnote m. It turns out that such spontaneous breaking mixes gauge bosons of SU(16) carrying different (B,L) quantum numbers and thereby induces proton decays such as p→$e^+\pi^\circ$ analogous to the cases of SU(5) and SO(10) where the violations are explicit. Symmetries like SU(16) do need, however, three mirror families (which are chiral and non-trivial under electroweak symmetry) to cancel triangle anomalies. Unlike in the 1970's, these mirror families are now constrained, however, by precision electroweak tests [60, 61].

60. M. Peskin and T. Takeuchi, *Phys. Rev. Lett.* **65**, 964 (1990); *Phys. Rev. D* **46**, 381 (1991).
61. G. Altarelli, R. Barbieri and S. Jadach, *Nucl. Phys. B* **369**, 3 (1992); G. Altarelli, R. Barbieri and F. Caravaglios, *Nucl. Phys. B* **405**, 3 (1993).
62. W. H. Furry, *Phys. Rev.* **56**, no. 12, p. 1184 (1939).
63. For a recent review on theory and experiments on neutrinoless double beta-decay, see e.g. S. Dell'Oro *et al.*, Advances in High Energy Physics Vol. 2016 (2016), ID 2162659, and references there in.
64. V. Kuzmin, *JETP Lett.* **12**, 228 (1970); S. L. Glashow, *Proc. Neutrino* **79** *Conf., Berg*, **2**, 518 (1979); R. N. Mohapatra and R. E. Marshak, *Phys. Rev. Lett.* **44**, 1316 (1980). For an overview of theory and phenomenology of $n - \overline{n}$ oscillation, see R.N. Mohapatra, *J. Phys. G* **36**, 104006 (2009).
65. For a concise review on "Baryon Number Violation" covering theory and experiments on both proton decay and $n - \bar{n}$ oscillation in the context of unified theories see the Snowmass Report (2013); arXiv:1311.5285 (hep-ph), Conveners: K.S. Babu and E. Kearn.
66. For lower limit on $\Gamma^{-1}(p \to e^+\pi^\circ)$: K. Abe *et al.* (SuperKamiokande Collaboration), *Phys. Rev. D* **95**, no. 1, 012004 (2017).
67. For lower limit on $\Gamma^{-1}(p \to \overline{\nu}K^+)$: K. Abe *et al.* (SuperKamiokande Collaboration), *Phys. Rev. D* **90**, no. 7, 072005 (2014); Talk presented on behalf of SuperKamiokande collaboration by M. Shiozawa at 34th Intl. Cosmic Ray Conference (ICRC 2016).
68. D. Chang, R. N. Mohapatra, and M. K. Parida. *Phys. Rev. Lett.*, **52**, 1072 (1984).
69. F. Gursey, P. Ramond and P. Sikivie, *Phys. Lett. B* **60**, 177 (1976).
70. For reviews, see e.g. P. Langacker, talk at Gatlinburg Conference, June 94, hep-ph 9411247, and P. Langacker and N. Polonsky, *Phys. Rev. D* **52**, 3081 (1995), and references there in.
71. LEP and SLD Average: On the weak angle: G. Gabbiani *et al.*, *Phys. Rep.* **427**, 257 (2006). For LEP measurement on $\alpha_3(m_z)$, see W. M. Yao *et al.* (Particle Data Group, *J. Phys. G* **33**, 1 (2006)).
72. D. Chang, R. N. Mohapatra, J. Gipson, R. E. Marshak and M. K. Parida, *Phys. Rev. D* **31**, 1718 (1985); R. N. Mohapatra and M. K. Parida, *Phys. Rev. D* **47**, 264 (1993) [hep-ph/9204234]; N. G. Deshpande, E. Keith and P. B. Pal, *Phys. Rev. D* **46**, 2261 (1993); S. Bertolini, L. Di Luzio and M. Malinsky, *Phys. Rev. D* **85**, 095014 (2012) [arXiv:1202.0807 [hep-ph]].
73. D. G. Lee, R. N. Mohapatra, M. K. Parida and M. Rani, *Phys. Rev. D* **51**, 229 (1995) [arXiv:hep-ph/9404238].
74. G. Altarelli and D. Meloni, *JHEP* **1308**, 021 (2013) [arXiv:1305.1001 [hep-ph]].

that the H_d of MSSM is a mixture of $\mathbf{10}_H^d$ and $\mathbf{16}_H^d$. This turns out to be the origin of non-trivial CKM mixings (See Ref. 24).

101. The flavor charge(s) of $\mathbf{45}_H(\mathbf{16}_H)$ would get determined depending upon whether $p(q)$ is one or zero (see below).

102. The basic presumption here is that effective dimensionless couplings allowed by $SO(10)/G(224)$ and flavor symmetries are of order unity [i.e., $(h_{ij}, g_{ij}, a_{ij}) \approx 1/3\text{-}3$ (say)]. The need for appropriate powers of (S/M) with $\langle S\rangle/M \sim M_{\rm GUT}/M_{\rm string} \sim (1/10 - 1/20)$ in the different couplings leads to a hierarchical structure. As an example, consider just one U(1)-flavor symmetry with one singlet S. The hierarchical form of the Yukawa couplings exibited in Eqs. (7) and (8) would be allowed, for the case of $p = 1$, $q = 0$, if ($\mathbf{16}_3$, $\mathbf{16}_2$, $\mathbf{16}_1$, $\mathbf{10}_H$, $\mathbf{16}_H$, $\mathbf{45}_H$ and S) are assigned U(1)-charges of (a, $a+1$, $a+2$, $-2a$, $-a - 1/2$, 0, -1). It is assumed that other fields are present that would make the U(1) symmetry anomaly-free. The value of "a" would get fixed by the presence of other operators (see later).

103. These effective non-renormalizable couplings can of course arise through exchange of (for example) **45** in the string tower, involving renormalizable $\mathbf{16}_i\overline{\mathbf{16}}_H\mathbf{45}$ couplings. In this case, one would expect $M \sim M_{\rm string}$.

104. K. S. Babu, J. C. Pati and P. Rastogi, "Tying in CP and flavor violations with fermion masses and neutrino oscillations", hep-ph/04010200, *Phys. Rev. D* **71**, 015005 (2005).

105. The expressions $V_{\rm cb}$ and θ_{23}^{ν} are given in Eq. (12) in terms of the simple square-root-formulas only to exhibit the qualitative reason based on symmetry (explained in the text) for the enhancement of θ_{23}^{ν} and suppression of $V_{\rm cb}$. The simple formulas do not really hold, however, when each square-root-term is rather large, as in θ_{23}^{ν}. The values quoted in Eq. (12) are based on exact numerical calculations.[24]

106. It has been noted[93] that small intrinsic (non-seesaw) masses $\sim$(few) $\times\, 10^{-3}$ eV of the LH neutrinos can arise quite plausibly through higher dimensional operators of the form: $W_{12} \supset \kappa_{12}\mathbf{16}_1\mathbf{16}_2\mathbf{16}_H\mathbf{16}_H\mathbf{10}_H\mathbf{10}_H/M_{\rm eff}^3$, without involving the standard seesaw mechanism, and in accord with the U(1) flavor symmetry if a $= 1/2$ (see Ref. 102). One can verify that such a term would lead to an intrinsic Majorana mixing mass term of the form $m_{12}^0 \nu_L^e \nu_L^\mu$, with a strength given by $m_{12}^0 \approx \kappa_{12}\langle\mathbf{16}_H\rangle^2(175\ \text{GeV})^2/M_{\rm eff}^3 \sim (1.5\text{-}6) \times 10^{-3}$ eV, for $\langle\mathbf{16}_H\rangle \approx (1\text{-}2)M_{\rm GUT}$ and $\kappa_{12} \sim 1$, if $M_{\rm eff} \sim M_{\rm GUT} \approx 2 \times 10^{16}$ GeV.[93] Such an intrinsic Majorana $\nu_e\nu_\mu$ mixing mass $\sim$ few$\times 10^{-3}$ eV, though small compared to $m(\nu_3)$, is still much larger than what one would generically get for the corresponding term from the standard seesaw mechanism. Now, the diagonal $(\nu_L^\mu \nu_L^\mu)$ mass-term, arising from the standard seesaw mechanism is expected to be $\sim (3 - 10) \times 10^{-3}$ eV for a natural value of $|y| \approx 1/20\text{-}1/15$, say.[24] Thus, taking the net values of $m_{22}^0 \approx (8 - 10) \times 10^{-3}$ eV, $m_{12}^0 \approx 3 \times 10^{-3}$ eV as above and $m_{11}^0 \ll 10^{-3}$eV, which are all plausible, we obtain $m_{\nu_2} \approx (7 - 9) \times 10^{-3}$ eV, $m_{\nu_1} \sim$ (1 to few) $\times\, 10^{-3}$ eV, so that $\Delta m_{21}^2 \approx (6 - 8) \times 10^{-5}$ eV^2 and $\theta_{12}^{\nu} \approx (20^\circ\text{–}35^\circ)$, in accord with the data, which show $\theta_{12}^{\nu} \approx 33^\circ$. (For review and notations see Refs. 107 and 108). Likewise, a non-seesaw contribution can be used to plausibly obtain $\theta_{13}^{\nu} \approx (2^\circ - 10^\circ)$ to be compared with the observed value of $\approx 9^\circ$. (See Refs. 107 and 108).

107. Particle Data Group Review: "Neutrino Mass, Mixing and Oscillations" by K. Nakamura and S. T. Petcov, Polg.lbl.gov/2016/reviews/rpp2016-rev-neutrinomixing.pdf, Oct. 2016. Best fit values quoted here are: $\sin^2\theta_{23} = 0.437$ or 0.569.

108. For a recent review of experimental and theoretical aspects of neutrino physics, the latter being in the context of unification models possessing flavor symmetries, see S.F. King, arXiv: 1701.04413, and references therein. The global fits quoted here, for

the case of normal hierarchy, yield (for one sigma range): $\theta_{23} \approx 41.4^\circ \pm 1.6^\circ, \theta_{12} \approx 33.2^\circ \pm 1.2^\circ, \theta_{13} \approx 8.45^\circ \pm 0.15^\circ$. The best fit (for normal ordering) yields: $\Delta m^2_{31} = (2.55 \pm 0.05) \times 10^{-3}\,\text{eV}^2; \Delta m^2_{21} = (7.45 \pm 0.25) \times 10^{-5}\,\text{eV}^2$.

109. K. S. Babu J. C. Pati and P. Ratogi, hep-ph/0502152, *Phys. Lett. B.* (2005); P. Rastogi, Ph.D. Thesis, Univ. of Maryland (2006).
110. S. M. Barr, *Phys. Lett. B* **112**, 219 (1982); J. P. Derendinger, J. E. Kim and D. V. Nanopoulos, *Phys. Lett. B* **139**, 170 (1984); I. Antoniadis, J. Ellis, J. Hagelin and D. V. Nanopoulos, *Phys. Lett. B* **194**, 231 (1987).
111. For reviews on different aspects of leptogenesis, see e.g. W. Buchmuller, R. D. Pecei and T. Yanagida, *Ann Rev. Nucl. Part. Sci.* **55**, 311 (2005); S. Davidson, E. Nardi and Y. Nir, *Phys. Rept.* **466**, 105 (2008); S. Blanchet and P. Di Bari, *New J. Phys.* **14**, 125012 (2012).
112. J. C. Pati, *Phys. Rev. D* **68**, 072002 (2003).
113. W. Buchmuler, P. Di Bari and M. Plumacher, hep-ph/0401240; G. F. Giudice, A. Notart, M. Raidal, A. Riotto and A. Strumia, hep-ph/0310123.
114. WMAP Collaboration, D. N. Spergel *et al.*, Astrophys. J. Suppl **148**, 175 (2003); astro-ph/0310723, yielded: $(Y_B)_{\text{WMAP}} \approx (8.7 \pm 0.4) \times 10^{-11}$. The latest results based on measurements of CMB anisotropies by the PLANCK Collaboration (quoted in the Particle Data Group Review (2016)) lead to: $Y_B \approx (8.65 \pm 0.10) \times 10^{-11}$.
115. S. Weinberg, *Phys. Rev. Lett.* **48**, 1303 (1982).
116. W. Buchmuler, *et al.*, and G. F. Giudice *et al.* (Ref. 113).
117. M. Kawasaki, K. Kohri, T. Moroi and A. Yotsuyanagi, *Phys. Rev. D* **78**, 065011, 2008; M. Kawasaki, K. Kohri and T. Moroi, astro-ph/0402990.
118. See R. Jeannerot, S. Khalil, G. Lazarides and Q. Shafi, *J. High Energy Physics* 012 (2000); hep-ph/0002151 and references there in. For a review on supersymmetric hybrid inflation, see e.g. G. Lazarides, arXiv: hep-ph/0607032. For a recent work on inflationary non-thermal leptogenesis based on a supersymmetric $G(2,2,4)$ model, see C. Pallis and N. Toumbas, arXiv: hep-ph/1207.3730.
119. Incorporating such an effective super potential in accord with the assignment of flavor charges suggested in Refs. 102 and 106 (for this purpose, Φ and $\overline{\Phi}$ are quivalent to $\overline{16}_{\text{H}}$ and 16_{H} respectively) would involve two additional signlets with appropriate flavor charges. One or both of those may have a VEV $\sim$ M.
120. J. Hisano, H. Murayama and T. Yanagida, *Nucl. Phys. B* **402**, 406 (1993) [hep-ph/9707279].
121. N. Sakai and T. Yanagida, *Nucl. Phys. B* **197**, 533 (1982); S. Weinberg, *Phys. Rev. D* **26**, 287 (1982).
122. H. Murayama and A. Pierce, *Phys. Rev. D* **65**, 055009 (2002) [hep-ph/0108104].
123. B. Bajc, P. Fileviez Perez and G. Senjanovic, arXiv:hep-ph/0210374; I. Gogoladze and A. Kobakhidze, *Phys. Atom. Nucl.* **60**, 126 (1997) [*Yad. Fiz.* **60N1**, 136 (1997)] [hep-ph/9610389].
124. J. C. Pati, "*Confronting The Conventional ideas on Grand Unification With Fermion Masses, Neutrino Oscillations and Proton Decay*", hep-ph/0204240; Proceedings of the ICTP, Trieste School 2001 and Proceedings of the DAE Meeting, Allahabad, India (2002).
125. Note that in the presence of a second SO(10)-singlet field $\mathbf{S'}$ carrying flavor charge of 2a+1/2, an effective coupling of the form $\mathbf{16_H}.\overline{\mathbf{16_H}}.\mathbf{S'}$ would be allowed preserving the U(1)-flavor symmetry introduced in Sec. 4. With $\mathbf{S'}$ having a VEV of GUT-scale, such a coupling would generate a mass-term $\mathbf{16_H}.\overline{\mathbf{16_H}}$ that enters into Fig. 4. The presence of this second singlet $\mathbf{S'}$ does not in any way affect the hierarchical pattern of effective couplings exhibited in Eqs. (7) and (9). I thank Qaisar Shafi for raising this point.

First of all, we claim that this problem is a much more elementary and important one than what the literature suggests, and that most of the suggested cures are not well enough thought through. What should be done, is to stretch as much as is possible what one can do using standard, mainstream physics. Then, finally, impose the demand that black holes should behave, as much as possible, as ordinary forms of matter. When doing this properly, this leads to amazing observations.

Only few people noted that the application of existing physical knowledge can bring us much further than any wild speculation. Only recently,[c] we discovered a way to calculate things that could have been applied decades earlier. Our observation was[22] that we can use a partial wave expansion to describe energy and momentum entering the black hole, and that, applying Einstein's equations, the information carried in by each partial wave is carried out again, in a partial wave with the same values of the quantum numbers ℓ and m.

This calculation seems to be as elementary as the calculation of the spectrum of a hydrogen atom using quantum mechanics. As in the hydrogen atom, the boundary conditions are of crucial importance, but in a black hole, these boundary conditions are quite counter intuitive.

What this calculation does, is to expose our problem more clearly than ever before, and now we can see what the answers must be. The general coordinate transformation relating the "inside" of a black hole with what an outside observer sees, must be a topologically nontrivial one.[23–25] Only then do the equations make sense.

This does remarkable things with the fabric of space and time itself, not noticed before. For one thing, space and time must be discrete. This was already suspected long ago, leading people to fantasize about noncommutative space and time. However, noncommutative geometry itself does not give the right equations, one has to do this more precisely. This should now be possible.

Yet the deepest mysteries are still there. The fact that black holes emit radiation at a finite temperature seems to be a strong indication that degrees of freedom near the horizon are discrete, so that space and time themselves will be discrete. We do not know, let alone understand, what this does to local Lorentz invariance. The question asked can be summarized as:

> If space and time are discrete, how does one distribute fields on space–time? How can we reconcile this with Lorentz invariance? In short: *what is Nature's book keeping system?*[26]

It seems that, at the Planck scale, all we really need to specify is: how is the information processed? At that scale, our world seems to be nothing more than an information processing machine. Many investigators now assume that this machine will be processing *quantum information*, or qubits, by some, continuous or discrete,

---

[c] Some of the remarks reported here in the proceedings are to be considered as *notes added in proof*; these concepts were not yet completely clear when the presentation was given in the conference.

version of Schrödinger's equation. This lecturer is suspecting a bolder scenario, which is that the information at the Planck scale will be no more than just classical information.[27] This information is apparently being processed in such a complex manner that, if we go to larger scales in space and time, only effectively quantum mechanical equations can tell us how the bulk of the information is processed at much larger scales of space and time.

Superstring theory does suggest to us what some of the answers to such questions will be like: a particle cannot be completely localized to one point in space. Even the black hole horizon will be 'fuzzy'. My point here is, that such answers are not precise enough. Fuzzy answers are often confused with sloppy answers. The precise equations are much more interesting. The early developments in quantum physics constitute important lessons here. The notion that a particle can be both a particle and a wave, seems sloppy at first, but once we had the Schrödinger equation, all sloppiness disappeared. We have to work out these problems precisely.

5. The Hierarchy Problem

As stated, the value found for the Higgs mass, around 125 GeV, is remarkably special, making the inner world just barely stable against scale transformations. Exactly at this mass value, the Higgs self-interactions stay practically constant as we go to very much tinier scales. This could indicate that there will be remarkably little structure in the desert between the TeV scale and the Planck scale, see Fig. 1.

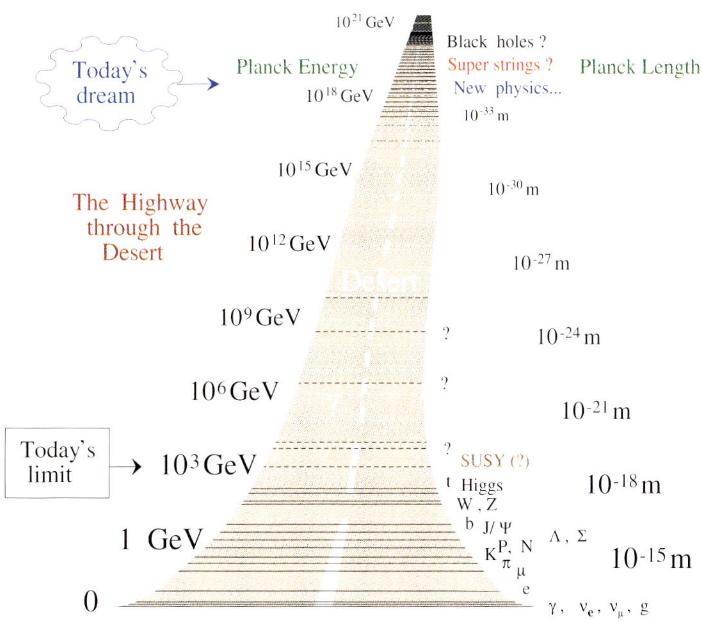

Fig. 1. The highway through the desert. Horizontal lines: known particles. Milestones: left, the energies per particle, in factors of 1000; right, the length scales, in factors 1000.

BCSPIN: 1989 and Thank You Note, 1994 School

- Bangladesh, China, SriLanka, Pakistan, India, Nepal.
- Salam present in the first school in 1989; also King of Nepal.
- BC(V)SPIN/Asian American Advanced Study Institute 2009.
- Held in Nepal, India, China, Vietnam, Mexico (2014).
- Supported by ICTP, NSF (USA), China, Mexico, Univ. of Delaware, Mitchell Foundation (Texas A&M), ...

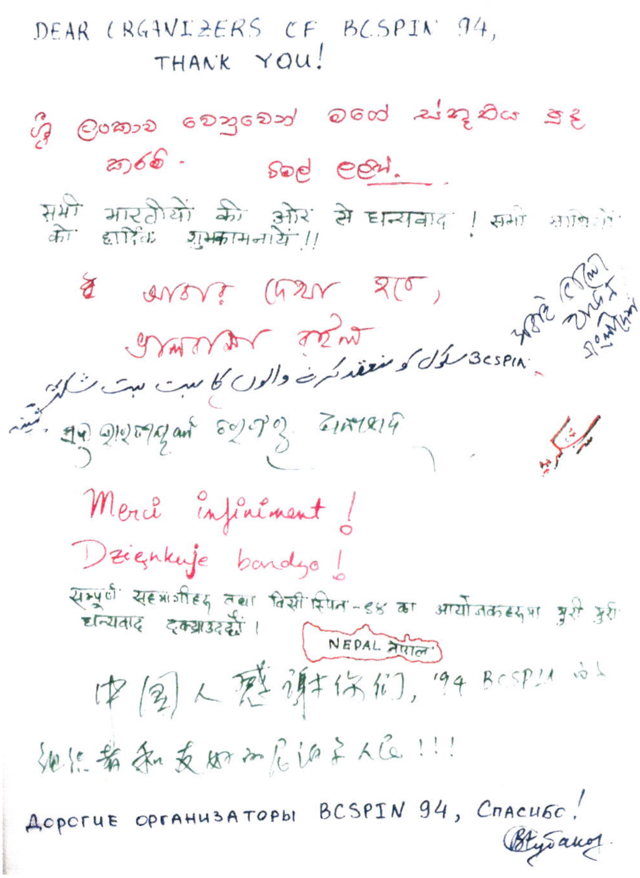
DEAR ORGANIZERS OF BCSPIN 94,
THANK YOU!

सभी भारतीयों की ओर से धन्यवाद ! सभी आयोजकों को हार्दिक शुभकामनायें !!

BCSPIN کو منعقد کرنے والوں کا بہت بہت شکریہ

Merci infiniment !

Dziękuje bardzo !

सम्पूर्ण सहभागीहरु तथा बिसी स्पिन-९४ का आयोजकहरुप्रति मुरी मुरी धन्यवाद दक्षाउदछौं !

NEPAL नेपाल

中国人感谢你们, '94 BCSPIN

Дорогие организаторы BCSPIN 94, Спасибо!

- Quark-Lepton Unification, lepton number as 4th color, electric charge quantization, neutrino mass, ... (with Jogesh Pati, JCP).
- Baryon number violation (with JCP).
- Superfields & R-symmetry (with John Strathdee).
- Kaluza–Klein Theories (JS, ...).

- Pati–Salam $\rightarrow SU(4)_c \times SU(2)_L \times SU(2)_R$: $(4,2,1) + (\overline{4},1,2)$

$$\begin{pmatrix} u & u & u & \nu_e \\ d & d & d & \ell \end{pmatrix}_{L,R} \Longrightarrow \text{16 chiral fields.}$$

  - SM neutrinos have non-zero masses.

- Georgi–Glashow $\rightarrow SU(5)$: $10 + \overline{5}$
  - 15 chiral fields.
  - Massless neutrinos.

b-τ YU in Supersymmetric $SU(4) \times SU(2)_L \times SU(2)_R$ (422)

(with Ilia Gogoladze, Rizwan Khalid, Shabbar Rizvi, Cem Un)

- $m_{16}, m_{H_i}, M_i, A_0, \tan\beta, \text{sign}(\mu)$.
- $m_{16} \equiv$ Universal soft SUSY breaking (SSB) sfermion mass.
- $m_{H_d,H_u} \equiv$ Universal SSB MSSM Higgs masses.
- $M_i \equiv$ SSB gaugino masses

$$M_1 = \frac{3}{5}M_2 + \frac{2}{5}M_3.$$

- $A_0 \equiv$ Universal SSB trilinear interaction.
- $\tan\beta = \frac{v_u}{v_d}$.
- $\mu \equiv$ SUSY bilinear Higgs parameter $\mu > 0$.

Random scans for the following parameter range (NUHM2):

$$\begin{array}{rll} 0 \leq & m_{16} & \leq 20\,\text{TeV}, \\ 0 \leq & M_2 & \leq 5\,\text{TeV}, \\ 0 \leq & M_3 & \leq 5\,\text{TeV}, \\ -3 \leq & A_0/m_{16} & \leq 3, \\ 0 \leq & m_{H_d} & \leq 20\,\text{TeV}, \\ 0 \leq & m_{H_u} & \leq 20\,\text{TeV} \\ 2 \leq & \tan\beta & \leq 60, \\ \mu > & 0, & m_t = 173.3\,\text{GeV}. \end{array}$$

	Point 1	Point 2	Point 3
m_0	1086.39	460.72	497.64
M_1	979.1	3313.58	3606.12
M_2	979.1	4579.38	4908.89
M_3	979.1	1414.89	1651.94
A_0	-3244.79	-1270.88	-1390.14
$\tan\beta$	28.49	15.41	16.47
μ	1853	176	746
m_h	124.06	124	124.1
m_H	1862	2856	3109
m_A	1850	2838	3088
$m_{H^\pm}$	1864	2857	3110
$m_{\tilde{\chi}^0_{1,2}}$	$424, 807$	$180, 182$	$759, 762$
$m_{\tilde{\chi}^0_{3,4}}$	$1845, 1847$	$1477, 3757$	$1620, 4032$
$m_{\tilde{\chi}^\pm_{1,2}}$	$810, 1850$	$188, 3754$	$780, 4023$
$m_{\tilde{g}}$	2180	3048	3515
$m_{\tilde{u}_{L,R}}$	$2239, 2174$	$3842, 2719$	$4253, 3118$
$m_{\tilde{t}_{1,2}}$	$1084, 1744$	$1039, 3394$	$1467, 3768$
$m_{\tilde{d}_{L,R}}$	$2240, 2166$	$3843, 2629$	$4254, 3025$
$m_{\tilde{b}_{1,2}}$	$1721, 1947$	$2524, 3436$	$2905, 3808$
$m_{\tilde{\nu}_1}$	1261	2980	3182
$m_{\tilde{\nu}_3}$	1098	2972	3164
$m_{\tilde{e}_{L,R}}$	$1265, 1144$	$2978, 1296$	$3181, 1407$
$m_{\tilde{\tau}_{1,2}}$	$719, 1107$	$1189, 2961$	$1276, 3156$
σ_{SI}(pb)	9.24×10^{-12}	1.79×10^{-10}	2.84×10^{-10}
σ_{SD}(pb)	2.46×10^{-09}	2.29×10^{-06}	2.36×10^{-07}
$\Omega_{CDM}h^2$	7.06	0.007	0.11
Δ_{EW}	827	15.4	134
Δ_{HS}	1110	51.3	181

	Point 1	Point 2	Point 3	Point 4	Point 5
m_{16}	12730	9839	17640	7477	11940
M_1	1172	1903	1462	1496	1700
M_2	1820	2881	2327	2335	2660
M_3	550	435.3	165	237	260
m_{H_d}, m_{H_u}	11720, 14690	5967, 7279	12890, 5640	6624, 1513	3111, 5478
$\tan\beta$	36.3	41.3	52.9	32.4	39.0
A_0/m_0	−2.07	−2.41	−2.62	−2.56	−2.63
m_t	173.3	173.3	173.3	173.3	173.3
μ	4957	9186	19086	8552	13149
$\Delta(g-2)_\mu$	0.82×10^{-11}	0.72×10^{-11}	0.28×10^{-11}	0.97×10^{-11}	0.45×10^{-11}
m_h	126.4	125.9	123.9	125	123.3
m_H	2262	2157	1799	7900	3058
m_A	2247	2144	1788	7849	3039
$m_{H^\pm}$	2264	2160	1802	7901	3061
$m_{\tilde{\chi}^0_{1,2}}$	641,1682	918, 2585	770, 2276	715, 2087	837, 2441
$m_{\tilde{\chi}^0_{3,4}}$	4973, 4974	9137, 9137	18924, 18924	8537, 8537	13101, 13101
$m_{\tilde{\chi}^\pm_{1,2}}$	1697, 4979	2604, 9133	2281, 18927	2104, 8534	2457, 13090
$m_{\tilde{g}}$	1625	1314	879	790	943
$m_{\tilde{u}_{L,R}}$	12743, 12860	9988, 9900	17708, 17538	7616, 7393	12019, 11977
$m_{\tilde{t}_{1,2}}$	689, 6131	1042, 4668	5577, 7056	781, 4077	901, 5263
$m_{\tilde{d}_{L,R}}$	12743, 12715	9988, 9853	17708, 17721	7617, 7525	12019, 11933
$m_{\tilde{b}_{1,2}}$	6234, 8566	4706, 5997	6884, 7646	4125, 5259	5293, 7047
$m_{\tilde{\nu}_1}$	12859	10035	17634	7562	12091
$m_{\tilde{\nu}_3}$	11262	8267	12950	6496	10076
$m_{\tilde{e}_{L,R}}$	12846, 12581	10027, 9814	17630, 17854	7554, 7623	12081, 11906
$m_{\tilde{\tau}_{1,2}}$	9129, 11263	5711, 8239	5525, 12875	5399, 6519	7366, 10045
σ_{SI}(pb)	0.71×10^{-13}	0.16×10^{-13}	0.70×10^{-14}	0.62×10^{-14}	0.27×10^{-13}
σ_{SD}(pb)	0.18×10^{-9}	0.19×10^{-11}	0.14×10^{-14}	0.41×10^{-12}	0.59×10^{-16}
$\Omega_{CDM}h^2$	0.13	0.86	0.45	0.09	0.123
R	1.06	1.18	1.04	1.19	1.09

preserves the number of supersymmetries. Further dimensional reduction leads to the unique maximal supergravity theories on nine and less dimensions. By a maximal supergravity we mean a theory that is invariant under a supersymmetry that has 32 component parameters. The dimensional reduction of the IIB supergravity theory leads to the same nine-dimensional theory; this must be the case as the nine-dimensional maximal supergravity theory is unique. Thus there is a mapping between the IIA and IIB supergravity theories on a circle. These relations inherit into corresponding superstring theories. The strong coupling limit of the IIA string theory can be thought of as defining an 11-dimensional theory whose low energy limit is the 11-dimensional supergravity theory.[21] The relations between the IIA and IIB superstring theories on circles is an example T duality transformation.

These ideas have come to be known as *M* theory but, as is rather clear, this is not a theory rather it is a set of relations between the different theories.

The problem with including branes as elements in the underlying theory is that unlike the string there is very little known about how to quantise branes. While there has been progress on formulating an action for multiple coincident M2 branes and the use of open strings has allowed us to understand some properties of D branes one does not even know the quantum states of a single brane and the problem of scattering amplitudes for branes is still very far from being solved. Thus progress in string theory has lead us into some kind of no mans land in that we realise that we understand very little about what the underlying theory could be.

In this review we will explain that all the maximal supergravity theories, that is, the low energy effects of the superstring theories, can be unified in a single theory that contains a very large new symmetry, the Kac–Moody algebra E_{11}. This theory is a nonlinear realisation of the semi-direct product of E_{11} with its vector representation denoted $E_{11} \otimes_s l_1$. This field theory is similar to that used to formulate the dynamics of pions mentioned above, but it differs in that it automatically contains a space–time as part of the group structure. This is a first step that one can hope may be used to determine some of the properties of the underlying theory of string and branes.

We begin by giving a review of the theory of nonlinear realisations in Sec. 3. Section 4 contains a short account of Kac–Moody algebras followed by Sec. 5 gives some of the historical motivation for the idea that the underlying theory of strings and branes has an E_{11} symmetry. Section 6 constructs the Kac–Moody algebra E_{11} and its representation which of most interest to us, the l_1, or vector, representation. Section 7 contains the construction of the nonlinear realisation of $E_{11} \otimes_s l_1$ nonlinear realisation and the 11-dimensional dynamics it predicts. Section 8 explains how the theories in D dimensions emerge from this nonlinear realisation and Sec. 9 shows that the nonlinear realisation of $E_{11} \otimes_s l_1$ is a unifying theory in that it contains the many different type II maximally supersymmetric theories. Section 10 is a discussion of the meaning of the results. This review is an expanded version of the lecture given in Singapore.

3. Nonlinear Realisations

As mentioned in the introduction the theory of nonlinear realisations was once well known but this knowledge has largely been lost, and worse still, been replaced by misunderstandings. As a result in this section we will review the theory of nonlinear realisation. The data required to specify a nonlinear realisations is a group G with a choice of subgroup H. The nonlinear realisation of a group G with local subgroup H is, by definition, constructed out of a group element $g \in G$ which is subject to the transformations

$$g \to g_0 g, \quad g_0 \in G, \quad \text{as well as} \quad g \to gh, \quad h \in H. \tag{3.1}$$

The group element $g_0 \in G$ is a rigid transformation, that is, it is a constant, while $h \in H$ is a local transformation, that is, like g it depends on the space–time that the theory possess. The space–time may be introduced by hand, as was the case for the original use of the nonlinear realisation used in particle physics, or it may be introduced as part of the construction by including corresponding generators that belong to the group G. This latter case is the one of interest to us in this paper. Clearly we can use the local transformation h to gauge away part of the group element g.

We now take the group G to have a particular form, that is, it is the semi-direct product of a group $\hat{G}$ with one of its representations l; we denote this by $\hat{G} \otimes_s l$. We denote the generators of $\hat{G}$ by R^α and for each element of the l representation we introduce a corresponding generators l_A. Then the algebra $\hat{G} \otimes_s l$ can be written as

$$\left[R^\alpha, R^\beta\right] = f^{\alpha\beta}{}_\gamma R^\gamma, \tag{3.2}$$

$$\left[R^\alpha, l_A\right] = -(D^\alpha)_A{}^B l_B. \tag{3.3}$$

The first equation is just the Lie algebra for $\hat{G}$ and in the second equation the matrix $(D^\alpha)_A{}^B$ is the matrix representation of the l representation. One can verify that it satisfies the Jacobi identity by virtue of this fact. The commutators of the l generators are restricted by the Jacobi identity. The simplest consistent choice is to take them to commute but we will leave them unspecified for the time being.

The reader is very familiar with the notion of the semi-direct product as the Poincaré group P in D dimensions can be written as $P = SO(1, D-1) \otimes_s T^D$ where T^D are the translations generators corresponding to the vector representation of $SO(1, D-1)$. If we denote the space–time translations by P_a and the Lorentz rotations by J_{ab}, with $a, b, \ldots = 0, 1, 2, \ldots, D-1$ then the Lorentz algebra is given by

$$[J_{ab}, J_{cd}] = \eta_{bc} J_{ad} - \eta_{ac} J_{bd} - \eta_{bd} J_{ac} + \eta_{ad} J_{bc} \tag{3.4}$$

while Eq. (3.3) is written, for this case, as

$$[J_{ab}, P_c] = -\eta_{ac} P_b + \eta_{bc} P_a. \tag{3.5}$$

The group element g of $\hat{G} \otimes_s l$ can be written in the form

$$g = e^{x^A l_A} e^{A_\alpha R^\alpha} \equiv g_l g_A, \tag{3.6}$$

where x^A and A^α parameterise the group element and in the second equation g_A and g_l involves the generators of $\hat{G}$ and the l representation respectively. We will interpret the x^A as the coordinates of a space–time and the A^α as fields that live on this space–time, that is, they depend on the coordinates x^A.

The dynamics of the nonlinear realisations is just a set of equations that are invariant under the transformations of Eq. (3.1). To understand why the nonlinear realisation leads to equations of motion one just has to realise that the group element g of Eq. (3.1) contains the fields of the theory which depend on the generalised space–time. As a result when one finds a set of quantities, constructed out of the group element g, that is, invariant under the transformations of Eq. (3.1) one is necessarily constructing an equation of motion for the fields of the theory. Hence the nonlinear realisation leads to dynamical equations for the fields which are either unique, or almost unique, provided one specifies the number of derivatives involved. As with every application of any symmetry one has to specify the number of space–time derivatives the action should contain. Nonlinear realisations are a bit different to the more familiar situation where one has some fields that transform linearly under a symmetry as in the case of the nonlinear realisation the symmetry and the fields are very closely linked and it is this that leads to the prediction of the dynamics in such a precise way.

We now consider three types of nonlinear realisation, one that leads just to a space–time, one that leads to fields that depend on a space–time that is introduced by hand and finally one that leads to a space–time and fields that depend on this space–time. We denote these as types I, II and III.

1. *Type I*

Let us first consider the case that the local subgroup $H = \hat{G}$ and in this case we can write the group element in the form $g = e^{x^A l_A}$ as the second factor in the group element involving the group $\hat{G}$ can be gauged away using the local H transformation of Eq. (3.1). Thus in this case we are just left with the coordinates x^A and there are no fields. If we take G to be the Poincaré group, that is, $\hat{G} = SO(1, D-1)$ and $H = SO(1, D-1)$ the group element is $g = e^{x^a P_a}$ and the transformations resulting from the rigid transformation are the Poincaré transformations of Minkowski space–time.

Another classic example is to take G to be the super-Poincaré group in four dimensions with one supercharge Q_α and $H = SO(1, D-1)$. The super-Poincaré group is a semi-direct product of the Lorentz group and its representation consisting of P_a and Q_α. We note that in this case the elements of the albeit reducible l representation no longer commute with themselves. The group element can be chosen to be of the form $g = e^{x^a P_a} e^{\theta^\alpha Q_\alpha}$ and the rigid transformations are those of superspace first found in the classic paper of Salam and Strathdee, Ref. 2 in the first section.

Type I nonlinear realisation contain no fields and are just the cosets G/H found in elementary mathematics books on group theory.

The Cartan forms for this type of nonlinear realisation are given by

$$\mathcal{V} = g^{-1}\, dg = dx^{\Pi}\, e_{\Pi}{}^{A}\, l_A + dx^{\Pi}\, \omega_{\Pi,\alpha} R^{\alpha}. \tag{3.7}$$

By studying the transformations of the objects $e_{\Pi}{}^{A}$ and $\omega_{\Pi,\alpha}$ under the nonlinear realisation one finds that they can be taken to be the vielbein and spin connection of the coset space G/H. It is this interpretation that encourages the use of the local indices $\Pi, \ldots$ rather than the tangent indices $A, \ldots$ according to whether they transform under local H transformations or rigid g_0 transformations induced from such transformations on the coordinates. The tangent space of the coset has tangent group H and it is easy to find that $e_{\Pi}{}^{A}$ and $\omega_{\Pi}{}^{\alpha}$ transform under the local group H as they should.

2. *Type II*

We now consider a second kind of nonlinear realisation which involves taking no generators in the l representation, that is $G = \hat{G}$ and H is a subgroup of G. The group element takes the form $g = e^{A_\alpha R^\alpha}$. So far we have no space–time but, by hand, we introduce a space–time with coordinates x_A simply by taking the fields A_α to depend on these coordinates. We note that the coordinates are dummy variables and, in this case, have no relation with the generators G. Local in this case means that the group element g and the local transformations h of Eq. (3.1) depend on the coordinates x_A. One can use the local symmetry to choose the group element to be of a particular form and so set to zero some of the fields A_α. Indeed the number of fields one can set to zero is the dimension of H leaving the dimension of G minus the dimension of H fields. A fact that is consistent with Goldstone's theorem.

Our problem is to find the dynamics that is invariant under the transformations of Eq. (3.1). The usual method is to construct the Cartan forms

$$\mathcal{V} = g^{-1}\, dg = P + Q, \tag{3.8}$$

where Q belongs to the Lie algebra of H and P contains only the remaining generators in G. Considering the rigid transformations of Eq. (3.1) we see that the Cartan forms are invariant under these transformations. However, under the local transformations they transform as

$$\mathcal{V}' = h^{-1}\mathcal{V}h + h^{-1}\, dh. \tag{3.9}$$

Clearly the P part of the Cartan form transforms covariantly, that is, as $P' = h^{-1}Ph$. If we demand that the action we seek has only two derivatives then it is of the form

$$\int dx_A \, \mathrm{Tr}(P^2), \tag{3.10}$$

where we have chosen the generators of G to be in a particular representation, indeed any matrix representation will do. The number of possible terms in the action one can write is determined by the way the adjoint representation of G decomposes into the representations of H. If there is only one representation in addition to the adjoint representation of H then the action is unique. The general theory for this type of nonlinear realisation was given in the classic papers.[22] A more extensive review of this type of nonlinear realisation can be found in Subsec. 13.2 of Ref. 23.

As we have mentioned above the maximal supergravity theories have some symmetries associated with the scalars. In fact the dynamics of the scalars in these supergravity theories is just the nonlinear realisation of the corresponding symmetry group. In particular the IIB supergravity theory has two scalars and their dynamics is the nonlinear realisation of $SL(2,R)$ with local subgroup $SO(2)$,[13] while the maximal supergravity theory in four dimensions has 70 scalars that belong to the nonlinear realisation of E_7 with local subgroup $SU(8)$.[17] In general the scalars in the maximal supergravity theory in D dimensions, for $D \leq 9$, belong to the nonlinear realisation $E_{11\mathrm{D}}$ with a local subgroup which is the maximal compact subgroup of $E_{11\mathrm{D}}$.

It was a type II nonlinear realisation that was used to account for the pion dynamics, discussed in the introduction, by taking $G = SU(2) \otimes SU(2)$ and H to be the diagonal $SU(2)$ subgroup. The low energy dynamics is uniquely determined.

3. *Type III*

Finally we give an account of the type of nonlinear realisation used in this talk. Now we consider no restriction and so $G = \hat{G} \otimes_s l$ and the local subgroup H is a subgroup of $\hat{G}$. The group element has the form of Eq. (3.6) and we find a space–time with coordinates which are in one to one correspondence with generators in the l representation. The fields A_α depend on the coordinates x_A and the group element transforms as in Eq. (3.1).

Since the generators in the l representation belong, by definition, to a representation of $\hat{G}$ we can write the transformations of Eq. (3.1) under the rigid g_0 belonging to $\hat{G}$ act as

$$g_l' = g_0 g_l g_0^{-1}, \quad g_A' = g_0 g_A. \tag{3.11}$$

An exception is when the rigid transformations $g_0 \in l$ and in this case they just give a shift the coordinates. While the local $h \in H$ transformations act as

$$g_l' = g_l\,, \quad g_A' = g_A h. \tag{3.12}$$

As a result the local subalgebra transformations only change the fields and leave the coordinates alone.

To construct the dynamics we consider the Cartan forms which now take the form

$$\mathcal{V} \equiv g^{-1}\,dg = \mathcal{V}_A + \mathcal{V}_l, \tag{3.13}$$

where

$$\mathcal{V}_A = g_A^{-1}\,dg_A \equiv dz^{\Pi}\,G_{\Pi,\underline{\alpha}}\,R^{\underline{\alpha}}, \tag{3.14}$$

belongs to the Lie algebra $\hat{G}$ and are the Cartan forms for $\hat{G}$, while the part that contains the generators of the l representation is given by

$$\mathcal{V}_l = g_E^{-1}(g_l^{-1}\,dg_l)g_E = g_E^{-1}\,dz \cdot lg_E \equiv dz^{\Pi}\,E_{\Pi}{}^{A}\,l_A. \tag{3.15}$$

While both $\mathcal{V}_E$ and $\mathcal{V}_l$ are invariant under rigid transformations, under local transformations of Eq. (3.5) they transform as the

$$\mathcal{V}_E \to h^{-1}\mathcal{V}_E h + h^{-1}\,dh \quad \text{and} \quad \mathcal{V}_l \to h^{-1}\mathcal{V}_l h. \tag{3.16}$$

Type III nonlinear realisations were not as well studied as type II in the old days. However, Isham, Salam and Strathdee worked out in detail the nonlinear realisation of the conformal group in four dimensions with the local subgroup being the Lorentz group.[24] Borisov and Ogivestsky considered the nonlinear realisation of $GL(4) \otimes_s T^4$.[25] In this case the dynamics was not unique but one could choose the undetermined coefficients so that it lead Einstein's gravity. A review of this calculation in D dimensions can be found in Subsec. 16.2 of Ref. 23 which also develops the theory of type III nonlinear realisation further as was done in the E_{11} papers referenced later on in this review. An early review which also contains a discussion of these type III nonlinear realisations can be found in Ref. 26.

4. Kac–Moody Algebras

In this section we will explain how Kac–Moody algebras were discovered[27,28] and by doing so give some insight into what they are. We will gloss over many important points, however, the reader can read a detailed and pedagogical account of Kac–Moody algebras in Chap. 16 of Ref. 23. Group theory emerged from the study of the roots of polynomial equations, however, physicists are more used to thinking of groups as sets of matrices. Given a group it was found that one could reconstruct the part connected to the identity by considering the Lie algebra. It also turns out that all finite-dimensional Lie algebras can be constructed from a subset of Lie algebras that are finite-dimensional and semi-simple and so we work just with these. The precise meaning of semi-simple can be found in Subsec. 16.1 of Ref. 23. The Lie algebra, denoted E, contains a set of commuting generators which we denote by H_i, $i = 1, 2, \ldots, r$ where r is by definition the rank of the algebra. This Abelian algebra is called the Cartan subalgebra. We can now diagonalise the remaining generators E_α with respect to the Cartan subalgebra, that is, we write the commutators of all the other generators in the Lie algebra E with those of the Cartan subalgebra subgroup $SO(1,3)$ where l is the vector representation of $SL(4)$. As we have just mentioned the gravity sector in the 11-dimensional supergravity theory arose from the $K^a{}_b$ generators which belong to the algebra $GL(11)$.

The algebra $\mathcal{A}^{11}$ that emerged from formulating the 11-dimensional supergravity theory as a nonlinear realisation was not a finite-dimensional algebra in the list of Cartan and nor was it a Kac–Moody algebra. This was to be expected as using this construction one only finds generators associated with the fields of the theory and this does not include generators for any local subalgebra. Indeed one should expect to find only the Borel subalgebra of some large algebra. However, if one demanded that this algebra $\mathcal{A}^{11}$ was contained in a Kac–Moody algebra then the smallest such algebra was E_{11}.[30] Motivated by this realisation it was conjectured that the $E_{11\mathrm{D}}$ symmetries were not a quirk of dimensional reduction but that the exceptional symmetries found in the lower-dimensional maximal supergravity theories were part of a vast E_{11} symmetry of the 11-dimensional theory.[30] The price to pay for changing the algebra used in the nonlinear realisation to E_{11} was that it lead to a theory that contains an infinite number of fields, only the first few of which were those of 11-dimensional supergravity.

It is instructive to examine how the above construction would have proceed for the four-dimensional maximal supergravity. This theory can also be formulated as a nonlinear realisation. To do this one introduces for each field of the theory a generator and adopts a suitable Lie algebra that they satisfy which can largely be found by requiring that the corresponding nonlinear realisation gives the equations of motion of four-dimensional maximal supergravity theory. In particular for the scalar fields one introduces generators, which carry no Lorentz indices, in the Borel subalgebra of E_7, however, we must also introduce the generators $K^a{}_b$ for the graviton and R^{aN} for the vectors. We can think of this as extending the nonlinear realisation of the scalars to include the other fields and in so doing so we must introduce generators that transform nontrivially under space–time transformations. We note that we only have some of the generators of the full algebra in particular we have only the Borel subalgebra of E_7 rather than the full E_7 algebra. Demanding that the algebra be extended to a Kac–Moody algebra leads to the E_7 algebra in the scalars sector, but the algebra E_{11} for the full theory.

In the above we have sidestepped the question of how we are to introduce space–time into the theory. Thinking about the gravity sector of the nonlinear realisation it is apparent that we should consider a type III nonlinear realisation, that is, include generators in the algebra which lead to the coordinates of space–time, rather than the type II nonlinear realisation used for the scalars. In the first papers on E_{11} one just introduced the space–time translation generators P_a even though it was clear that this could only be part of the solution as it was not an E_{11} covariant introduction. The correct way to introduce space–time is to introduce generators corresponding to a representation of E_{11}, the l_1 representation, which generalises the space–time translations to include an E_{11} multiplet of generators, and take the algebra used in the nonlinear realisation to be the semi-direct product of E_{11} and this l_1 representation.[31]

The above method of proceeding has an analogy with the original use of non-linear realisations in particle physics. The analogue of pion dynamics is the maximal supergravity theories as these are thought to contain the low energy dynamics of strings and branes. The algebra $SU(2) \otimes SU(2)$ is replaced with E_{11}. Of course the theory of pion dynamics was finally understood after the introduction of quarks and the $SU(3)$ gluon gauge theory. However, as we explained in the introduction this approach to pion dynamics played a key role in unravelling the correct theory, that is, the standard model.

6. The E_{11} Algebra, its Vector Representation and the $E_{11} \otimes_s l_1$ Algebra

The Dynkin diagram of the Kac–Moody algebra E_{11} is given by

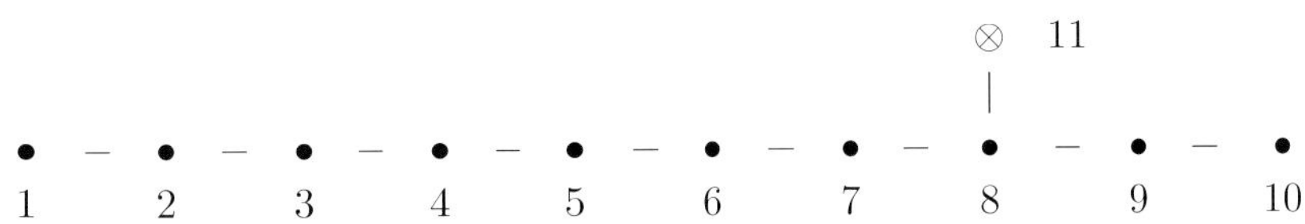

As with any Kac–Moody algebra we do not know all the generators of E_{11}. However, one can gain some understanding of the E_{11} algebra by considering the decomposition of E_{11} into representations of $SL(11)$. In the Dynkin diagram this corresponding to deleting node eleven so as to leave the algebra $SL(11)$. This is the meaning of the cross in the E_{11} Dynkin diagram. The decomposition leads to a set of generators which belong to representations of $SL(11)$ and can be classified according to a level. The level is the number of the number of up minus down $SL(11)$ indices divided by three and it is preserved by the commutators of E_{11}. We will denote the generators of E_{11} by $R^{\underline{\alpha}}$ and those of the l_1 representation by l_A. The positive level generators are given by[30]

$$K^a{}_b,\ R^{a_1a_2a_3},\ R^{a_1a_2\cdots a_6} \quad \text{and} \quad R^{a_1a_2\cdots a_8,b}, \ldots, \tag{6.1}$$

where the generators $R^{a_1a_2a_3}$ and $R^{a_1a_2\cdots a_6}$ are totally antisymmetric in their indices, while the next generator obey the constraints $R^{a_1a_2\cdots a_8,b} = R^{[a_1a_2\cdots a_8],b}$ and $R^{[a_1a_2\cdots a_8,b]} = 0$. The indices $a, b, \ldots = 1, 2, \ldots, 11$. The generators $K^a{}_b$ are those of $GL(11)$ and have level zero, the other generators have levels $1, 2, 3, \ldots$. The negative level generators are given by

$$R_{a_1a_2a_3},\ R_{a_1a_2\ldots a_6},\ R_{a_1a_2\cdots a_8,b}, \ldots, \tag{6.2}$$

where the last generator obey an analogous constraint.

The E_{11} algebra can be constructed using the Serre procedure but it is much easier to construct it level by level using the known generators at a given level,

Fig. 1. We opened for thee a manifest victory.

although brief, was very strong. He was to me the role model of a dedicated scientist pioneering in helping fellow scientists, especially those who come from developing countries. For his insight and kindness I am eternally grateful.

2. Volume Quantization

The work I am reporting in this presentation is the result of a long-term collaboration with Alain Connes over a span of twenty years starting in 1996.[3–8] In the latest work on volume quantization we were joined by Slava Mukhanov.[9,10] On the inner fluctuations of the Dirac operator over automorphisms of the noncommutative algebra times its opposite, we were joined by Walter von Suijlekom.[11–13] A detailed analysis of what appears in this article will appear somewhere else.[14]

At very small distances, of the order of Planck length 1.6×10^{-35} m, we expect the nature of space–time to change. It is then natural to ask whether there is a fundamental unit of volume in terms of which the volume of space is quantized. The present volume of space is of the order of 10^{60} Planck units,[5] but at the time of the big bang, this volume must have been much smaller. To explore this idea, we start with the observation that it is always possible to define a map Y^A, $A = 1, \ldots, n+1$, from an n dimensional manifold M_n to the n-sphere S^n so that $Y^A Y^A = 1$.[15] This map has a degree, the winding number over the sphere, which is an integer. This is defined in terms of an n-form ω_n, the integral of which is a topological invariant

$$\omega_n = \frac{1}{n!} \epsilon_{A_1 A_2 \cdots A_{n+1}} Y^{A_1} dY^{A_2} \cdots dY^{A_{n+1}}. \tag{1}$$

If we equate ω_n with the volume form

$$v_n = \sqrt{g} dx^1 \wedge dx^2 \cdots \wedge dx^n \tag{2}$$

so that

$$\int_{M_n} v_n = \int_{M_n} w_n = \deg(Y) \in \mathbb{Z} \tag{3}$$

then the volume of the manifold M_n will be quantized[15] and given by an integer multiple of the unit sphere in Planck units. This hypothesis, however, has topological obstruction. For the equality to hold, the pullback of the volume form is a four-form that does not vanish anywhere, and thus the Jacobian of the map Y^A does not vanish anywhere and is then a covering of the sphere S^n. However, since the sphere is simply connected, the manifold M_n must be a covering of the sphere S^n.[23,10] This implies that the manifold must be disconnected, each component being a sphere. This gives a bubble picture of space, where every bubble of Planckian volume has the topology of a sphere S^n. This is not an attractive picture because one would have to invent a mechanism for condensation of the bubbles at lower energies. To rescue this idea, we first rewrite the proposal (3) in a different form. Let D_0 be the Dirac operator on M_n given by

$$D_0 = \gamma^a e_a^\mu \left(\partial_\mu + \frac{1}{4} \omega_\mu^{bc} \gamma_{bc} \right) \tag{4}$$

where e_a^μ is the (inverse) vielbein and ω_μ^{bc} is the spin-connection. Notice that in momentum space, the Dirac operator could be identified with momenta p_a, Feynman slashed with the Clifford algebra spanned by γ^a. In analogy, introduce a new Clifford algebra Γ^A such that

$$\{\Gamma_A, \Gamma_B\} = 2\kappa \delta_{AB}, \quad \kappa = \pm 1, \quad A = 1, \ldots, n+1 \tag{5}$$

and slash the coordinates Y^A with Γ_A

$$Y = Y^A \Gamma_A, \quad Y = Y^*, \quad Y^2 = 1 \tag{6}$$

then a compact way of writing equation (3) is given by

$$\langle Y [D_0, Y]^n \rangle = \gamma, \qquad n = \text{even} \tag{7}$$

where γ is the chirality operator on M_n, $\gamma = \gamma_1 \cdots \gamma_n$ and $\langle\ \rangle$ denotes taking the trace over the Clifford algebra spanned by Γ^A. In this form, the quantization condition is a generalization of the Heisenberg commutator for momenta and coordinates $[p, x] = -i\hbar$. It is also identical to the Chern character formula in noncommutative geometry, which is a special case of the orientability condition with idempotent elements. This suggests to consider the above proposal for a noncommutative space defined by a spectral triple $(\mathcal{A}, \mathcal{H}, D)$ together with reality operator J and chirality γ.[16–18] Here $\mathcal{A}$ is an associative $*$ algebra with involution and unit element, $\mathcal{H}$ a Hilbert space, D is a self-adjoint operator with bounded spectrum for $(D^2 + 1)^{-1}$. The chirality operator commutes with the algebra $\mathcal{A}$, $\gamma a = a \gamma$, $\forall a \in \mathcal{A}$.

The following properties are assumed to hold

$$J^2 = \epsilon, \quad JD = \epsilon' DJ, \quad J\gamma = \epsilon'' \gamma J, \quad \epsilon, \epsilon', \epsilon'' \in \{-1, 1\} \tag{8}$$

which defines a KO dimension (mod 8) of the noncommutative space. As an example, for a Riemannian manifold $\mathcal{A} = C^\infty(M)$, $\mathcal{H} = L^2(S)$, D is the Dirac operator (4), γ is the chirality, J is the charge conjugation operator.

The operator J sends the algebra $\mathcal{A}$ into its commutant $\mathcal{A}^o$

$$[a, Jb^* J^{-1}] = 0, \quad \forall a, b \in \mathcal{A} \tag{9}$$

where $Ja^* J^{-1} \in \mathcal{A}^o$, and thus the left action and the right action acting on elements of the Hilbert space commute.

Going back to the volume quantization condition, the slashed coordinates Y when acted on with the J become $Y' = JYJ^{-1}$ which commutes with it $[Y, Y'] = 0$. If Y is slashed with the Clifford algebra (5) where $\kappa = 1$ so that Y' will correspond to the Clifford algebra with $\kappa = -1$. It is essential to have a volume quantization condition involving both Y and Y'. To do this, let $e = \frac{1}{2}(Y + 1)$ so that $e^2 = e$ and similarly $e' = \frac{1}{2}(Y' + 1)$ with $e'^2 = e'$. The product $E = ee'$ also satisfies $E^2 = E$ which implies that the composite coordinate $Z = 2E - 1$ satisfies $Z^2 = 1$. This suggests that we modify our volume quantization condition (7) to become[9]

$$\langle Z[D_0, Z]^n \rangle = \gamma, \quad n = \text{even}. \tag{10}$$

For dimensions $n = 2$ and $n = 4$ this relation splits into two pieces, one is a function of Y and the other a function of Y':[10]

$$\langle Y[D_0, Y]^n \rangle + \langle Y'[D_0, Y']^n \rangle = \gamma, \quad n = 2, 4. \tag{11}$$

For $n = 6$ there are mixing terms between Y and Y' and the relation (10) does not factorize. Thus the only realistic case where we take the volume quantization condition (10) to hold is for manifolds of dimension $n = 4$. In what follows we restrict our considerations to dimensions $n = 4$, where we will find out the special importance of the number 4.

3. Noncommutative Space

From now on we specialize to dimension $n = 4$. In this case, (11) holds and with this condition, we prove that if M_4 is an oriented four-manifold then a solution of the two sided equation (10) exists and is equivalent to the existence of the two maps Y and $Y' : M_4 \to S^4$ such that the sum of the two pullbacks $Y^*(\omega) + Y'^*(\omega)$ does not vanish anywhere and

$$\text{vol}(M) = ((\deg Y) + (\deg Y'))\text{vol}(S^4). \tag{12}$$

The proof is difficult because in four dimensions, the kernel of the map Y is of codimension 2. Details are given in Ref. 10. Fortunately, for this relation to hold, the only conditions on the manifold M_4 is the vanishing of the second Steifel–Whitney class w_2, which is automatically satisfied for spin manifolds, and that vol(M) should be larger than four units.[10] In this setting, the four dimensional manifold emerges as a composite of the inverse maps of the product of two spheres of Planck size. The manifold M_4 which is folded many times in the product, unfolds to macroscopic size. The two different spheres, associated with the two Clifford algebras can be considered as quanta of geometry which are the building blocks to generate an arbitrary oriented four dimensional spin-manifold. We can show that the manifold M_4, the two spheres with their maps Y, Y' and their associated Clifford algebras define a noncommutative space which is the basis of unification of all fundamental interactions, including gravity.

To study this noncommutative space, we first note that the Clifford algebras with $\kappa = 1$ and $\kappa = -1$ are given by[19]

$$\mathrm{Cliff}(+,+,+,+,+) = M_2(\mathbb{H}), \tag{13}$$
$$\mathrm{Cliff}(-,-,-,-,-) = M_4(\mathbb{C}) \tag{14}$$

and thus the algebra $\mathcal{A}_F$ of the finite space is

$$\mathcal{A}_F = M_2(\mathbb{H}) \oplus M_4(\mathbb{C}). \tag{15}$$

The finite Hilbert space $\mathcal{H}_F$ is then the basic representation $(4,4)$ where the first 4 is acted on by the matrix elements $M_2(\mathbb{H})$ and the second 4 is acted on by the matrix elements $M_4(\mathbb{C})$. It is tantalizing to observe that the same finite algebra (15) was obtained by classifying all finite algebras of KO dimension 6, required to avoid mirror fermions. The maps Y and Y' are functions of the coordinates x^μ. Since $Y^2 = 1$ composing words from the elements of the algebra $M_2(\mathbb{H})$ and Y of the form $a_1 Y a_2 Y \cdots a_i Y$, $\forall i$, and similarly for Y' will generate the algebra

$$\mathcal{A} = C^\infty(M_4, M_2(\mathbb{H}) \oplus M_4(\mathbb{C})) \tag{16}$$
$$= C^\infty(M_4,) \otimes (M_2(\mathbb{H}) \oplus M_4(\mathbb{C})). \tag{17}$$

The associated Hilbert space is then

$$\mathcal{H} = L^2(S) \otimes \mathcal{H}_F \tag{18}$$

and the Dirac operator is

$$D = D_0 \otimes 1 + \gamma_5 \otimes D_F \tag{19}$$

where D_F is a self-adjoint matrix operator acting on $\mathcal{H}_F$. The reality operator J_F acting on the finite algebra $\mathcal{A}_F$ satisfies

$$J_F(x,y) = (y^*, x^*) \tag{20}$$

so that the reality operator J will be

$$J = C \otimes J_F \tag{21}$$

four-dimensional Lorentzian manifold. Then three dimensional space volume will be quantized, provided that the field X that maps the real line have a gradient of unit norm $g^{\mu\nu}\partial_\mu X \partial_\nu X = 1$. It is known that this condition when satisfied gives a modified version of Einstein gravity with integrating functions giving rise to mimetic dark matter.[22,23] All of this could be considered as a first step towards quantizing gravity. What is done here is the analogue of phase space quantization where the Dirac operator plays the role of momenta and the maps Z play the role of coordinates. The next step would be to study consequences of this new idea on the quantization of fields which are functions of either coordinates Z or Dirac operator D_A. All of this and other ideas will be presented elsewhere.

To conclude, the influence of Salam on my work which started with supersymmetry, supergravity and their applications continued to string theory, topological gravity and noncommutative geometry, cannot be underestimated. It is my honor and privilege to have followed his footsteps. It was my good fortune to have him as my thesis advisor at Imperial College and postdoctoral mentor at ICTP.

Acknowledgments

I would like to thank Alain Connes for enjoyable collaboration during the last twenty years. I would also like to thank Slava Mukhanov for his friendship and collaboration. The joint work with Walter van Suijlekom opened a new avenue in developing our program. This work is supported in part by the National Science Foundation under Grant No. Phys-1202671 and Phys-1518371.

References

1. A. Salam and J. Strathdee, On superfields and Fermi-Bose symmetry, *Phys. Rev. D* **11**, 1521 (1975).
2. J. Pati and A. Salam, Lepton number as the fourth color, *Phys. Rev. D* **10**, 275 (1974).
3. A. H. Chamseddine and A. Connes, The spectral action principle, *Comm. Math. Phys.* **186**, 731 (1997).
4. A. H. Chamseddine, A. Connes and M. Marcolli, Gravity and the standard model with neutrino mixing, *Adv. Theo. Math. Phys.* **11**, 991–1089 (2007).
5. A. H. Chamseddine and A. Connes, The uncanny precision of the spectral action, *Comm. Math. Phys.* **293**, 867–897 (2010).
6. A. H. Chamseddine and A. Connes, Why the Standard Model, *J. Geom. Phys.* **58**, 38–47 (2008).
7. A. H. Chamseddine and A. Connes, Noncommutative geometry as a framework for unification of all fundamental interactions including gravity, *Fortsch. Phys.* **58**, 553 (2010).
8. A. H. Chamseddine and A. Connes, Reselience of the spectral standard model, *JHEP* **1209**, 104 (2012).
9. A. H. Chamseddine, A. Connes and V. Mukhanov, Quanta of geometry: Noncommutative aspects, *Phys. Rev. Lett.* **114**, 091302 (2015).
10. A. H. Chamseddine, A. Connes and V. Mukhanov, Geometry and the quantum: Basics, *JHEP* **12**, 098 (2014).

11. A. H. Chamseddine, A. Connes and W. van Suijlekom, Inner fluctuations in noncommutative geometry without the first order condition, *J. Geom. Phys.* **73**, 222 (2013).
12. A. H. Chamseddine, A. Connes and W. van Suijlekom, Beyond the spectral standard model: Emergence of Pati-Salam unification, *JHEP* **11**, 132 (2013).
13. A. H. Chamseddine, A. Connes and W. van Suijlekom, Grand unification in the spectral Pati-Salam models, *JHEP* **2511**, 011 (2015).
14. A. H. Chamseddine, A. Connes and V. Mukhanov, in preparation.
15. W. Greub, S. Halperin and R. Vanstone, *Connections, Curvature and Cohomology*, Volumes 1–3, and in particular pages 347–351, Volume 2 (sphere maps).
16. A. Connes, *Noncommutative Geometry* (Academic Press, 1994).
17. A. Connes, A short survey of noncommutative geometry, *J. Math. Phys.* **41**, 3832–3866 (2000).
18. A. Connes and H. Moscovici, The local index formula in noncommutative geometry, *Geom. Funct. Anal.* **5**, 174–243 (1995).
19. H. Lawson and M. Michelson, *Spin Geometry* (Princeton University Press, 1989).
20. A. Connes, Noncommutative geometry and the standard model with neutrino mixing, *JHEP* **0611**, 081 (2006).
21. J. Barrett, A Lorentzian version of the noncommutative geometry the standard model of particle physics, *J. Math. Phys.* **48**, 012303 (2007).
22. A. H. Chamseddine and V. Mukhanov, Mimetic dark matter, *JHEP* **1311**, 135 (2013).
23. A. H. Chamseddine, V. Mukhanov and A. Vikman, Cosmology with mimetic matter, *JCAP* **1406**, 017 (2014).

by an elastic membrane,

$$T_{\mu\nu} = \mathrm{diag}\left[P, \left[P - \frac{1}{2}\ell\delta(r-\ell)\right] r^2, \left[P - \frac{1}{2}\ell\delta(r-\ell)\right] r^2 \sin^2\theta, 3M(4\pi\ell^3)^{-1}\right], \quad (8)$$

one finds for the external $B(r)$ function

$$\begin{aligned} B(r) = 1 &- \frac{\kappa^2 M}{8\pi\gamma r} + \frac{\kappa^2 e^{-m_2 r}}{\gamma r}\left\{\frac{M}{2\pi\ell^3}\left[\frac{\ell\cosh(m_2\ell)}{m_2^2} - \frac{\sinh(m_2\ell)}{m_2^3}\right]\right. \\ &\left. - P\left[\frac{\sinh(m_2\ell)}{m_2^3} - \frac{\ell\cosh(m_2\ell)}{m_2^2} + \frac{\ell^2\sinh(m_2\ell)}{3m_2}\right]\right\} \\ &- \frac{\kappa^2 e^{-m_0 r}}{2\gamma r}\left\{\frac{M}{4\pi\ell^3}\left[\frac{\ell\cosh(m_0\ell)}{m_0^2} - \frac{\sinh(m_0\ell)}{m_0^3}\right]\right. \\ &\left. - P\left[\frac{\sinh(m_0\ell)}{m_0^3} - \frac{\ell\cosh(m_0\ell)}{m_0^2} + \frac{\ell^2\sinh(m_0\ell)}{3m_0}\right]\right\} \end{aligned} \quad (9)$$

depending on the M, P and ℓ source parameters but limiting to the point-source result (7) as $\ell \to 0$.

3.2. *Frobenius asymptotic analysis*

Proceeding on to confront the full nonlinear theory, if one restricts attention to solutions that are asymptotically flat as one approaches spatial infinity, the linearized analysis of Section 3.1 will serve to characterize correctly the behavior of solutions to the full theory as $r \to \infty$, but one needs to complement this with an analysis of solutions at small radii as well. One may use the Frobenius technique of asymptotic analysis of the field equations near the origin through study of the *indicial equations* for behavior as $r \to 0$.[9] Let

$$\begin{aligned} A(r) &= a_s r^s + a_{s+1} r^{s+1} + a_{s+2} r^{s+2} + \cdots, \\ B(r) &= b_t r^t + b_{t+1} r^{t+1} + b_{t+2} r^{t+2} + \cdots \end{aligned}$$

and analyze the conditions necessary for the lowest-order terms in r of the field equations $H_{\mu\nu} = 0$ to be satisfied. This gives the following results,[15] for the general α, β theory:

$$(s,t) = (1,-1) \quad \text{with 5 free parameters} \quad (10a)$$

$$(s,t) = (0,0) \quad \text{with 3 free parameters} \quad (10b)$$

$$(s,t) = (2,2) \quad \text{with 6 free parameters.} \quad (10c)$$

Of the three types of $r \to 0$ asymptotic behavior (10), the (10b) type $(0,0)$ solutions are fully regular at the origin, and are the analogues of the 2-parameter family of flat spacetimes in Einstein theory (where the integration constants correspond to the independent scale coefficients in g_{tt} and g_{rr}). This is the "vacuum" family of solutions, which needs no sources. Consider now the coupling of a spherically symmetric solution to an "egg-shell" spherically symmetric δ-function source with a small radius ϵ surrounding the $r = 0$ origin. Inside the shell, the solution can only be of the (0,0) nonsingular type, needing no source. Now suppose that outside the shell one has a solution that would be of (2,2) type if one continued it all the way in to $r = 0$. Count parameters in this setup: 3 inside + 6 outside = 9 initially. However, from the two third-order field equations (4,5) there are 6 continuity and 'jump' conditions across the δ-function source. So one really has $9 - 6 = 3$ parameters left free. These three so-far unfixed parameters are just what is needed to impose two boundary conditions at spatial infinity to eliminate the rising exponential solutions, plus one 'trivial' parameter that is fixed by requiring $g_{00} \to -1$ as $r \to \infty$.

Thus one concludes that the exterior $(2,2)$ solution fits precisely with a positive-energy distributional source in the full non-linear theory. Exterior (1,−1) and (0,0) solutions would, however, be *overdetermined* according to the parameter counts of Eqs. (10a,10b). So coupling to a positive-energy distributional source works properly just for the $(2,2)$ family. This may be compared with the behavior of the scalar Green function for a $(\Box\Box - m^2\Box)$ kinetic operator in $D = 4$. Although a pure $1/r$ term is a solution to the 'vacuum' equation $(\Box\Box - m^2\Box)\phi = 0$ ignoring sources, it is not the one that couples properly to a positive δ-function source: for a static $-k\delta^3(x)$ source, one instead obtains a solution $\frac{k}{4\pi}(\frac{1}{r} - \frac{1}{r}\exp(-mr))$. In order to obtain just a pure $\frac{1}{r}$ solution, one would need to couple to a non-positive source involving a mixed combination of $\delta^3(x)$ and $\Box\delta^3(x)$.

4. Numerical Solution Families

The analysis of Section 3 establishes the general structure of asymptotically flat spherically symmetric solutions at spatial infinity and as one approaches the $r \to 0$ origin. Finding out how solutions behave in between these asymptotic regions calls for numerical analysis. We shall find a variety of types of behavior: horizonless $(2,2)$ solutions, wormholes and Schwarzschild and non-Schwarzschild black hole solutions with horizons.

4.1. $(2,2)$ *solutions without horizons*

The egg-shell distributional source analysis of Section 3.2 suggests that $(2,2)$ solutions may have a special rôle among the families of spherically symmetric solutions of the quadratic curvature theory (1). In order to learn more about them, one needs to apply numerical methods. We now omit further discussion of sources and consider just the 'vacuum' solutions to (2) with $T_{\mu\nu} = 0$. The task is to match up solutions at spatial infinity with $(2,2)$ asymptotic behavior approaching the $r = 0$ origin. For asymptotically flat solutions with nonzero spin-two Yukawa coefficient $C^{2-} \neq 0$ in the $r \to \infty$ asymptotic region, one numerically finds $(2,2)$ family solutions that have no horizon in the $m_2 = m_0$ theory[17] and in the $R - C^2$ theory,[15] as shown in Fig. 1.

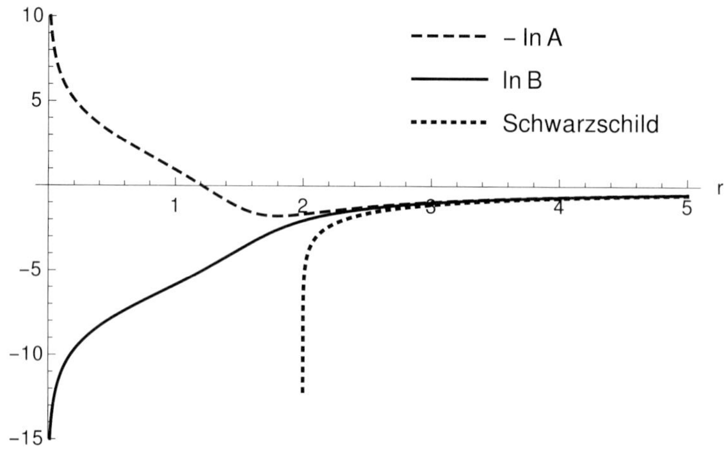

Fig. 1. Horizonless solution in $R + C^2$ theory, behaving as r^2 in both $A(r)$ and $B(r)$ as $r \to 0$.

Such $(2, 2)$ horizonless solutions asymptotically approach the Schwarzschild solution for large r, but differ strikingly in what would have been the inner-horizon region. Although there is a curvature singularity at the origin in the $(2, 2)$ class of solutions (e.g. for this class, one has $R_{\mu\nu\rho\sigma}R^{\mu\nu\rho\sigma} = 20a_2^{-2}r^{-8} + \cdots$), this is a *timelike* singularity, unlike the *spacelike* singularity of the Schwarzschild solution.

4.2. *Wormholes*

Another solution type has the character of a "wormhole" between different sheets of spacetime. Such solutions can have either sign of $M \sim -C^{20}$ and either sign of the falling Yukawa coefficient C^{2-}. As an example, one finds a solution with $M < 0$ in the $R - C^2$ theory. In this solution, $f(r) = 1/A(r)$ reaches zero at a point where $B(r) = a_0^2 > 0$. Making a coordinate change $r - r_0 = \frac{1}{4}\rho^2$, one then has

$$ds^2 = -\left(a_0^2 + \frac{1}{4}B'(r_0)\rho^2\right)dt^2 + \frac{d\rho^2}{f'(r_0)} + \left(r_0^2 + \frac{1}{2}r_0\rho^2\right)d\Omega^2$$

which is $\mathbb{Z}_2$ symmetric in ρ and can be interpreted as a "wormhole" passage between $\rho > 0$ and $\rho < 0$ spacetime regions, with the $r < r_0$ region excluded from spacetime.

5. Solutions with Horizons

Solutions with horizons require a different approach because the presence of a horizon complicates the numerical analysis coming in from spatial infinity. The presence of a horizon together with asymptotic flatness yields a substantial simplification of the relevant equations however, owing to a "no-hair" theorem[15,18] for the trace of the field equation (2).

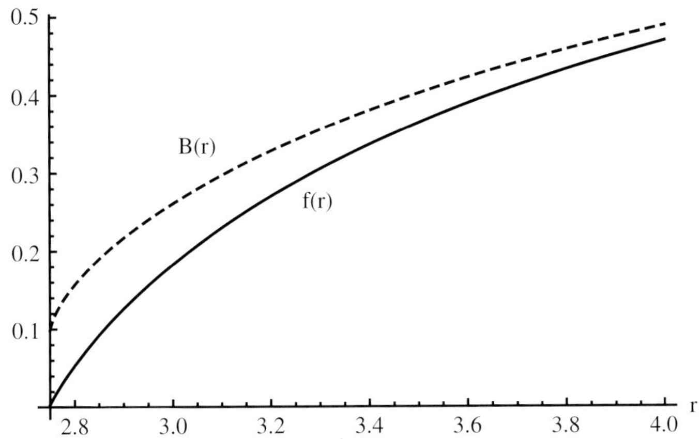

Fig. 2. "Wormhole" solution: $(r) = 1/A(r)$ reaches zero at a point where $B(r) = a_0^2 > 0$.

5.1. *No-hair theorems and horizons*

For $\beta > 0$ (i.e. for non-tachyonic $m_0^2 > 0$), take the trace of the $H_{\mu\nu} = 0$ field equation to obtain

$$\left(\Box - \frac{\gamma}{6\beta}\right) R = 0 \,. \tag{11}$$

Consider a 3D spatial slice of spacetime at a fixed time t on which h_{ab} is the 3D metric and $N = -t^a t^b g_{ab}$ is the (norm)^2 of the timelike Killing vector t^a that is orthogonal to the slice. Multiply Eq. (11) by $N^{\frac{1}{2}} R$, integrate with $\int d^3x\sqrt{h}$ over this 3D spatial slice and prepare for an integration by parts by separating off a total derivative, thus obtaining

$$\int d^3x\sqrt{h}[D^a(N^{\frac{1}{2}} R D_a R) - N^{\frac{1}{2}}(D^a R)(D_a R) - m_0^2 N^{\frac{1}{2}} R^2] = 0 \,, \tag{12}$$

where D_a is a 3D covariant derivative on the spatial slice.

From (12) with $m_0^2 > 0$ and provided the boundary terms arising from the total derivative give vanishing contributions, one finds that one must have $R = 0$ throughout the extra-horizon region. The boundary at spatial infinity gives a vanishing contribution provided $R \to 0$ as $r \to \infty$, while the inner boundary at a horizon null-surface gives a zero contribution since $N = 0$ there.

One conclusion that immediately follows from the trace no-hair theorem is that for solutions that have an asymptotic $\frac{1}{4}\exp(-m_0 r)$ scalar Yukawa correction to the Schwarzschild $\frac{1}{r}$ behavior as $r \to \infty$, there can be no horizon, because such a Yukawa correction implies $R \neq 0$.

The no-hair theorem significantly simplifies the numerical analysis of solutions because requiring $R = 0$ means calculations can effectively be done in the $R - C^2$ theory with $\beta = 0$, in which the field equations, thankfully, can be reduced to a system of two second-order equations.

5.2. *Schwarzschild and non-Schwarzschild black holes*

It is apparent from the full vacuum field equations (2) with $T_{\mu\nu} = 0$ that any spacetime with $R_{\mu\nu} = 0$ is also a solution in the quadratic curvature theory (1). So the Schwarzschild solution is automatically a solution to this theory as well. The question then arises whether there are any other black-hole type solutions to (2), i.e. other asymptotically flat solutions with horizons that differ from Schwarzschild.

Counting parameters in an expansion around a horizon and subject to the $R = 0$ condition, one finds just 3 free parameters. This is the same count as for the $(1, -1)$ family in the expansion around the origin when also subjected to the $R = 0$ condition. So asymptotically flat solutions with a horizon must belong uniquely to the $(1, -1)$ solution family, which contains the Schwarzschild solution itself. The Schwarzschild solution is characterized by two parameters: the mass M of the black hole, plus the trivial g_{00} normalization parameter at spatial infinity. So in the higher-derivative theory, there is just one "non-Schwarzschild" $(1, -1)$ parameter.

In Section 4.2 we saw that it is possible to have a zero of $f(r) = 1/A(r)$ at a point r_0 where one still has $B(r_0) > 0$. The converse is not possible, however. Use the metric parametrization

$$ds^2 = -B(r)dt^2 + \frac{dr^2}{f(r)} + r^2(d\theta^2 + \sin^2\theta d\phi^2)\,, \tag{13}$$

and consider $B(r)$ vanishing linearly in $r - r_0$ for some r_0. Analysis of the field equations then shows that one must also have $f(r)$ linearly vanishing as $r \to r_0$, and so one has a horizon. One can accordingly make near-horizon expansions

$$\begin{aligned} B(r) &= c\left[(r - r_0) + h_2\,(r - r_0)^2 + h_3\,(r - r_0)^3 + \cdots\right], \\ f(r) &= f_1\,(r - r_0) + f_2\,(r - r_0)^2 + f_3\,(r - r_0)^3 + \cdots \end{aligned} \tag{14}$$

and one finds that the parameters h_i and f_i for $i \geq 2$ can be solved-for in terms of r_0 and f_1. For the Schwarzschild solution, one has $f_1 = 1/r_0$, so it is convenient to parametrize the deviation from Schwarzschild using a "non-Schwarzschild" parameter δ with

$$f_1 = \frac{1+\delta}{r_0}. \tag{15}$$

Considering variation of this non-Schwarzschild parameter away from the $\delta = 0$ Schwarzschild value, it is clear that changing it generally has to do something to the solution at infinity. For a solution assumed to have a horizon and holding $R = 0$, the only thing that can happen initially is that a rising exponential is turned on, i.e. asymptotic flatness is lost at spatial infinity. So, for asymptotically flat solutions with a horizon in the near vicinity of the Schwarzschild solution, the only spherically symmetric static solution generally is the Schwarzschild solution itself.

This conclusion is formalized[15] by considering infinitesimal variations of a solution away from Schwarzschild and proving a no-hair theorem for the linearized equation governing the variation. This can successfully be done for coefficients α

that are not too large (i.e. for spin-two masses m_2 that are not too small). One concludes that the Schwarzschild black hole is at least in general isolated as an asymptotically flat solution with a horizon.

Now the question arises: what happens when one moves a finite distance away from Schwarzschild in terms of the $(1, -1)$ non-Schwarzschild parameter? Does the loss of asymptotic flatness persist, or does something else happen, with solutions arising that cannot be treated by a linearized analysis in the deviation from Schwarzschild?

This can be answered numerically.[14] The task becomes one of finding values of $\delta \neq 0$ for which the generic rising exponential behavior as $r \to \infty$ is suppressed. What one finds is that there does indeed exist an asymptotically flat family of non-Schwarzschild black holes which crosses the Schwarzschild family at a special horizon radius r_0^{Lich}. For $\alpha = \frac{1}{2}$, one finds the following phases of black holes:

The crossing point $r_0^{\text{Lich}} \simeq 0.876$ between these two black-hole families is key to a further understanding of the overall structure of the black-hole solutions to the quadratic curvature theory (1).[16]

6. The Lichnerowicz Equation and Implications for Stability

Now let us study in more detail the point where the new black hole family crosses the classic Schwarzschild solution family. We can study solutions in the vicinity of a Schwarzschild solution by looking at infinitesimal variations of the higher-derivative equations of motion (2) around an initially Ricci-flat background. For the $\delta R_{\mu\nu}$ variation of the Ricci tensor away from a background with $R_{\mu\nu} = 0$ one obtains

$$\gamma(\delta R_{\mu\nu} - \frac{1}{2}g_{\mu\nu}\,\delta R) + 2\left(\beta - \frac{1}{3}\alpha\right)(g_{\mu\nu}\Box - \nabla_\mu\nabla_\nu)\delta R$$
$$-2\alpha\Box\left(\delta R_{\mu\nu} - \frac{1}{2}g_{\mu\nu}\,\delta R\right) - 4\alpha\, R_{\mu\rho\nu\sigma}\,\delta R^{\rho\sigma} = 0. \tag{16}$$

Restricting attention to asymptotically flat solutions with horizons, however, we know from the trace no-hair theorem that $R = 0$ so $\delta R = 0$ and the $\delta R_{\mu\nu}$ equation then simplifies, upon recalling that $m_2^2 = \frac{\gamma}{2\alpha}$, to

$$\left(\Delta_L + m_2^2\right)\delta R_{\mu\nu} = 0, \tag{17}$$

where the Lichnerowicz operator is given by

$$\Delta_L\,\delta R_{\mu\nu} \equiv -\Box\delta R_{\mu\nu} - 2R_{\mu\rho\nu\sigma}\,\delta R^{\rho\sigma}\ . \tag{18}$$

Restricting attention to the $m_2^2 > 0$ non-tachyonic case, one sees that static black hole solutions deviating from Schwarzschild must have a $\lambda = -m_2^2$ negative Lichnerowicz eigenvalue for $\delta R_{\mu\nu}$.

In a study of the instability of the Euclideanized Schwarzschild solution in Einstein theory, Gross, Perry and Yaffe found that there is just one normalizable negative-eigenvalue mode of the Lichnerowicz operator for deviations from

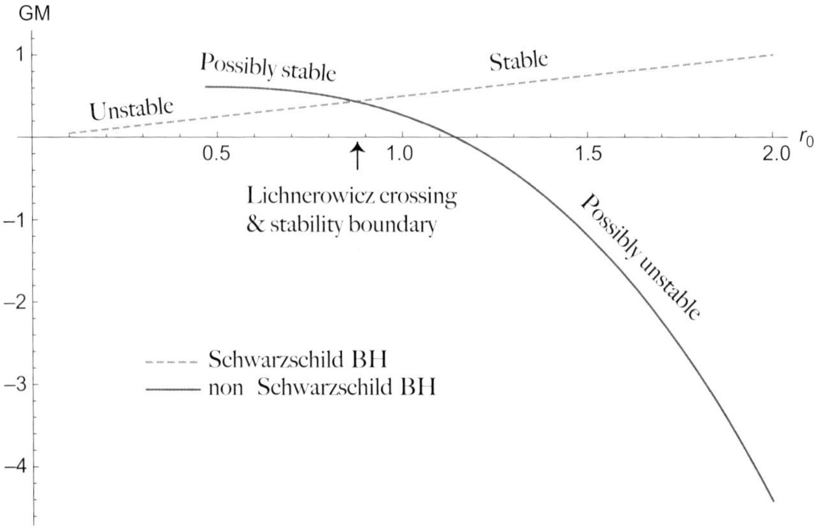

Fig. 4. Suggested classical stability regimes. The dashed line denotes Schwarzschild black holes and the solid line denotes non-Schwarzschild black holes.

this trajectory the possibility of a change in solution phase may once again be suggested[16] by the existence of an intermediate static perturbation that one encounters at mass $M = M_{\max}$. In this case, the static perturbation from the non-Schwarzschild family would consist simply of solutions lying along the straight-line Schwarzschild family of Fig. 3. This suggestion is reinforced also by consideration of the relative thermodynamic specific heats. Both for the Schwarzschild black holes and the non-Schwarzschild black holes the specific heats are negative,[14] but as one crosses $M_{\max}$ there is a changeover from the Schwarzschild black holes having the more negative specific heat for $M < M_{\max}$ to the non-Schwarzschild black holes having the more negative specific heat for $M > M_{\max}$.

Accordingly, a suggestion of possible classical stability structure with respect to S-wave instabilities may be made as shown in Fig. 4.

7. Remembering Professor Abdus Salam

In the years since I met Professor Abdus Salam in 1977, fundamental physics has made significant progress on many fronts: not least in the conclusive confirmation of the Standard Model of elementary particle physics, for which Professor Salam was awarded the Nobel Prize, together with Steven Weinberg and Sheldon Glashow, in 1979. Over the years, I remember many conversations with Professor Salam on questions relating to the quantization of gravity and on its ramifications. Always in my mind I remember clearly his enthusiasm for these deep issues. And always

I remember the twinkle in his eye; the spirit with which he approached fundamental physics is with us still.

References

1. G. 't Hooft and M. J. G. Veltman, One loop divergencies in the theory of gravitation, *Annales de l'Institut Henri Poincaré: Section A, Phys. Theor.* **20**, 69 (1974).
2. K. S. Stelle, Renormalization of higher derivative quantum gravity, *Phys. Rev. D* **16**, 953 (1977).
3. A. Salam and J. A. Strathdee, Remarks on High-energy stability and renormalizability of gravity theory, *Phys. Rev. D* **18**, 4480 (1978). doi:10.1103/PhysRevD.18.4480
4. E. S. Fradkin and A. A. Tseytlin, Renormalizable asymptotically free quantum theory of gravity, *Nucl. Phys. B* **201**, 469 (1982). doi:10.1016/0550-3213(82)90444-8
5. I. G. Avramidi and A. O. Barvinsky, Asymptotic freedom in higher derivative quantum gravity, *Phys. Lett.* B **159**, 269 (1985). doi:10.1016/0370-2693(85)90248-5
6. S. Weinberg, Critical phenomena for field theorists, in *Understanding the Fundamental Constituents of Matter*, ed. A. Zichichi [Lectures presented at 1976 Erice International School of Subnuclear Physics (Plenum, NY, 1978)].
7. M. Reuter, Nonperturbative evolution equation for quantum gravity, *Phys. Rev. D* **57**, 971 (1998). doi:10.1103/PhysRevD.57.971 [hep-th/9605030].
8. M. R. Niedermaier, Gravitational fixed points from perturbation theory, *Phys. Rev. Lett.* **103**, 101303 (2009). doi:10.1103/PhysRevLett.103.101303
9. K. S. Stelle, Classical gravity with higher derivatives, *Gen. Rel. Grav.* **9**, 353 (1978). doi:10.1007/BF00760427
10. B. Whitt, Fourth order gravity as general relativity plus matter, *Phys. Lett. B* **145**, 176 (1984). doi:10.1016/0370-2693(84)90332-0
11. A. A. Starobinsky, A new type of isotropic cosmological models without singularity, *Phys. Lett. B* **91**, 99 (1980). doi:10.1016/0370-2693(80)90670-X
12. V. F. Mukhanov and G. V. Chibisov, Quantum fluctuations and a nonsingular universe, *JETP Lett.* **33**, 532 (1981) [*Pisma Zh. Eksp. Teor. Fiz.* **33**, 549 (1981)].

13. J. Martin, C. Ringeval and V. Vennin, Encyclopaedia inflationaris, *Phys. Dark Univ.* **5–6**, 75 (2014). doi:10.1016/j.dark.2014.01.003 [arXiv:1303.3787 [astro-ph.CO]].
14. H. Lu, A. Perkins, C. N. Pope and K. S. Stelle, Black holes in higher-derivative gravity, *Phys. Rev. Lett.* **114**, 171601 (2015). doi:10.1103/PhysRevLett.114.171601 [arXiv:1502.01028 [hep-th]].
15. H. L, A. Perkins, C. N. Pope and K. S. Stelle, Spherically symmetric solutions in higher-derivative gravity, *Phys. Rev. D* **92**, 124019 (2015). doi:10.1103/PhysRevD.92.124019 [arXiv:1508.00010 [hep-th]].
16. H. L, A. Perkins, C. N. Pope and K. S. Stelle, article in preparation.
17. B. Holdom, On the fate of singularities and horizons in higher derivative gravity, *Phys. Rev. D* **66**, 084010 (2002). doi:10.1103/PhysRevD.66.084010 [hep-th/0206219].
18. W. Nelson, Static solutions for 4th order gravity, *Phys. Rev. D* **82**, 104026 (2010). doi:10.1103/PhysRevD.82.104026 [arXiv:1010.3986 [gr-qc]].
19. D. J. Gross, M. J. Perry and L. G. Yaffe, Instability of flat space at finite temperature, *Phys. Rev. D* **25**, 330 (1982). doi:10.1103/PhysRevD.25.330
20. B. Whitt, The stability of Schwarzschild black holes in fourth order gravity, *Phys. Rev. D* **32** 379 (1985). doi:10.1103/PhysRevD.32.379
21. B. F. Schutz and C. M. Will, Black hole normal modes: A semianalytic approach, *Astrophys. J.* **291**, L33 (1985). doi:10.1086/184453
22. R. Gregory and R. Laflamme, Black strings and p-branes are unstable, *Phys. Rev. Lett.* **70**, 2837 (1993). doi:10.1103/PhysRevLett.70.2837 [hep-th/9301052].
23. Y. S. Myung, Stability of Schwarzschild black holes in fourth-order gravity revisited, *Phys. Rev. D* **88**, 024039 (2013). doi:10.1103/PhysRevD.88.024039 [arXiv:1306.3725 [gr-qc]].
24. C. de Rham, G. Gabadadze and A. J. Tolley, Resummation of massive gravity, *Phys. Rev. Lett.* **106**, 231101 (2011). doi:10.1103/PhysRevLett.106.231101 [arXiv:1011.1232 [hep-th]].
25. E. Babichev and A. Fabbri, Instability of black holes in massive gravity, *Class. Quant. Grav.* **30**, 152001 (2013). doi:10.1088/0264-9381/30/15/152001 [arXiv:1304.5992 [gr-qc]].
26. R. Brito, V. Cardoso and P. Pani, Massive spin-2 fields on black hole spacetimes: Instability of the Schwarzschild and Kerr solutions and bounds on the graviton mass, *Phys. Rev. D* **88**, 023514 (2013). doi:10.1103/PhysRevD.88.023514 [arXiv:1304.6725 [gr-qc]].
27. H. S. Reall, Classical and thermodynamic stability of black branes, *Phys. Rev. D* **64**, 044005 (2001). doi:10.1103/PhysRevD.64.044005 [hep-th/0104071].
28. S. S. Gubser and I. Mitra, The Evolution of unstable black holes in anti-de Sitter space, *JHEP* **0108**, 018 (2001). doi:10.1088/1126-6708/2001/08/018 [hep-th/0011127].
29. S. Hollands and R. M. Wald, Stability of black holes and black branes, *Commun. Math. Phys.* **321**, 629 (2013). doi:10.1007/s00220-012-1638-1 [arXiv:1201.0463 [gr-qc]].

ICTP: From a Dream to a Reality in 50+ Years

F. Quevedo

Abdus Salam International Centre for Theoretical Physics,
Trieste, 34151, Italy
and
DAMTP, University of Cambridge, UK
fquevedo@ictp.it

For more than 50 years, the Abdus Salam International Centre for Theoretical Physics has fostered the growth and sustainability of physics and mathematics in the developing world, benefitting hundreds of thousands of scientists. What began as a dream by its founder, Pakistani Nobel Laureate Abdus Salam, has become a first-rate international research hub connecting scientists from all corners of the globe. As the social and economic situations in many developing countries has shifted, ICTP has responded with the creation of relevant research and training programmes that continue to boost science in disadvantaged parts of the world. Today, ICTP remains a beacon of hope for scientists who aspire to greatness.

Keywords: Abdus Salam; ICTP; developing countries.

1. Introduction

The right to pursue science should be universal, regardless of a country's economic or technological status. For more than 50 years, the Abdus Salam International Centre for Theoretical Physics (ICTP), based in Trieste, Italy, has enforced this right for scientists in developing countries, ensuring that they have access to the same resources and opportunities enjoyed by their counterparts in wealthier parts of the world. It is then a pleasure and an honour for me to participate in this celebration of Abdus Salam's 90th birthday representing ICTP.

Created during the Cold War in the heart of Europe, a continent separated at the time by the Iron Curtain, ICTP provided a rare line of communication between scientists from the East and West. Also, ICTP emerged as a focal point of cooperation between the North and South, aiming to help scientists from developing countries overcome their isolation and contribute to state-of-the-art research in physics and mathematics. While details have changed with time, the basic relevance of the Centre remains the same.

ICTP is an international organization, the first and leading global institution for scientific research and education with an emphasis on supporting science in developing countries. It was created by an agreement between the Italian government and the International Atomic Energy Agency (IAEA), and is now run by a tri-partite

How the Centre implements its mandate and maintains its level of research and training excellence is described in the following paragraphs.

3. In Pursuit of Research and Educational Excellence

ICTP strives to have a proper balance between pure and applied science. Following the original spirit of its founders, research and education on basic sciences are at the core of its activities. The rigorous training in basic sciences that young students receive at ICTP prepares them not only to do research in their specialized fields but also, and more importantly, to be versatile problem solvers, a quality that is highly needed in the challenging, dynamic world we live in. Judging from the success of our alumni, ICTP's efforts seem to be effective: many of the scientists educated here have later occupied prominent roles in society, including successful researchers, heads of departments, rectors of universities, ministers of science and technology, heads of international organizations and even one prime minister.

ICTP has kept pace with these changes by broadening its research activities to include applied subjects, such as climate change, that have more direct and relatively shorter-term repercussions on society. We have also developed a Training and Research in Italian Laboratories (TRIL) programme, that brings thousands of developing world scientists and engineers to train in more than 400 Italian laboratories, with clear benefits to both the hosting institutes and the TRIL participants. This is allowing ICTP to have an even greater impact in the scientific landscape of developing countries.

3.1. *ICTP research*

ICTP's unique strength lies in its ability to bring together large numbers of gifted scientists from developing and developed countries to participate in joint research. From its early focus on theoretical high energy physics, the Centre's research areas have evolved in response to the needs of physicists and mathematicians from the developing world, and now include:

High Energy, Cosmology and Astroparticle Physics (HECAP): HECAP was created at the start of ICTP in 1964, and includes the research field of Abdus Salam. The core staff of some 20 permanent researchers and postdoctoral fellows is studying some of the most exciting areas in physics today, from string theory to physics at large energy colliders, and from neutrino phenomenology to alternative cosmologies.

Condensed Matter and Statistical Physics (CMSP): Created in 1974, this group has expanded and broadened over the years. With more than 25 members, including permanent staff, postdoctoral fellows and staff associates, research in the CMSP section spans some of the most exciting areas of theoretical condensed matter physics, including strongly correlated systems, quantum computation, physics of nanostructures, physics of many-body quantum systems at or far from equilibrium, the computer simulations of fluids and solids with atomistic, molecular and electronic structure methods, and the design of new materials for renewable energy applications.

Mathematics: Created in 1986, this group covers most aspects of pure and some applied mathematics. With about 20 members, including permanent staff, staff associates, distinguished staff associates, and postdoctoral fellows, the Mathematics section builds connections to groups and institutes in developing countries through a variety of activities. Currently its members focus on algebraic geometry, commutative algebra, differential geometry, dynamical systems, analysis and number theory.

Earth System Physics (ESP): Formed in 1998, this group has become a world leader on the physics of the Earth, including climate change and geophysics. The ESP section studies a wide spectrum of the Earth system, from its fluid components (oceans and the atmosphere) to the planet's interior. This section maintains a range of models and datasets and coordinates the Regional Climate research NETwork (RegCNET), encompassing over 600 participants worldwide.

Applied Physics: Created in the 1990s, this section encompasses several areas of physics of direct relevance to developing countries, from wireless communication, ICT (information and communications technology) for development and satellite navigation to X-ray imaging, electronics, optics and lasers and turbulent fluid flows. The areas are, in fact, among the activities for which the demand in developing countries is enormous and growing. With a varying group of researchers, depending on the relevant project, the Applied Physics section actively cooperates with developing-country scientists in education and joint projects of relevance to their region.

Quantitative Life Sciences (QLS): Created in 2011, this is the youngest — and so far smallest — research section at ICTP, but is growing in demand thanks to the great boom of quantitative methods in biology and all life sciences in general. By studying the attacks of malignant cancer cells, the flocking behaviour of starlings, and finding ways to manage huge datasets, ICTP's QLS scientists are uncovering the underlying physics in the broad domain of life sciences that encompasses disciplines ranging from molecular and cell biology to terrestrial and oceanic ecology, and economics and quantitative finance. This group currently has some 10 researchers among permanent staff, staff associates who have been helping to build the group and postdoctoral fellows.

ICTP's visiting researchers are immediately caught up in the culture of the Centre. Strategically placed blackboards welcome them to linger over physics and math problems, while bulletin boards announce a feast of upcoming seminars and colloquia. ICTP research is further strengthened by the Centre's infrastructure, including its recent investment in high performance computing equipment and its highly regarded Marie Curie Library.

Research at ICTP has always been at the highest standards, with many successes over the years. For example, research at ICTP has contributed directly, or indirectly, to at least four Nobel Prizes. These include:

- ICTP founder Abdus Salam's award in 1979 for his contribution to the theory of the unified weak and electromagnetic interaction between elementary particles;
- the heavy involvement of Filippo Giorgi, head of ICTP's Earth System Physics section, with the Intergovernmental Panel on Climate Change, which shared the 2007 Nobel Peace Prize;
- ICTP's contribution, led by Bobby Acharya of the Centre's High Energy Physics section, to CERN's ATLAS experiment, which helped confirm the existence of the Higgs Boson that had been theorized by François Englert and Peter W. Higgs, who shared the 2013 Nobel Prize in Physics;
- the theoretical work on neutrino oscillation by ICTP scientist Alexei Smirnov and colleagues, which was confirmed by one of the recipients of the 2015 Nobel Prize in Physics, Arthur McDonald. Professor Smirnov has also received several awards (Sakurai prize, Einstein medal) for his work on neutrinos and particularly for solving the solar neutrino problem. He is the 'S' in the famous MSW effect.

Many other important scientific contributions started or have been developed at ICTP (superfields, superspace, supermembranes, grand unification, proton decay, large extra dimensions, effective field theory of inflation, Car-Parrinello density functional theory, the Quantum ESPRESSO software for electronic structure calculations, regional climate model RegCM, etc.) due to its unique position of hosting permanent top scientists and being a hub where scientists from all countries meet.

ICTP has also identified research areas of potential expansion for the future, which are of timely importance, especially for developing countries, and complement the current research efforts at ICTP. These include renewable energies and high performance computing (HPC). For the latter, over the past five years great gains have been made in ICTP's HPC capabilities, which are important both as a service to ongoing ICTP research in several areas and in order to provide a master's degree in a rapidly growing and increasingly integral field. Starting in 2013, ICTP and SISSA (the International School for Advanced Studies) began collaboration on the installation of an HPC cluster, which opened in 2014. Also, new to ICTP is an 18-month master's degree programme in HPC, also in conjunction with SISSA.

3.2. *ICTP training and education*

As an international crossroad of scientific excellence, ICTP offers a unique environment for scientists at all stages of their careers to advance their knowledge in physics and mathematics. Each year more than 5,000 scientists from about 160 countries pass through ICTP, taking advantage of the Centre's worldwide reputation for truly outstanding workshops, conferences and advanced educational programmes that explore topics at the cutting edge of physics and mathematics.

Some scientists stay longer, perhaps as participants of the Centre's Associates Scheme, which supports visits of several months at a time over a three-year period; or, as students in ICTP's Postgraduate Diploma Programme, a year-long, intense course of study to prepare young scientists from the developing world for graduate study.

From the Centre's Postgraduate Diploma Programme, an intense, year-long course of study that gives young scientists from developing countries the boost they need for acceptance into doctoral programmes anywhere in the world, to the Centre's Associates Scheme, which supports visits of several months at a time over a three-year period, ICTP provides a lifeline for a lifetime of learning.

Today, ICTP alumni can be found in 188 countries around the world, serving as science ambassadors in their home countries and sharing their knowledge with new generations of scientists. The full spectrum of ICTP's training and education programmes is described here.

3.2.1. *ICTP degree programmes*

ICTP is investing in scientific capacity building by supporting the studies of students from developing countries who qualify to enrol in the Centre's joint masters and doctorate programmes. ICTP's seven degree programmes offer instruction in fields ranging from physics and mathematics to medical physics and high performance computing. Currently around 100 postgraduate students from all continents are enrolled in ICTP's joint programmes, bringing light to the already active ICTP academic life. The complete list of degree programmes is as follows:

- Joint Master in Physics (with University of Trieste)
- Master of Arts in Economics (with University of Turin)
- Master of Complex Systems (with consortium of European universities)
- Masters in Medical Physics (with University of Trieste)
- Masters in High Performance Computing (with SISSA)
- PhD in Physics and Mathematics (with SISSA)
- PhD in Earth Science and Fluid Mechanics (with University of Trieste)

3.2.2. *ICTP Postgraduate Diploma: The start of an educational journey*

ICTP recognizes that many students from developing countries lack the vigorous educational training needed to succeed in doctoral studies. Since 1991, the Centre's Postgraduate Diploma Programme has addressed this need by offering an intense, 12-month, pre-PhD course of study for talented young science students who have limited possibilities to pursue advanced studies in their home countries. More than 700 students from 75 countries have graduated from the programme; of these, 75% have gone on to attain doctoral degrees.

Currently 4 institutions have successfully applied to be official ICTP partner institutions in terms of becoming UNESCO category 2 institutes. These include ICTP-SAIFR and MCTP, plus ICTP-AP (Asian Pacific), an institute that is planned to start in Beijing at the University of the Chinese Academy of Science (UCAS), and ICTP-EAIFR (East Africa) at the university of Kigali, Rwanda. Each new institute is expected to adapt to the needs of the region in order to improve the level of research and international scientific collaboration in the region following the example of ICTP. One centre in Turkey, ICTP-ECAR, is also carrying activities following the ICTP format.

4.3. *Science dissemination*

ICTP helps scientists from around the world to follow all ICTP conferences and courses by livestreaming many of its activities on its YouTube channel. Also, the Centre's Science Dissemination Unit (SDU) has recorded and published online all Postgraduate Diploma Programme courses, as well as many conferences and workshops, using their automated EyA system for the webcasting of physics and mathematics. As of 2015, ICTP's nearly 14,000 hours of online Diploma Programme lectures had received more than one million unique visitors, around 50% of whom come from India, China and Africa. These numbers suggest that the lectures, taught in English by ICTP scientists, are a useful learning resource for students in developing countries. Some parts of the world, however, are restricted both by language barriers and bandwidth constraints. To tackle these issues, SDU has implemented the project "Didactica para el Desarrollo" with educational scientific lectures in different languages. Other SDU services include the development, implementation and management of open source applications, especially in support of science and education in developing countries via mobile science and learning platforms, and the provision of grants and training for the low-cost production of scientific contents by institutions and scholars.

4.4. *Physics without frontiers*

Physics Without Frontiers is a new ICTP outreach programme targeting the far reaches of the developing world to inspire and engage undergraduate and masters physics students. Founded by ICTP high energy physics postdoctoral fellow Kate Shaw, the programme currently embarks on physics roadshows, organizing one-day, intensive masterclasses at universities. Students have the opportunity to analyse real data from the LHC at CERN, which Shaw has been involved with as a member of the INFN Udine/ICTP ATLAS group. In 2015, Shaw brought the wonder of particle physics to Palestine, Lebanon and Nepal, reaching over 260 students. The programme also brings high level masters courses to universities; in particular, the programme has organized a masters course in particle physics in Palestine these past three years with remote teaching and face-to-face tutorials.

5. A Clear Future

ICTP's efforts would not be possible without a solid structure tested throughout fifty years of experience: 30 permanent scientists who are supported by 100 temporary researchers and long-term visitors, 500 visitors daily, for a total of about 6000 visits each year. I want to emphasize the fact that ICTP, in contrast to many other international organizations and funding agencies, is an institution that is run by scientists for scientists. It is exactly this relationship of trust and harmony, mediated by the common language of science, which is our great strength.

Despite the many efforts of ICTP and other institutions, the scientific gap between developed and least developed countries is increasing at a worrisome level. We are now entering a new phase in which ICTP, with the collaboration of others, should take a leading role in promoting better working conditions for scientists in the developing world in order to close this gap.

ICTP's mission remains as important, if not more, than it was at the time of its inception. ICTP is itself a success story, but the relatively small size of our centre limits our ability to have a greater impact in the developing world. Establishing coherent collaborations with other institutions sharing our goals can substantially enhance the impact of our mission, and help to ensure that budding scientists, no matter what the economic and political situation of their native countries, have the opportunity to nurture their ambitions in an environment conducive to the highest levels of scientific knowledge and discovery. Furthermore, ICTP can be considered as an early model for "science diplomacy" since it has allowed scientists from countries in conflict to know and communicate with each other in the universal language of science.

Today, ICTP continues its role as an international crossroad for those who are fascinated with physics or who marvel at the beauty of mathematics. Curiosity and the will to explore the mysteries of the universe are still at the very heart of what motivates the Centre. Its inclusive approach spreads knowledge and peace to every corner of the world, to every country despite adverse learning or research conditions. ICTP remains a lifeline for scientists that after 50 years is stronger than ever. Abdus Salam's dream continues.

Salam's Dream and Dynamic Changes in Chinese Condensed Matter Physics: A Personal Perspective

Lu Yu
Institute of Physics, Chinese Academy of Sciences

Professor Abdus Salam deeply believed that 'scientific thought is the common heritage of all mankind' and he had the dream that the developing world should benefit and could contribute substantially to that heritage, on par with the developed world.

Professor Salam was the wise man and mentor in my professional career. I was fortunate to be able to spend almost 17 years at the International Centre for Theoretical Physics (ICTP) in Trieste, a unique institution in the world, created and guided by him, the Mecca for scientists from the developing world, and was working directly under his supervision for 10 years. I could witness how he devoted his wisdom, energy, heart and soul to materialize the dream of the developing world; I would also witness how his dream is coming true, at least partially, in some parts of the South, including China, although the path is not straightforward, and is full of challenges and difficulties.

Half a century ago, modern condensed matter physics was almost non-existent in China. The reason was simple — before 1949, the basis of science was very weak, because of war and lagging far behind economy. After 1949, everything needed a boost; a few well-trained physicists returned to China from the West, for example Kun Huang who was the best known Chinese condensed matter physicist, Xide Xie, and others. But they were too few, and because of the Cold War, China was totally isolated from the rest of the World. That was the beginning — a humble beginning.

On the other hand, in the past 30 years, especially since the beginning of the 21st century, Chinese condensed matter physics entered the fast track. A number of outstanding young physicists from China with cutting edge research achievements now have global recognition. I will give a few examples later on to illustrate this 'quantal' transition.

1. Humble Beginning

In the 50 s and 60 s, the way out of total isolation in China was 'to learn from the Soviet Union' (SU). At that time, China changed its entire higher education system, switching from the American system to the Russian one, which was strongly delegation in 1972 to attend the Annual Meeting of the Canadian Association of Physics, where he could hear M. E. Fisher's talk on Ken Wilson's work. Although he could not catch the essence of the lecture, he could at least catch a few keywords, which were really helpful. After he returned we rushed to the library and were very fortunate to find that although the journals were not displayed at all during the Cultural Revolution, they were still received, being left unpacked and neglected in the corner. We found with a big shock that during the 'Cultural Revolution' in China, a genuine revolution was taking place in the studies of the phase transitions and critical phenomena in the World. The strong feeling of lagging far behind urged us to take immediate actions: we started again the intensive process of self- and mutual-education with hunger and thirsty; we read the important papers one-by-one and discussed them in detail at group seminars. It was not an easy job: the stacked lecture notes had thickness of 30–40 cm, but it was a genuinely exciting and enlightening time.

The main outcome was two-fold. First, we accumulated enough materials for systematic lectures in that area and for a semi-popular book in Chinese on the phase transitions and critical phenomena. The lecture notes on phase transitions and renormalization group were published along with other lectures in a combined volume *Progress in Statistical Physics* by *Science* press. Amazingly, that volume was reprinted later in Taipei using the traditional (un-simplified) Chinese characters, without authors' names (as an evidence of the abnormal relations between the two sides of the Taiwan strait at that time). The semi-popular book on phase transitions and critical phenomena was very well received by the Chinese scientific community, in that all copies of the 3 earlier editions were sold out quickly, and the fourth one just came out of print. Many young colleagues told us later that they have benefited a lot from reading that book.

Second, with Bailin Hao, we were able to calculate the critical exponents for the continuous phase transitions to the high orders. It was not just to learn but also to do something when we read the important papers one by one. In particular, we studied carefully a paper by Elihu Abrahams and Toshihiko Tsuneto on the skeleton graph expansion for calculating the critical exponents where they claimed to be able to push calculations to the high orders. We thought to check that with our own calculations and eventually found the correct way to proceed. These calculations are rather sophisticated. When we finished the calculations in 1973, we felt very happy, but were discouraged at the same time, when we saw in *Physics Letters* a two-page brief report with no derivations by E. Brezin *et al.* announcing similar results. The two groups were using completely different technologies: They used the Callan–Symanzik equation from the quantum field theory, while we explored the skeleton graph expansion, based on the renormalization group arguments. We also calculated the specific heat exponent directly without resolving to the scaling relations, so we decided to submit it for publication anyway. It appeared in the Chinese physics journal two years later, even without a title/abstract in English. So it was not known outside China at that time.

Another important event during the Cultural Revolution, which had a big impact on my career, happened in 1975: An American delegation of Solid State Physics visited China for almost a month. The delegation was led by Charles Slichter from University of Illinois, and it included 4 Physics Nobel Laureates: John Bardeen, Nicolaas Bloembergen, Ivar Giaever and Bob Schrieffer. The delegation spent 3 full working days at IoP and obtained quite a good picture of the actual situation. I happened to be the interpreter of Bob Schrieffer's lecture on solitons and could present our joint work with Hao on the critical exponent calculation. Our American colleagues were very much impressed as seen from their comments in the published official report (*Solid State Physics in the People's Republic of China*, National Academy of Sciences, Washington, 1976). 'This study used renormalization-group theory and diagrammatic analysis, methods similar to those used in the most advanced contemporary work in the West and in the Soviet Union ... these studies were the most conspicuous exception because everywhere in China physicists usually try to do practical work, a classical or semi-classical phenomenology useful for such things.'

To tell just a small episode: during Bob's lecture at IoP, I reminded him, beyond my interpreter's duty, when the 'key word' of his talk — 'Soliton', escaped from his mind temporarily. For me that was nothing special, as I read his paper in advance. However, Bob was quite pleased and told the story to several friends, most probably also including Stig Lundqvist from Sweden, who would later invite me to join the staff of the ICTP in Trieste, Italy.

3. Opening-up and Recovery

In 1978 my fate changed radically, like all intellectuals in China. I was allowed to travel and went on my first trip abroad after 17 years. That year I had the fortune to attend the Solvay Conference in Brussels on 'Order and Fluctuations in Equilibrium and Non-equilibrium Statistical Mechanics', where I met Phil Anderson, Leo Kadanoff, Mike Fisher and many others. Our international colleagues were very warm and friendly to us. They were curious about how we could survive the Cultural Revolution. The fact that we could still do some work under those conditions, as seen from our papers in Chinese, has won sympathy and support for further scientific exchanges.

A year later, I went to Harvard to visit Bert Halperin's group for more than a year. When we were sent to the countryside in late 60s, in a movement officially named 're-education by farmers', 're-education' meant for me a complete stop of the research work. Just to the opposite, my visit to Harvard was an inspiring experience of genuine 're-education' to gain fresh feeling of frontline research. Bert jokingly called me his 'senior postdoc'. 'Senior' is true, as I am 4 years older than him, but I do not have a PhD at all.

After Harvard I went to visit Bob Schrieffer at UC Santa Barbara. There I got involved in research on the conducting polymers, strongly influenced by the

the record transition temperature for iron-arsenic superconductors up to now. All this happened within a period of a few weeks, during which people worked day and night. *Science* magazine once commented: 'New superconductors propel Chinese physicists to the forefront.' Zhao and colleagues have received the National Prize in Natural Sciences of the first rank in China which has been vacant for several years, and many other awards, quite deservingly.

I would consider these achievements to be not merely 'intermittent bubbles', but rather the cumulative result of sustained governmental support of basic research over the past decades. Without the government's increasing investment and the persistent efforts of researchers in this area, it would not have been possible for the Chinese team to reach this pioneering position quickly in 2008. Over the past decades, the total support for superconductivity research in China was in fact higher than that in the US and Europe.

Realizing the progress in China, the US began actively pursuing collaborations with the Chinese scientists. In 2010, Harold Weinstock of US Air Force Office of Scientific Research (AFOSR) proposed and co-sponsored with Zhongxian Zhao the first 'China-US Bilateral Meeting on Exploration of Superconductivity Materials' in Beijing. Two more meetings were held in Santa Barbara and Hong Kong in subsequent years. In addition, the US Department of Energy (DOE) also proposed collaboration with China, and up to now three joint workshops already took place. This is quite unusual, since over the preceding decades the US has always been exporting knowhow to Chinese high-energy physics. But in the case of superconductivity, China is on an equal footing in the collaboration. Harriet Kung of the DOE once said that she hopes this collaboration might lead to a shared Nobel Prize for Chinese and US scientists.

Iron-based superconductors is not the unique area where the Chinese scientists are among the main driving forces. Similar situation occurs again in the research on topological insulators, quantum anomalous Hall effect, Weyl semi-metals, etc.

The study of topological properties began with the discovery of the quantum Hall effect in the 1980s. Studies of topological properties opened a new area in condensed-matter physics, where the concepts of the Landau's Fermi liquid theory no longer apply. Scientists began to search for new theoretical paradigm to describe these systems.

Many scientists have made seminal contributions in this area. For example, Shou-cheng Zhang of Stanford University successfully predicted the existence of two-dimensional materials for topological insulators, which were soon confirmed by experimentalists in Germany. Zhang has initiated very fruitful collaborations with Mainland Chinese scientists. Calculations made by Zhong Fang and Xi Dai from CAS Institute of Physics, in collaboration with Zhang, predicted the most important three-dimensional (3D) topological insulators in the family of Bi_2Se_3 compounds. Soon afterwards they realized that these 3D topological insulators, when doped with magnetic materials to break the time-reversal symmetry, would be an ideal material for observing the quantum anomalous Hall effect (QAHE). After this work was published, many scientists around the world began to compete to be the first to observe this phenomenon. In 2013, after growing and measuring over 1000 samples, the team led by Qikun Xue at Tsinghua University succeeded in producing rather complicated (quaternary compounds) topological insulators with 3D magnetically ordered chromium (CrBiSbTe), and observed a clear signal of QAHE.

Looking back, the path to this discovery was not straightforward. In the beginning, Duncan Haldane of the Princeton University predicted that one may find the anomalous Hall effect in a honeycomb lattice like graphene that breaks the time-reversal symmetry. But that could not be realized at the time. Later, several different proposals were made to observe the QAHE, including various combinations involving topological insulators. Although Zhong Fang and colleagues predicted the 3D material for potential discovery, it was still very difficult to synthesize and study the predicted materials. There were many failures, but experimentalists trusted that the theoretical calculations were correct. The success of this project is a good demonstration of a new paradigm in condensed matter physics: Theorists work in very close collaboration with sample growers, making precise predictions of what is the most promising material in which to look for new phenomena, and with experimentalists advising them on what kind of results they should anticipate.

The measurement of the QAHE requires extremely high precision at low temperatures, and it happens that only Li Lv's group at the IoP could cool the electron temperature to 4 mK. This is clearly top technology in the world. Although the QAHE does not require such ultra-low temperature, the ultra-low noise level was the key element for the high-quality data. There was an interesting episode in the story. When the paper by Xue's team was submitted to *Science*, the reviewers originally did not believe it, but finally became convinced when the original raw data were shown to them.

Topological materials is now a hot area around the Globe. Besides quantum anomalous Hall effect, the latest advancement in China is in the area of topological semi-metals. In these 3D Dirac systems, the Fermi level is located right at the Dirac cone top, but usually electrons of opposite chiralities are degenerate. By breaking the translational or other symmetries, this degeneracy is removed, and the Weyl fermions with right and left chiralities become independent of each other, leading to a number of exotic phenomena like giant magnetoresistance (a counterpart of the chiral anomaly in particle physics). In designing, making and precisely characterizing the Weyl semi-metals, the Chinese team again took the lead, through fruitful collaborations between theoreticians, materials scientists and experimentalists. In fact, the achievements of this Chinese team, along with the parallel, independent contribution of the American team was selected as one of the 10 Top Breakthroughs in physics of 2015 by *Physics World* of the European Physical Society, as well as one of the 8 Highlights of 2015 by the American Physical Society.

Looking at the picture on a bigger scale: 30 years ago when the Third World Academy of Sciences (TWAS), was created by Professor Abdus Salam, the developing countries, which made up 80% of the world's population, were producing only 5% of the world's scientific publications. At that time, the Research and Development (R&D) budget in the developing countries was only 0.25% of their GDP (Gross Domestic Product), whereas in the developed countries, it accounted for 2.5%. In actual monetary terms, it is 2 versus 100 billion USD. According to a UNESCO report, in 2010, the scientific output from seven developing countries, which already comprised two-thirds of the developing world, was already at 32%, of which China alone contributed 10.6%, second place after the United States.

Recently, the gap between China and the US has been further narrowed. In 1991, China was spending 0.7% of the GDP ($413 billion in total) on R&D, while in 2014 it was 2.09% of the GDP ($11 trillion in total). As for the scientific output, the ratio of China vs US publications has been raised from 1/5 in 2005 to 3/4 in 2014. Not just the quantity. The 'Nature Index' is a typical indicator of high-quality scientific publications. The representative journals for that index have been selected by two independent panels of experts. That index for China has been increased by 37% from 2012 to 2014, while for the US it has been decreased by 4% for the same period.

Of course, China still has a long way to go to build up a solid basis for a sustainable development for science and technology, especially for the disruptive innovations: It requires fundamental changes in the education systems and cultivation of a healthy, productive culture for that purpose. It should require the efforts of many generations to come. Condensed matter physics in China is not an exception.

6. Some Remarks

Surely, the dramatic changes in Chinese condensed matter physics for the last 30 years did not come out of blue. Several important factors have already been mentioned above. To recapitulate the following points worth mentioning:

(1) The development of the economy is the most important prerequisite. The scientific knowledge can be learned by exchange, while the technical know how can be only bought: there is no 'free lunch'. Without the rapid growth of economy initiated by the opening-up policy of Deng Xiaoping all dreams of catching up the advanced nations would still remain utopia for China.
(2) The sustainable support from government for research and education, in particular for the basic science, is the key driving force for the growth of research. For the last 20 years the budget of the Natural Science Foundation of China has been growing at the annual rate of 10–15%. In 1986 it was about $10 million, whereas now it is about $3 billion, nearly a half of the US NSF budget.
(3) In parallel to funding, the manpower is the crucial factor for the scientific research. The 'brain drain' is the biggest problem for the developing countries. Recently in China we are pleased to see the 'backflow' of well-educated young people via various schemes of recruiting talents. The respectable living conditions, sufficient funding for creative research work and intellectual freedom to

pursue one's own research agenda, are the most attractive incentives for their coming back.

(4) Scientific exchange is a 'MUST' for any substantial progress. We greatly enjoy, appreciate and benefit enormously from the open exchange available nowadays. It was even more valuable when we were in isolation and at the beginning of the opening-up. We Chinese have the tradition of remembering those who offered the helping hand in the difficult time, like during the 'Cultural Revolution', as in the Chinese saying: 'Don't forget those who dug the well when you drink the water.'

(5) The formation of the research community in the developing countries, as emphasized by Professor Salam, is extremely important. Since 2002 the 'Beijing Forum on High-Temperature Superconductivity' has been taking place every year (with one year of interruption due to the SARS outbreak), attracting a large number of international scientists in this area. These meetings have established an effective format of communication that consists of mainly discussions, and served a useful function in elevating the research and shaping up the research community.

Professor Salam said at the TWAS Inauguration Conference in 1985: 'In recent times, in this adventure of discovery on the frontier, the South has not been able to play a commensurate role ... principally because of lack of opportunity. This, however, is not a situation which young men and women from the Third World will accept. They enviously, and deservedly, long to participate in this exciting adventure of scientific creation on equal terms.' Some people doubted at that time whether scientists from the developing countries could indeed contribute 'on equal terms'.

Now I would like to say that Professor Salam's dream has, at least partially, come true, and I would tell him to rest in peace!

Bob Schrieffer with Lu Yu at IoP in 1975.

Prof. Abdus Salam, My Teacher and Mentor: The Role of ICTP in Africa

Francis Kofi A. Allotey
President, AIMS-GHANA

It is with great pleasure, honor and privilege that I participate in this meeting to celebrate the life and achievements of Professor Abdus Salam. I thank the organizers for the kind invitation.

The presentation will be categorized as follows:

- Prof. Salam, Mentor and Teacher
- Prof. Salam, ICTP and Development of Physics and Mathematics in Africa
- Prof. Salam, a candidate for the post of Director General of UNESCO
- Epilogue

Prof. Salam, Mentor and Teacher

I first met Professor Salam in 1959 at Imperial College as a graduate student in the Department of Mathematics, then located at the Huxley Building, South Kensington. He was a Professor of Applied Mathematics and Head of Applied Mathematics Section of the Department of Mathematics.

I attended Prof. Salam's lectures on Group Theory and Fundamental Particles Physics. It was a very popular course for almost all postgraduate students and young lecturers in Theoretical Physics at the various Colleges in the University of London. His lectures were brilliant and insightful and contained many innovative ideas, particularly his classification of the then known fundamental particles using group theoretical methods. He was confident and inspiring and always wore black gown. Considering the large number of students who attended his lectures, he always used the largest auditorium in the Huxley Building which was always full to capacity and still had other students standing during lectures. I was the only African student attending his course. He became very much interested in my academic work and welfare. His wife invited my wife several times for afternoon tea in their home.

(Photo credit: ICTP Photo Archives)

25 years Staff Loyal Service Celebration, ICTP, Trieste, 1992.

Prof. Salam was religious and a devoted Muslim. He was encouraged and inspired by his maternal uncle who was an educationist and a pioneer of the Ahmadiyya missionary in Ghana, specifically, Saltpond, a town where I was born and grew up. It was his uncle who founded the Ahmadiyya Muslim Education Service in Ghana. The land for the construction of the Mosque and the Islamic school in the town was donated by the Catholic Church.

Though, Professor Salam was a devout Muslim, he never discussed his religious beliefs with his students, colleagues and employees. He was very kind and a philanthropist and spent his fund assisting some high schools in Africa. In the 1960's, Prof. Salam sent one of his graduate students from Pakistan to Ghana to teach physics and mathematics in a high school for two years before commencing his Ph.D. at Imperial College.

Prof. Salam and Prof. Harry Jones, the Head of Mathematics Department at Imperial College assisted me with references when I was leaving for my Ph.D. at Princeton University.

After my Ph.D. and returned to Ghana, Prof. Salam wrote to me in 1967 that he was introducing a new programme; Condensed Matter Physics at International Centre for Theoretical Physics (ICTP), a Centre that he had established in 1964 and he wanted me to participate. I readily accepted the invitation and since 1967, I have been visiting ICTP regularly without a break as: Participant, Associate, Senior Associate, Course Director and since 1996, as a member of the ICTP Scientific Council.

Prof. Salam in 1992 nominated me to represent him on the Scientific Council of the International Institute for Theoretical and Applied Physics at the Iowa State University, Ames, USA, which was established on the model of the ICTP.

Prof. Salam, ICTP and Development of Physics and Mathematics in Africa

ICTP was the first institution to play a crucial role in positively developing Mathematics and Physics in Africa through human capacity development in various areas of physics and mathematics on the continent. This has helped African scientists keep active in research while staying in their home countries.

Every year, almost 6000 scientists visit ICTP for advanced training and research and out of which are about 600 from Africa. The Centre also organizes several training courses, workshops and conferences in Africa. Many African physicists study and do research in Italian laboratories under ICTP's Training and Research in Italian Laboratories (TRIL) programmes.

There is also an ICTP/IAEA Sandwich programme for Ph.D. in physics and mathematics and related areas which African physicists and mathematicians have benefited from, known as the Sandwich Training Education Programme (STEP).

In this programme, a student who has registered for a PhD in a developing country is chosen for three visits to ICTP for a total of about 18 months, usually, having a co-advisor in ICTP or in an ICTP-affiliated institution. These students are awarded their PhD from their home institutions. The idea behind the STEP programme is to minimize brain drain while providing opportunity for collaboration and access to top facilities.

ICTP has several affiliated centres, projects and networks, including Laser Atomic Molecular-Network (LAM-N) in Africa and has helped train several senior

African physicists and mathematicians who are occupying significant positions in Africa. Each region of Africa is represented and one can find associates and former associates of the centre. These elites play important roles in designing the scientific policy of their respective countries and many hold important academic and research positions.

There are associates who have become ministers or secretaries in scientific affairs and presidents of universities. Almost all the Presidents of Physical Societies in Africa have been associated with ICTP. The immediate past Commissioner for Science and Technology of the African Union was an ICTP associate and a former Director of an ICTP-affiliated center in his home country, the Republic of Benin.

Abdus Salam ICTP/KNUST Microprocessor Laboratory Course, Kumasi, 2003.

The idea to form a Society for African Physicists originated at ICTP on Friday 26th August 1983, when thirty-five African Scientists, visiting ICTP from various parts of Africa held a meeting and resolved to form the Society of African Physicists and Mathematicians (SAPAM) due to the following concerns:

- The poor state of physics and mathematics in Africa,
- Lack of cohesion and functional links among African scientists,
- Great scientific and technological gap between the industrialized and developing countries particularly in Africa, and
- Aware that mathematics and physics are the basis for the creation of modern industries.

The idea to form SAPAM received an overwhelming support from Prof. Salam and he agreed to host the formal inaugural meeting at ICTP in October 1984 during ICTP's 20th Anniversay Celebration. During the inauguration, a pan-African Symposium on the "State of Physics and Mathematics in Africa" was organized under the patronage of Prof. Salam with me as a Chairman of the planning committee.

Dr. G. Andreotti, then Italian Foreign Minister and former Prime Minister of Italy attended the inaugural meeting. Dr. Andreotti stressed the importance of science, technology and innovation for sustainable economic development and he was happy about the inauguration of SAPAM.

Opening Ceremony of 20th Anniversary of ICTP and Inauguration of SAPAM, 1984.

He said and I quote "...the populations affected by food and health problems must be helped, but they must also face these problems within a wider context, which would give them the tools for self development. Thus, there is (besides the immediate direct help) another kind of help nonetheless essential from the point of our civil responsibility: the formation of researchers, technicians and experts whose competence is the base of the economic development of less developed countries and will ensure the good use of long term investments prerequisite for any civic progress."

Dr. Andreotti substantially increased the amount requested by Professor Salam from the Italian government for the ICTP and this enabled ICTP to establish the Office of External Activities and TWAS. He asked other development partners and donor agencies to assist Africa in the area of Science, Technology and Innovation.

Cape Town, South Africa, it was resolved that SAPAM should be transformed and be known as the African Physical Society (AfPS).

It was recognized that a professional society was needed to become an advocate for Physics and physicists at the African Union, in governments of the 54 African countries, amongst the Universities, research institutions, corporate bodies, schools and in the African general public; a society that organizes meetings, conducts professional development workshops, suggests standard of professional conduct and provides information and does all the things that professional associations do.

At the iThemba Laboratory meeting in Cape Town South Africa, the official publication of AfPS known as African Review of Physics was inaugurated with the editorial office located at ICTP, Trieste, Italy.

It is a free on-line peer-reviewed international journal dedicated to publishing in all branches of experimental and theoretical physics with emphasis on originality and relevance to basic understanding of contemporal physics and related interdisciplinary fields.

The AfPS was formally launched on the 12th January 2010 under the distinguished patronage of His Excellency, Maitre Abdulaye Wade, then the President of the Republic of Senegal. There were 110 African Physicists from 21 African Countries and all national physical societies in Africa were present.

LAM-N and AfPS Members with President Wade, 2010.

It is worth mentioning here that African Physical Society played a leading role through the government of Ghana for the UNESCO Executive Board to endorse 2015 as the International Year of Light (IYL) and then for the United Nations to proclaim 2015 as International Year of Light and Light based technologies.

The African Physical Society has received a request from the International Commission for Acoustics to play a similar leading role in seeking an endorsement from UNESCO for the United Nations to proclaim 2019 as an International Year of Sound.

Presidents, Italian Physical Society and AfPS, 45th Anniversary Celebration of ICTP, Trieste, November 2010.

The initiative to establish an African Academy of Sciences (AAS) originated at ICTP in 1985. It was during the inauguration of The World Academy of Sciences (TWAS) when Professor Salam organized Regional meetings for the scientists participating in the inauguration of TWAS. Prof. Salam took part in the African Regional meeting as an observer. It was during the meeting that the idea to form AAS evolved. He gave his strong moral and financial support to make AAS a reality.

Prof. Salam, a candidate for the Post of Director General of UNESCO

With the huge success of ICTP and his strong belief that science was a common human heritage, Prof. Salam wanted to apply the ICTP model to UNESCO, particularly the "S" (Science) part, hence his quest for the post of Director General of UNESCO. He wanted to establish ICTP Centers in many developing countries and placed science technology and innovation (STI) on the high agenda in developing countries.

Prof. Salam strongly believed that the development gap between countries in the North and those in the South was basically a manifestation of science and technological gap. Nothing else, neither different cultural values or different perception or religious thoughts nor different systems of economy or governance can explain why the North to the exclusion of the South can master this globe of ours and beyond.

Prof. Salam made me his campaign manager to vie for African votes. We spent about ten days soliciting votes for him at the UNESCO Headquarters in Paris. Despite the massive support for him from the developing countries, particularly African countries, his country Pakistan did not endorse his candidature to the UNESCO Executive Board. His candidature was therefore not placed at the General Conference for the election to the post of Director General of UNESCO.

Epilogue

Professor Abdus Salam would have been very pleased to know that his vision of establishing ICTP centres in developing countries was being realized with the granting by the UNESCO General Conference in November 2015, a UNESCO Category II status to four institutions in Brazil, China, Mexico and Rwanda which will bear the name ICTP.

Some years after the death of Professor Salam, while I was attending a Board meeting of International Atomic Energy Agency (IAEA) Board of Governors in Vienna, Austria, of which Prof. Salam was also a former Governor), a Pakistani Official approached me.

He informed me that though Pakistan, a Muslim country, does not approve the mounting of human images and busts, it would, however, not object if Prof. Salam's bust would be mounted at IAEA, similar to that of Prof. Homi Bhabha, Prof. Marie Curie-Sklodowska, Prof. Otto Hahn and Prof. Igor Kurchatov.

He further informed me that he would ask ICTP to follow it up and that Pakistan would bear the cost involved.

I wish to propose to ICTP to follow this up.

At this stage, I wish to apologize for being personal. Through ICTP, I have been able to contribute to physics, mathematics and international science policy and development while still working in Ghana. As I mentioned earlier, since 1967, I have been visiting ICTP regularly without a break as; Participant, Associate, Senior Associate, Course Director and since 1996, as a member of the ICTP Scientific Council. I have been Governor at IAEA, Vice President of International Union of Pure and Applied Physics (IUPAP), President of African Physical Society, Vice President African Academy of Sciences, a co-author, UN Blue Book Series No1(1981), Comprehensive Study on Nuclear Weapons. I was a member of 12 experts, "called the 12 wise men" that advised UN on "IAEA Beyond the Year 2000".

In Ghana, I have been a Chairman of Ghana Atomic Energy Commission, Pro Vice Chancellor of Kwame Nkrumah University of Science and Technology, President of Ghana Academy of Arts and Sciences, Chairman, Council for Scientific and Industrial Research and President, African Institute for Mathematical Sciences (AIMS), Ghana.

I played a leading role through the government of Ghana in the process that led UN to declare 2015, International Year of Light and Light Based Technologies. I am a Chartered Fellow of the British Computer Society and in 2012, the UK Institute of Physics conferred on me an Honorary Fellow.

To commemorate my contribution to science, the government of Ghana issued a postage stamp with my portrait in 2006.

Postage stamp with my portrait in 2006.

I am what I am in the world of physics due to Prof. Salam's mentorship and inspiration of which I am very grateful.

Thank you for your kind attention.

Action Principles for Hydro- and Thermo-Dynamics

Christian Fronsdal

Department of Physics and Astronomy,
University of California Los Angeles, CA 90095-1547, USA

For all its brilliant success stories, theoretical physics is actually in a lamentable state. The best way to highlight the situation and at the same time point out that it is not difficult to do better — using action principles — is to present several examples.

1. First Example. Gas Mixtures

This example is chosen, not so much for any intrinsic importance as for its simplicity, which allows us to come to a basic point as directly as possible. It is an elementary example of the improved predictive power of a theory that is formulated as an action principle.

Gaseous mixtures present several interesting problems, each of which has been successfully analyzed, but with unrelated methods.

1.1. *The speed of sound*

The theory of propagation of sound forms a very interesting chapter in the evolution of hydrodynamics and thermodynamics. The first attempt at predicting the speed of sound in our atmosphere may have been that of Newton. Among the assumptions that he made to arrive at an estimate a crucial one was that the temperature was not involved; precisely, that the temperature was fixed and did not oscillate along with the density and the pressure. Newton's estimate was within 20 percent of the measured value and could be considered a partial success. Then Laplace hit the nail on the head with his proposal to the effect that it was not a fixed value of the temperature that characterized the development, but a fixed value of the entropy. This is something that deserves to be remembered, for it was a very early example of adiabatic, non-equilibrium thermodynamics.

The speed of sound varies from one gas to another, even if we limit observations to normal, atmospheric conditions, and in a (homogeneous) mixture it depends on the concentration. A theory that is based on the interpretation of the gas as a collection of particles is summarized in the report of an experiment (Lofqvist *et al.*, 2003).[1] It leads to the following formula,

$$c^2 = \frac{\tau + \kappa}{\tau + 1}\left(1 + \frac{\tau + \kappa}{n_1\tau + n_2\kappa}\right), \tag{1}$$

where n_1, n_2 are the adiabatic indices, $\tau = \rho_1/\rho_2$ and κ is the ratio of molecular weights. This formula is remarkably successful. It could be said that it is also remarkably naive, for the mixture has additional degrees of freedom and more than one mode of vibration should be expected (as is indeed the case). But it does no good to quarrel with success and indeed that is not the intention. To get to the point we must consider another problem.

1.2. *Critical behavior*

This is about the phenomenon of liquification of gases, especially mixtures. The basic tool of analysis has been, for more than a century, the proposal by van der Waals of a relation between density, temperature and pressure that applies to a regime that encompasses that of condensation/evaporation. For a pure gas

$$p = \frac{\mathcal{R}T}{V-b} - \frac{a}{V^2}, \quad a, b \text{ constant}.$$

The standard approach has been to consider a mixture as a compromise between two pure gases. One simply replaces the fixed parameters a, b by functions of the concentration, most often polynomials

$$a(c) = \sum_n a_n c^n, \quad b(c) = \sum b_n c^n, \quad c = \rho_1/\rho.$$

This introduces a number of free parameters, success is guaranteed and, consequently, meaningless.

1.3. *To improve the predictive power of these theories*

It is proposed to develop a formulation of the physics that is based on an action principle; one that introduces a single Lagrangian that must be made responsible for both — in fact all — aspects of gaseous mixtures. And in the special case under discussion that is not difficult.

An action principle for hydrodynamics was invented — or at any rate reported — by Fetter and Walecka in their 1980 textbook.[2] It is not well known among physicists and engineers. For a pure gas the action is

$$A = \int d^3x \mathcal{L}, \quad \mathcal{L} = \rho(\dot{\Phi} - \vec{\nabla}\Phi^2/2) - W[\rho].$$

The independent field variables are the density ρ and the velocity potential Φ; the velocity is

$$\vec{v} = -\vec{\nabla}\Phi.$$

This is fully fledged Hamiltonian dynamics; ρ and Φ is a pair of conjugate canonical variables.

The functional $W[\rho]$ is closely related to the thermodynamic internal energy, but in the present context it is usually a simple expression, the most popular being the polytropic $W[\rho] \propto \rho^\gamma$.

Variation of the velocity potential leads to the equation of continuity and variation of the density gives the Bernoulli equation; together, these two make up the full set of equations that govern simple hydrodynamics.

This theory, with a choice of $W[\rho]$, is sufficient to calculate the speed of rectilinear propagation c of sound in the gas. In the case of a polytrope, it is

$$c^2 = \frac{dp}{d\rho}\Big|_S \propto \rho^{\gamma-1}.$$

The value is usually given as $\sqrt{\gamma \mathcal{R} T}$, but this requires thermodynamics.

An extension of the Lagrangian formulation of simple hydrodynamics was proposed in 2014 (actually in 2011, but publication was difficult). It is very simple,

$$\mathcal{L} = \rho(\dot{\Phi} - \vec{\nabla}\Phi^2/2) - f - sT.$$

Here f and s are the densities of free energy and entropy. In the case of an ideal gas

$$f(\rho, T) = \mathcal{R}T \ln \frac{\rho}{T^n}, \quad s = \rho S,$$

with the specific entropy S constant. Variation of the temperature gives the adiabatic relation

$$\mathcal{R}\left(\ln \frac{\rho}{T^n} - n\right) + S = 0$$

that can be used to eliminate the temperature in favor of ρ and the constants n, S, which leads to the polytropic condition with $\gamma = 1 + 1/n$.

We need the Lagrangian for the mixtures, and the natural expression to try first is the simple sum,

$$\mathcal{L} = \rho_1(\dot{\Phi}_1 - \vec{\nabla}\Phi_1^2/2) - f_1 - s_1 T + \rho_2(\dot{\Phi}_2 - \vec{\nabla}\Phi_2^2/2) - f_2 - s_2 T.$$

Additivity of energy and entropy is a governing principle in thermodynamics.

Variation of the temperature now gives a single adiabatic condition,

$$\frac{\partial}{\partial T}(f_1 + f_2) + s_1 + s_2 = 0.$$

This can be used to eliminate the temperature; it explains why there is no theory of mixtures in hydrodynamics.

The success of this method is not immediate, for it predicts a set of two different sound modes, in contradiction with experience. The remedy is an interaction between the gases, and in Lagrangian dynamics the natural way to do that is to add a term to the free energy density

$$f_1 + f_2 \rightarrow f_1 + f_2 + f_{\text{int}}.$$

The simplest choice is

$$f_{\text{int}} = a(\rho_1 \rho_2)^k \quad a, k \text{ constant}.$$

This model has been applied to several different mixtures. The result is that, if $k = 1/2$, then for sufficiently large value of the parameter a it leads to the same

This resembles the relation derived by Einstein and Laub. But their theory has only the irrotational velocity $\vec{v}$, with its essential property of shifting under Lorentz transformations. Instead, it is the vortex vector $\dot{\vec{X}}$ that appears here; it is invariant under small Lorentz transformations. Their theory also overlooks the pressure term.

The equation seen here is much the same as that used in magnetohydrodynamics, or plasma theory, where the importance of the pressure term is recognized, but with the important difference that the crucial term is missing in this context because the current is said to be zero.

The principal point to be made in this connection is that the theory of electromagnetism in solids and in liquids need to have a conserved current for the potential to couple to. We have taken a step in that direction but the interpretation of the field $\dot{\vec{X}}$ and the parameter κ are not yet clear, especially in the context of the electron gas. The approach presented here is effective in describing certain experiments (e.g. Tolman (1910),[9] Tolman and Stewart (1915)[10]), but it is not yet known how to deal effectively with the Wilson — and — Wilson experiment.

3. Third Example. Astrophysics

Astrophysics, the study of the internal constitution of stars, was started by Homer Lane (1870)[11] who, as the title of his paper announced, proposed to apply the knowledge gained in terrestrial laboratories to the study of stellar interiors. He actually tried to treat the interior of the sun as an ideal gas, a suggestion that met with quite unexpected success in the hands of Emden and Eddington. What was achieved was an adaptation of classical hydrodynamics to a new domain. Gravitation played a role as an external force but in most respects the context was familiar.

Then came General Relativity. It's very creation was a search for an Action, and it was entirely successful. In its range of applications it's relatively simple mathematical structure survived generalization through many orders of magnitude, outwards to unimaginable distances and inwards in accuracy. In the early years General Relativity was brought into contact with other physical systems, contact that was and is necessary to judge its relevance, by coupling it to other field theories. This was easy and natural, for the prevailing material theories, in the first instance Maxwell's electrodynamics, are also relativistic field theories. A relativistic field theory is needed because the Bianchi identity demands it, it is the integrability condition for Einstein's equations.

But then Tolman (1935)[12] suggested an alternative strategy. He was an eminent physicist who had started his career doing pioneering experiments in electrodynamics[9,10] and had gone on to become an expert on thermodynamics, electrodynamics and relativity. It is probably not known if Tolman tried to develop the action principle approach to General Relativity. What I know is that several people tried during the 1960s (Taub, Schutz, Bardeen). Other attempts may have been overlooked. As far as I know, there is no recorded history of any serious attempt to formulate the interaction of Einstein's theory of General Relativity with ordinary fluids (as opposed to streams of non-interacting particles), there was no relativistic hydrodynamics. (Hydrodynamics without the equation of continuity is not hydrodynamics.)

Tolman's approach was based on his understanding of the energy–momentum tensor. This is an object that originates in Relativistic Lagrangian Field Theories. It is a useful tool within that context, but used on its own it is a cumbersome object. Tolman approached Einstein's equations,

$$G_{\mu\nu} = 8\pi G T_{\mu\nu}$$

by identifying, correctly, the right hand side an energy–momentum tensor, but he did not discover a suitable relativistic Lagrangian field theory from which to get such an object. The difficulty has nothing to do with relativity, special or general, but simply the fact that ordinary hydrodynamics did not possess an action principle formulation.

Perhaps most people will agree on this: if hydrodynamics had been available as a Lagrangian field theory 100 years ago, then a relativistic version would quickly have been found, and in this case it would have been the favored place to look for the tensor to place on the right hand side of Einstein's equation.

But since this path was not open, history took another turn, and astrophysics became a phenomenological theory with very limited predictive power.

At about the same time a development began to take place that may be not unrelated: it became fashionable to make imaginative statements about stellar developments that were only weakly supported by sound scientific reasoning and observations. It has been said that astrophysicists became more and more sure of themselves, though always ready to adapt to new observations. The prediction concerning the maximum size of a White Dwarf (Chandrasekhar, 1935)[13] was very successful but many others were not. Some papers from the 1940s are still being quoted as having "shown" that certain kinds of stars must collapse and then explode. The studies on the basis of which such statements were being made involved laborious numerical calculations of extremely limited range. Today one can complete such calculations in seconds, but one is much more circumspect in expressing any opinion about the future development of this or that stellar system.

In giving up the search for a mathematically consistent formulation of General Relativity we lowered the expectations as well as the criteria that are important in physics research. The brand of "relativistic thermodynamics" that resulted gave up its very soul when it abandoned the equation of continuity as well as the respect for mathematical rigor that must attend the development of a theory where integrability conditions are paramount. For a more technical assessment see my paper in arXiv:1106.2271.[14]

The relativistic version of the Action Principle of Fetter and Walecka introduced the Lagrangian density[15]

$$\mathcal{L} = \rho\left(\frac{1}{2}g^{\mu\nu}\psi_{,\mu}\psi_{\nu} - c^2\right) - W[\rho]). \tag{5}$$

Is Left-Right Symmetry the Key?

Goran Senjanović
GSSI, L'Aquila, Italy
ICTP, Trieste, Italy
goran@ictp.it

Abdus Salam challenged the chiral gauge nature of the Standard Model by paving the road towards the Left-Right symmetric electro-weak theory. I describe here the logical and historical construction of this theory, by emphasising the pioneering and key role it played for neutrino mass. I show that it is a self-contained and predictive model with the Higgs origin of neutrino mass, in complete analogy with the SM situation regarding charged fermions.

Keywords: Left-Right symmetry; neutrino mass; Higgs mechanism.

1. Prologue

Fifties were a great era for weak interactions. In 1956 neutrino existence was fully established by the Cowan-Reines experiment and, following the Lee-Yang bombshell, near maximal parity violation was discovered by Wu *et al.* and Lederman *et al.* experiments. In 1957 enters Salam, and independently from Landau, suggests a chiral neutrino. Soon after emerged the celebrated V-A theory that would pave the way for the Standard Model, the electro-weak gauge theory developed by Glashow, Salam and Weinberg in the sixties. The crucial ingredients were chirality, gauge principle and the Higgs mechanism, and now that the Higgs boson was found and seems to agree with the SM predictions, we can say that the rest is history.

I will get to hear of Salam as a founder and director of the International Centre for Theoretical Physics in Trieste towards the end of my pre-graduate days in Beograd in early seventies. For most Yugoslavs Trieste at that time was only a shopping mecca of the West; however for a few of us it was becoming a mecca of physics. I found it beautiful that in a small town at the border with the Balkans Salam would build a place that was supposed to bridge science and scientists from all worlds, especially developing ones. ICTP had no PhD program though and so, following the footsteps of my older brother, I went to do my PhD at the City College of New York.

CCNY was an exciting place in the seventies. Bob Marshak, one of the fathers of the V-A theory, had become a president and started to build a strong physics research from scratch. And the times in particle physics were great. The whole world of light and excitement opened through the advancement of asymptotic freedom and renormalisability of the spontaneously broken gauge theories. There were fundamental theories and new techniques to learn and you could be a part of history in making. I fell in love with the Standard electro-weak model but hated its left-right asymmetry which was supposed to remain for all seasons.

And here Salam entered directly into my life in physics, when I ran across the Pati-Salam theory which liked the world to be left-right symmetric. Rabi Mohapatra and Jogesh Pati worked out the minimal Left-Right model, with the conclusion however that Left-Right symmetry had to be broken explicitly. I found that hard to swallow and soon enough, Mohapatra and I wrote a paper showing that parity could be broken spontaneously which completed a theory that was to last to this day as a serious candidate for the physics beyond the Standard Model.

There was a dark cloud on the horizon though: the theory was predicting massive neutrino against the common wisdom of the day. The trouble was the smallness of neutrino mass, hard to understand with neutrino being a Dirac particle, just like the electron. It took some time, but with the advent of the seesaw mechanism and the solar neutrino puzzle, neutrino mass became a blessing. The essential point was the existence of the right-handed neutrino which in the seesaw picture becomes a heavy neutral Majorana lepton and in turn makes the light neutrino Majorana too. The end result is lepton number violation, through neutrinoless double beta decay at low energies, and the production of same sign charged di-leptons at hadronic colliders, as Wai-Yee Keung and I showed in the early eighties.

Disclaimer. The physics discussed here has been recently reviewed in more depth in Ref. 1 where one can also find a more complete set of references. This short review contains personal bias and cannot do justice to the topics in question.

2. General View

It is clear from the Prologue that the LR symmetric theory is rather old, so why talk about it today? The answer is twofold. First, finally the LHC has a potential of observing it which makes it more timely than ever. Second, there have been two fundamental developments in recent years that make the theory self-contained and predictive: (i) the testable Higgs origin of neutrino mass,[2] in complete analogy with the Standard Model case of charged fermions; (ii) the analytic expression for the right-handed quark mixing matrix,[3] a challenge that lasted some forty years. Moreover, there has been a furry of activity devoted to the LHC potential and the low energy processes such as neutrinoless double beta decay, lepton flavor violation and such.

I give the main results at the outset in order to ease the pain for the casual reader and to motivate her to keep reading on.

(i) In the SM the Higgs boson decay rates are completely determined by the masses of particles in question. This is crux of the Higgs mechanism, completed by Weinberg[5] and GIM.[6] In particular, the one-to-one correspondence between masses and Yukawa couplings of charged fermions allows one to predict the Higgs boson decays into fermion anti–fermion pairs

$$\Gamma(h \to f\bar{f}) \propto m_h \frac{m_f^2}{M_W^2}. \tag{1}$$

This is what it means to understand the origin of particle masses. One can worry why the masses are what they are, but this question, if it is ever to be answered, comes after one establishes their Higgs-Weinberg origin.

It is in this sense that the LR symmetric model is the theory of neutrino mass, as will be discussed in Sec. 6. In direct analogy with (1) one can predict[2] the Higgs decay into light and heavy neutrinos, or better, the decay of heavy right-handed neutrino N (when it is heavier than the Higgs) into the Higgs and light neutrino. As an illustration, I give here the relevant expression[4] for a simplified case described in Sec. 6

$$\Gamma(N_i \to h\nu_j) \propto \delta_{ij}\ m_{\nu_i} \frac{m_{N_i}^2}{M_W^2}. \tag{2}$$

This would be hard to observe, needles to say; however, there is an experimentally more accessible decay channel of right-handed neutrino N into the W boson and charged lepton[2,4]

$$\Gamma(N_i \to W\ell_j) \propto V_{ij}^2 m_{\nu_i} \frac{m_{N_i}^2}{M_W^2}, \tag{3}$$

where V is the PMNS leptonic mixing matrix.

In the general case the above expressions look more complicated, but all the essential features are caught here. One has a complete analogy with the Standard Model situation regarding the charged fermions, only now one has to know the (Majorana) masses and mixings of left and right handed neutrinos separately. More about it below.

(ii) The right-handed quark mixing matrix V_R has a simple approximate form[3] as a function of the usual left-handed CKM matrix V_L

$$(V_R)_{ij} \simeq (V_L)_{ij} - i\epsilon \frac{(V_L)_{ik}(V_L^\dagger m_u V_L)_{kj}}{m_{d_k} + m_{d_j}} + O(\epsilon^2) \tag{4}$$

where ϵ is a small unknown expansion parameter. It can be shown that the left and right mixing angles are almost the same, and right-handed phases depend only on V_L and ϵ. A determined reader should go to Sec. 7 for more details and for some immediate consequences of (4) regarding the right-handed mixing angles and phases.

The rest of this short review is organised as follows. I first discuss the salient features of the theory in the next section, and then try to give a historical development that took one to the seesaw based version of the model. Thus, in Sec. 4 I go through the original version of theory that had Dirac neutrinos and struggled explaining why their masses were so small. Section 5 is devoted to the modern version of the theory based on the seesaw mechanism with naturally light Majorana neutrinos. Next, I go over the issues (i) and (ii) above in Secs. 6 and 7, respectively, before offering an outlook for the future. I end with an epilogue, in order to make the presentation not only LR but also top-bottom symmetric.

3. Generic Features

The minimal LR symmetric theory is based on the $SU(2)_L \times SU(2)_R \times U(1)_{B-L}$ gauge group, augmented by the symmetry between the left and right sectors.[7–9]

Quarks and leptons are then completely LR symmetric

$$q_{L,R} = \begin{pmatrix} u \\ d \end{pmatrix}_{L,R}, \qquad \ell_{L,R} = \begin{pmatrix} \nu \\ e \end{pmatrix}_{L,R}. \tag{5}$$

Clearly, the LR symmetry says that if there is a LH neutrino, there must be the RH one too and neutrino cannot remain massless. A desire to cure the left-right asymmetry of weak interactions lead automatically to neutrino mass.

The formula for the electromagnetic charge becomes

$$Q_{em} = I_{3L} + I_{3R} + \frac{B-L}{2}, \tag{6}$$

which trades the hard to recall hyper-charge of the SM for $B-L$, the physical anomaly-free global symmetry of the SM, now gauged. Both LR symmetry and the gauged $B-L$ require the presence of RH neutrinos.

LR symmetries. It is easy to verify that the only realistic discrete LR symmetries, preserving the kinetic terms, are $\mathcal{P}$ and $\mathcal{C}$, the generalised parity and charge-conjugation respectively, supplemented by the exchange of the left and right $SU(2)$ gauge groups (for a recent discussion and references, see Ref. 10).

Higgs sector. The analog of the SM Higgs doublet is now a bi-doublet[7,8]

$$\Phi = \begin{bmatrix} \phi_1^0 & \phi_2^+ \\ \phi_1^- & -\phi_2^{0*} \end{bmatrix} \tag{7}$$

in order to provide masses for charged fermions. This amounts to two $SU(2)_L$ doublets, but one of them ends up being very heavy and effectively decouples from low energies.[9] In analogy with their charged partners neutrinos get Dirac mass.

4. Classic Era

So far so good. But what fields should be used for the large scale of symmetry breaking?

In the original version[7,8] one opted for $B-L=1$ LH and RH doublets, i.e. the doublets under $SU(2)_L$ and $SU(2)_R$ groups respectively. It seemed a logical choice, a LR extension of the SM Higgs doublet. Looking back, it is hard to understand why for some years no alternative was studied, since there was nothing special about this choice. After all, we had already used the $SU(2)_L$ doublets, in a form of a bi-doublet, in order to give masses to charged fermions. The large scale Higgs sector determines the ratio of new gauge boson masses, W_R and Z_R but why in the world should it mimic the SM situation with W and Z? It is both instructive and amusing how one gets sidetracked and confused at the beginnings.

In any case, one sat down to show that parity could be broken spontaneously.[8,9] The symmetry of the potential tells you immediately that there are only two possibilities: (i) same vevs for LH and RH doublets, unacceptable; (ii) one of the two vevs vanishing, as required by experiment. It was easy to show that there was the stable minimum with only the RH doublet vev, and the task of breaking the theory down to the SM was achieved. The trouble was neutrino mass, hard to understand[11] why so small with neutrino being a Dirac particle, just like the electron.

5. Modern Era

The theory was prophetic in predicting neutrino mass so early, years before experiment, but it seemed to fail to account for its smallness. It turned out that the problem was not LR symmetric gauge group, but simply the choice of the heavy Higgs sector. This let to the with a version of the theory based on the seesaw mechanism.[12–14]

The main point was to choose the right Higgs in order to make RH neutrino a heavy Majorana lepton, so that it could mix with the LH one and give it in turn a tiny mass. All that it required was to substitute doublets by the appropriate triplets. In this way, the theory leads naturally to a Majorana neutrino and lepton number violation (LNV). In our paper,[13] Mohapatra and I emphasised this, and argued that the resulting LNV process, the neutrinoless double decay,[15] could be easily generated by new RH sector, and not by small Majorana neutrino mass. Yet, it is often claimed to this day that this process is a direct probe of neutrino Majorana mass; this is simply wrong. By the way, the idea of new physics being possibly behind the neutrinoless double beta decay dates back all the way to the late fifties.[16]

Some years later Wai-Yee Keung[17] and I made a case for an analog high-energy Lepton Number Violating process, the production and the subsequent decay of the RH neutrino. We realised that due to its Majorana nature, the RH neutrino, once produced on-shell would decay equally into a charged lepton and anti-lepton. This allows to test and measure directly its Majorana nature, not just indirectly through low energy effective processes. Also, besides the usual LNV conserving final state, one would have direct LNV in the form of the same sign charged di-leptons and (two) jets. This turns out to be a generic property of any theory that leads to Majorana neutrino and has become over the years the paradigm for LNV at hadronic colliders, and today both CMS and ATLAS are looking into it.

Fig. 4. Professors Abdus Salam (right) with Zulfiqar Ali Bhutto (left) and Munir Ahmed Khan (centre) at the Multan science conference in 1972.

with ICTP in Miramare. I fondly remember, among the faculty, Paolo Budinich, Giuseppe Furlan, Luciano Bertocchi, and Nello Paver. It was also there that I heard for the first time lectures from some of the great Italian physicists Eduardo Amaldi, Nicola Cabbibo, Sergio Fubini, Luciano Maiani, and Tullio Regge, among others. Likewise, we often had seminar speakers and visitors from the UK, including Paul Dirac and Dennis Sciama. The stay at the ICTP opened for me, like for thousand other visitors from developing countries, an entirely new world of science. The two-year stay at the ICTP was a continuous intellectual feast, which used to reach its climax in the summer, with extensive colleges and scientific meetings, some of which were held at the picturesque Parco di Miramare at the Adriatic Sea.

The extended stay at the ICTP and numerous subsequent visits gave me an opportunity to see Abdus Salam at work from close quarters. By the time that the neutral weak currents predicted by the Glashow–Salam–Weinberg model were discovered in 1973 by the Gargamelle experiment at CERN,[5] electroweak unification was already a done deal for Salam, and he had moved on to the next level of unification. Among other visitors, there was Jogesh Pati from the university of Maryland, who spent several summers and had a longer stay at the ICTP. He collaborated actively with Prof. Salam, and from this period came some of the remarkable Pati-Salam series of papers which are still very relevant from the present day perspective. These include, among others, the paper on the unified Lepton-Hadron symmetry,[6] followed by the one entitled: Lepton number as the fourth colour,[7] which introduced what is now popularly known as the Pati–Salam SU(4) group, with the leptoquarks, which are again *en vogue* as discussed later, and the paper entitled: Is Baryon number conserved?[8] A little later, in early 1974, came the remarkable paper with John Strathdee on supersymmetry in which a systematic method for constructing the Wess-Zumino super-gauge transformation was exhibited.[9] These are some of the papers in the period 1972–74, which found a large following and made a big impact on particle physics research. Pati–Salam papers led to a surge of theoretical interest in finding more realistic grand unified theories, culminating in the Georgi–Glashow $SU(5)$[10] and subsequently the $SO(10)$ groups,[11] which are still popular templates in discussing deeper theoretical issues. I am sure that if proton decay had been experimentally established, Pati and Salam, being among the very first providing a model for baryon non-conservation, would be the front contenders for a Nobel Prize in physics. A full compendium of Prof. Salam's scientific papers can be seen in the volume I edited with Chris Isham, Tom Kibble and Riazuddin.[12]

After finishing my post-doc at the Stevens Institute of Technology in Hoboken, New Jersey, I returned to ICTP in the summer of 1975. In the intervening two years, the waters flowing down the Indus had become very muddy, which also impacted Salam' s mood and outlook. It is my impression that Salam's direct involvement in the scientific advisory work in Bhutto administration had cooled down considerably. His parting of ways with Bhutto was not very perceptible in the beginning, but their relations definitely came to a break in 1974 due to a legislation pushed by Bhutto in the parliament, which discriminated Salam and the Ahmadi sect to which he belonged. Salam was a devout Muslim, and this became more evident after the 1974 Anti-Ahmadi resolution adopted in Pakistan. To underscore his religious credentials, he started to sign letters with Muhammad Abdus Salam, and grew a beard as a token of his spiritual identity. Salam was particularly dismayed by Bhutto, who despite his western liberal education and exposure to great institutions, such as Oxford and Harvard, had bowed to the street pressure in Pakistan betraying his own party's secular and democratic credentials, which culminated in the politically expedient action of ex-communicating Ahmadis from public life. Salam had obviously opposed this development, as he was of the firm opinion that the resolution pushed by Bhutto in the parliament would put Pakistan on a religious fundamental trajectory never to return, but he couldn't do much to reverse it. Subsequent events in Pakistan proved Salam's prophecy. Also, Bhutto was engulfed in the rising tide of religious fundamentalism, and his playing footsie with the reactionary political outfits and religious zealots became his undoing. He was overthrown in a military coup by General Zia-ul-Haque and was hanged subsequently.

After the Nobel prize for Physics n 1979,[13–15] which Salam shared with Glashow and Weinberg, ICTP substantially enlarged its activities, in some of which I also participated and this association went well over a decade. Let me summarise my personal impressions about Salam that I gained in this period. In my opinion, there were three dominant strains and themes that moulded Salam's personality. First and foremost, he was a physicist at the forefront, destined to unearth the laws of nature. This intellectual pursuit was conducted with highest professional standards and impeccable honesty. He practiced this during his entire professional life, and this was also the advice that he gave to his students and others around him. Salam was also a cherished and much sought-after thesis supervisor and he guided a large number of students and collaborators, who in the course of their professional life became leaders in their own right. Some of them are present at this meeting.

As a scientific advisor, Salam wore many different hats and played very diverse roles in this capacity. To name a few, he worked closely with the IAEA, Vienna, and in that capacity he convinced the IAEA board to set up ICTP in Trieste, which he directed for almost thirty years and which helped thousands of physicists and other scientists from developing countries in coming out of their scientific isolation. In the meanwhile, the idea behind ICTP has caught on, and several such centres have now emerged in various parts of the world. This is being discussed by several speakers at this meeting. As a member of the scientific policy committee of CERN, he pushed the CERN proton–antiproton collider, leading to the discoveries of the $W^{\pm}$ and Z bosons, as recounted at this meeting by Carlo Rubbia. As the chief scientific advisor to the presidents of Pakistan, Salam was principally responsible for setting up the scientific infrastructure in the initial stages of this country. With the passage of time Salam converted to pacifism — a declared anti-nuclear activist who wanted to banish the nuclear weapons from our world. He was also an active participant and supporter of the Pugwash conferences, which played an important role in facilitating dialogues among adversaries during the cold war and in nuclear disarmament.

The third strain in Salam's life was his role he voluntarily opted to play as the world's best-known Muslim scientist. In that capacity, he travelled very widely, particularly after he became a Nobel Laureate, collecting awards and honorary degrees. In return, he lectured untiringly, insisting on the need of higher education and scientific research. This, however, fell on deaf ears, as such activities were considered as the indulgence of a few and hence largely futile in those parts of the world. But, he never gave up. Salam was also deeply conscious of his rich intellectual heritage from the times long past, as he had a certain affinity for scientists and philosophers of those days, among them Abu Jaffer Al-Khwarizmi, the founder of Algorithm, Abu Raihan al-Biruni, an all-round genius of his times, and Abu Ali Ibn Sina (a.k.a. Avicenna), known for his medical research, to name a few, in whose rows he rightly thought that he belonged. To his dismay, his dream of scientific renaissance in Islamic countries never became a reality.

Let me close this section by thanking Abdus Salam for all that he did for the scientific uplift of developing countries and personally for helping me in pursuing a career as a physicist. His legacy, the standard model of particle physics, still holds

Fig. 5. Professor Abdus Salam with Queen Silvia of Sweden at the 1979 Nobel Banquet (Courtesy: Nobelprize.org).

sway, despite minute experimental scrutiny, as I describe below in the context of flavour physics.

2. Rare B-decays in the Standard Model*

In the second part of my write-up, I will review some selected topics in rare B decays. The interest in studying these decays is immense. This is due to the circumstance that these decays, such as $b \to (s,d)\gamma$, $b \to (s,d)\ell^+\ell^-$, ..., are flavour-changing-neutral-current (FCNC) processes, involving the quantum number transitions $|\Delta B| = 1, |\Delta Q| = 0$. In the SM, they are not allowed at the tree level, but they do take place at the quantum (loop) level and are governed by the GIM (Glashow–Iliopoulos–Maiani) mechanism,[16] which imparts them sensitivity to higher masses, (m_t, m_W), from the loops. As a consequence, they determine the CKM (Cabibbo–Kobayashi–Maskawa)[17] weak interaction matrix elements. Of these, the elements in the third row, V_{td}, V_{ts} and V_{tb} are of particular interest. While $|V_{tb}|$ has been measured in the production and decays of the top quarks in hadronic collisions,[18] the first two are currently not yet directly accessible. In the SM, these CKM matrix elements have been indirectly determined from the B^0–$\bar{B}^0$ and B_s^0–$\bar{B}_s^0$ mixings. Rare B-decays provide independent measurements of the same quantities.

In theories involving physics beyond the SM (BSM), such as the 2-Higgs doublet models or supersymmetry, transitions involving the FCNC processes are sensitive to the masses and couplings of the new particles. Precise experiments and theory

*This part has significant overlap with the article entitled "Rare B-meson decays at the crossroads," *Int. J. Mod. Phys. A*, Vol. 31, No. 23 (2016) 1630036.

are needed to establish or definitively rule out the BSM effects. Powerful calculation techniques, such as the heavy quark effective theory (HQET)[19] and the soft collinear effective theory (SCET)[20–24] have been developed to incorporate power $1/m_c$ and $1/m_b$ corrections to the perturbative QCD estimates. In exclusive decays, one also needs the decay form factors and a lot of theoretical progress has been made using the lattice QCD[25] and QCD sum rule techniques,[26–29] often complementing each other, as they work best in the opposite q^2-ranges. Rare B-decays have enjoyed great attention in the past and they continue to so in the ongoing experiments in flavour physics. The current experimental frontier is now the large hadron collider (LHC), which will be soon joined by Belle II, which is expected to start taking data at KEK in 2018.

2.1. *The standard candle of rare B decays:* $B \to X_s\gamma$

The experimental era of rare B-decays started in 1993 with the measurement of $B \to K^*\gamma$ by the CLEO collaboration at the Cornell e^+e^- collider,[30] followed in 1995 with the measurement of the inclusive decay $B \to X_s\gamma$.[31] The photon energy spectrum in this process was already calculated in 1990 by Christoph Greub and me,[32] which came in handy for the CLEO measurements[33] shown in Fig. 6 (left frame), which is compared with the theoretical prediction.[32] Since then, a lot of experimental and theoretical effort has gone in the precise measurements and in performing higher order perturbative and non-perturbative calculations. As a consequence, $B \to X_s\gamma$ has now become the standard candle of FCNC processes, with the measured branching ratio and the precise higher order SM-based calculation providing valuable constraints on the parameters of BSM physics. The impact of

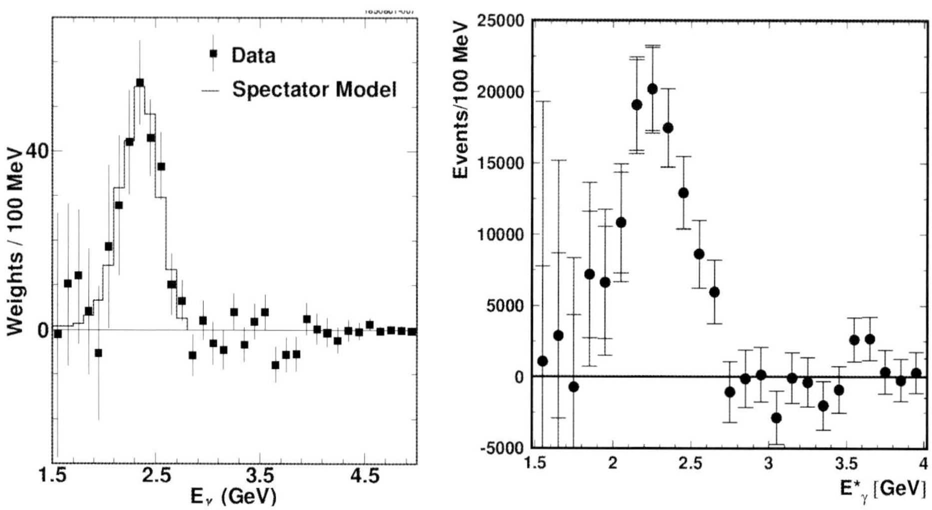

Fig. 6. Photon energy spectrum in the inclusive decay $B \to X_s\gamma$ measured by CLEO (left frame)[33] and Belle (right frame).[34]

the B-factories on this measurement can be judged the scale in Fig. 6 (right frame), which is due to the Belle collaboration.[34]

The next frontier of rare B-decays involves the so-called electroweak penguins, which govern the decays of the inclusive processes $B \to (X_s, X_d)\ell^+\ell^-$ and the exclusive decays such as $B \to (K, K^*, \pi)\ell^+\ell^-$. These processes have rather small branching ratios and hence they were first measured at the B-factories. Inclusive decays remain their domain, but experiments at the LHC, in particular, LHCb, are now at the forefront of exclusive semileptonic decays. Apart from these, also the leptonic B-decays $B_s \to \mu^+\mu^-$ and $B_d \to \mu^+\mu^-$ have been measured at the LHC.

I will review some of the key measurements and the theory relevant for their interpretation. This description is anything but comprehensive, for which I refer to some recent excellent references[35–38] and resources, such as HFAG[39] and FLAG.[25]

2.2. *Inclusive decays* $B \to X_s\gamma$ *at NNLO in the SM*

The leading order diagrams for the decay $b \to s\gamma$ are shown are shown in Fig. 7, including also the tree diagram for $b \to u\bar{u}s\gamma$, which yields a soft photon. The first two diagrams are anyway suppressed due to the CKM matrix elements, as indicated. The charm and top quark contributions enter with opposite signs, and the relative contributions indicated below are after including the leading order (in α_s) QCD effects. A typical diagram depicting perturbative QCD corrections due to the exchange of a gluon is also shown.

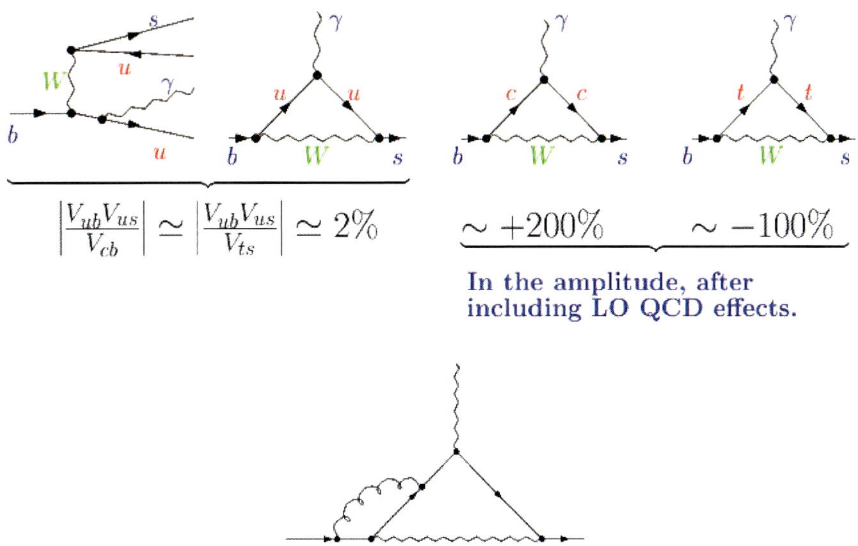

Fig. 7. Examples of the leading electroweak diagrams for $B \to X_s\gamma$ from the up, charm, and top quarks. A diagram involving a gluon exchange is shown in the lower figure.

The QCD logarithms $\alpha_s \ln M_W^2/m_b^2$ enhance the branching ratio $\mathcal{B}(B \to X_s\gamma)$ by more than a factor 2, and hence such logs have to be resummed. This is done using an effective field theory approach, obtained by integrating out the top quark and the $W^\pm$ bosons. Keeping terms up to dimension-6, the effective Lagrangian for $B \to X_s\gamma$ and $B \to X_s\ell^+\ell^-$ reads as follows:

$$\mathcal{L} = \mathcal{L}_{QCD\times QED}(q,l) + \frac{4G_F}{\sqrt{2}} V_{ts}^* V_{tb} \sum_{i=1}^{10} C_i(\mu) O_i$$

$(q = u, d, s, c, b, \quad l = e, \mu, \tau)$

$$O_i = \begin{cases} (\bar{s}\Gamma_i c)(\bar{c}\Gamma_i' b), & i = 1, (2), \quad C_i(m_b) \sim -0.26\ (1.02) \\ (\bar{s}\Gamma_i b)\Sigma_q(\bar{q}\Gamma_i' q), & i = 3,4,5,6, \quad |C_i(m_b)| < 0.08 \\ \frac{e m_b}{16\pi^2}\bar{s}_L\sigma^{\mu\nu}b_R F_{\mu\nu}, & i = 7, \quad C_7(m_b) \sim -0.3 \\ \frac{g m_b}{16\pi^2}\bar{s}_L\sigma^{\mu\nu}T^a b_R G^a_{\mu\nu}, & i = 8, \quad C_8(m_b) \sim -0.16 \\ \frac{e^2}{16\pi^2}(\bar{s}_L\gamma_\mu b_L)(\bar{l}\gamma^\mu(\gamma_5)l), & i = 9, (10) \quad C_i(m_b) \sim 4.27\ (-4.2). \end{cases}$$

Here, G_F is the Fermi coupling constant, V_{ij} are the CKM matrix elements, O_i are the four-Fermi and dipole operators, and $C_i(\mu)$ are the Wilson coefficients, evaluated at the scale μ, which is taken typically as $\mu = m_b$, and their values in the NNLO accuracy are given above for $\mu = 4.8$ GeV. Variations due to a different choice of μ and uncertainties from the upper scale-setting $m_t/2 \leq \mu_0 \leq 2m_t$ can be seen elsewhere.[35]

There are three essential steps of the calculation:

- Matching: Evaluating $C_i(\mu_0)$ at $\mu_0 \sim M_W$ by requiring the equality of the SM and the effective theory Green functions.
- Mixing: Deriving the effective theory renormalisation group equation (RGE) and evolving $C_i(\mu)$ from μ_0 to $\mu_b \sim m_b$.
- Matrix elements: Evaluating the on-shell amplitudes at $\mu_b \sim m_b$.

All three steps have been improved in perturbation theory and now include the next-to-next-to-leading order effects (NNLO), i.e., contributions up to $O(\alpha_s^2(m_b))$. A monumental theoretical effort stretched well over a decade with the participation of a large number of theorists underlies the current theoretical precision of the branching ratio. The result is usually quoted for a threshold photon energy to avoid experimental background from other Bremsstrahlung processes. For the decay with $E_\gamma > 1.6$ GeV in the rest frame of the B meson, the result at NNLO accuracy is[40,41]

$$\mathcal{B}(B \to X_s\gamma) = (3.36 \pm 0.23) \times 10^{-4}, \tag{1}$$

where the dominant theoretical uncertainty is non-perturbative.[42] This is to be compared with the current experimental average of the same[39]

$$\mathcal{B}(B \to X_s\gamma) = (3.43 \pm 0.21 \pm 0.07) \times 10^{-4}, \tag{2}$$

where the first error is statistical and the second systematic, yielding a ratio 1.02 ± 0.08, providing a test of the SM to an accuracy better than 10%.

The CKM-suppressed decay $B \to X_d\gamma$ has also been calculated in the NNLO precision. The result for $E_\gamma > 1.6$ GeV is[40]

$$\mathcal{B}(B \to X_d\gamma) = (1.73^{+0.12}_{-0.22}) \times 10^{-5}. \tag{3}$$

This will be measured precisely at Belle II. The constraints on the CP asymmetry are not very restrictive, but the current measurements are in agreement with the SM expectation. For further details, see HFAG.[39]

2.3. *Bounds on the charged Higgs mass from $\mathcal{B}(B \to X_s\gamma)$*

As the agreement between the SM and data is excellent, the decay rate for $B \to X_s\gamma$ provides constraints on the parameters of the BSM theories, such as supersymmetry and the 2 Higgs-doublet models (2HDM). In calculating the BSM effects, depending on the model, the SM operator basis may have to be enlarged, but in many cases one anticipates additive contributions to the Wilson coefficients in the SM basis. In the context of $B \to X_s\gamma$, it is customary to encode the BSM effects in the Wilson coefficients of the dipole operators $C_7(\mu)$ and $C_8(\mu)$, and the constraints from the branching ratio on the additive coefficients ΔC_7 and ΔC_8 then takes the numerical form[40]

$$\mathcal{B}(B \to X_s\gamma) \times 10^4 = (3.36 \pm 0.23) - 8.22\Delta C_7 - 1.99\Delta C_8. \tag{4}$$

To sample the kind of constraints that can be derived on the parameters of the BSM models, the 2HDM is a good case, as the branching ratio for the decay $B \to X_s\gamma$ in this model has been derived to the same theoretical accuracy.[44] The Lagrangian for the 2HDM is

$$\mathcal{L}_{H^+} = (2\sqrt{2}G_F)^{1/2}\Sigma_{i,j=1}^3 \bar{u}_i(A_u m_{u_i} V_{ij} P_L - A_d m_{d_j} V_{ij} P_R) d_j H^* + h.c., \tag{5}$$

where V_{ij} are the CKM matrix elements and $P_{L/R} = (1 \mp \gamma_5)/2$. The 2HDM contributions to the Wilson coefficients are proportional to $A_i A_j^*$, representing the contributions from the up-type A_u and down-type A_d quarks. They are defined in terms of the ratio of the vacuum expectation values, called $\tan\beta$, and are model dependent.

- 2HDM of type-I: $A_u = A_d = \frac{1}{\tan\beta}$,
- 2HDM of type-II: $A_u = -1/A_d = \frac{1}{\tan\beta}$.

Table 1. Measurements [HFAG 2014][39] and SM-based estimates of $\mathcal{B}(B \to (K^*, \rho)\gamma)$ and $\mathcal{B}(B_s \to \phi\gamma)$ in units of 10^{-5}, and the ratio $\mathcal{B}(B_s \to \phi\gamma)/\mathcal{B}(B^0 \to K^{*0}\gamma)$.

Decay Mode	Expt. (HFAG)	Theory (SM)
$B^0 \to K^{*0}\gamma$	4.33 ± 0.15	4.6 ± 1.4
$B^+ \to K^{*+}\gamma$	4.21 ± 0.18	4.3 ± 1.4
$B_s \to \phi\gamma$	3.59 ± 0.36	4.3 ± 1.4
$B_s \to \phi\gamma/B^0 \to K^{*0}\gamma$	0.81 ± 0.08	1.0 ± 0.2
$B^0 \to \rho^0\gamma$	$0.86^{+0.15}_{-0.14}$	0.65 ± 0.12
$B^+ \to \rho^+\gamma$	0.98 ± 0.25	1.37 ± 0.26

decays $B \to K^*\gamma$ and $B_s \to \phi\gamma$ are concerned, the partial NNLO theory is still the state-of-the-art. Their branching ratios as well as the ratio of the decay rates $\mathcal{B}(B_s \to \phi\gamma/\mathcal{B}(B \to K^*\gamma)$ are given in Table 1, together with the current experimental averages.[39] The corresponding calculations for the CKM-suppressed decays $B \to (\rho, \omega)\gamma$ are not yet available to the desired theoretical accuracy, due to the annihilation contributions, for which, to the best of my knowledge, no factorisation theorem of the kind discussed above has been proven. The results from a QCD-Factorisation based approach[48] for $B \to \rho\gamma$ are also given in Table 1 and compared with the data. The exclusive decay rates shown are in agreement with the experimental measurements, though theoretical precision is not better than 20%. Obviously, there is need for a better theoretical description, more so as Belle II will measure the radiative decays with greatly improved precision. I will skip a discussion of the isospin and CP asymmetries in these decays, as the current experimental bounds[39] are not yet probing the SM in these observables.

3. Semileptonic $b \to s$ Decays $B \to (X_s, K, K^*)\ell^+\ell^-$

There are two $b \to s$ semileptonic operators in SM:

$$O_i = \frac{e^2}{16\pi^2}(\bar{s}_L\gamma_\mu b_L)(\bar{l}\gamma^\mu(\gamma_5)l), \qquad i = 9, (10)$$

Their Wilson Coefficients have the following perturbative expansion:

$$C_9(\mu) = \frac{4\pi}{\alpha_s(\mu)}C_9^{(-1)}(\mu) + C_9^{(0)}(\mu) + \frac{\alpha_s(\mu)}{4\pi}C_9^{(1)}(\mu) + \cdots,$$

$$C_{10} = C_{10}^{(0)} + \frac{\alpha_s(M_W)}{4\pi}C_{10}^{(1)} + \cdots.$$

The term $C_9^{(-1)}(\mu)$ reproduces the electroweak logarithm that originates from the photonic penguins with charm quark loops, shown below.[58]

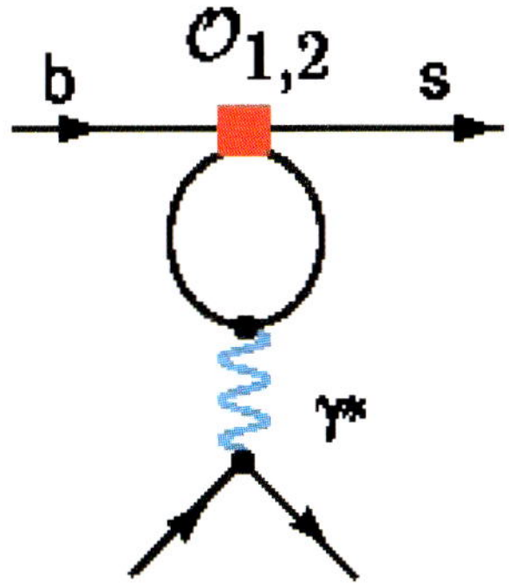

The first two terms in the perturbative expansion of $C_9(m_b)$ are

$$C_9^{(0)}(m_b) \simeq 2.2; \quad \frac{4\pi}{\alpha_s(m_b)} C_9^{(-1)}(m_b) = \frac{4}{9} \ln \frac{M_W^2}{m_b^2} + \mathcal{O}(\alpha_s) \simeq 2.$$

As they are very similar in magnitude, one needs to calculate the NNLO contribution to get reliable estimates of the decay rate. In addition, leading power corrections in $1/m_c$ and $1/m_b$ are required.

3.1. *Semileptonic decays $B \to X_s \ell^+\ell^-$*

A lot of theoretical effort has gone into calculating the perturbative QCD NNLO, electromagnetic logarithms and power corrections.[58–62] The B-factory experiments Babar and Belle have measured the dilepton invariant mass spectrum $d\mathcal{B}(B \to X_s\ell^+\ell^-)/dq^2$ practically in the entire kinematic region and have also measured the so-called Forward–Backward lepton asymmetry $A_{FB}(q^2)$.[63] They are shown in Fig. 10, and compared with the SM-based theoretical calculations. Note that a cut of $q^2 > 0.2$ GeV^2 on the dilepton invariant squared mass is used. As seen in these figures, two resonant regions near $q^2 = M^2_{J/\psi}$ and $q^2 = M^2_{J/\psi'}$ have to be excluded when comparing with the short-distance contribution. They make up what is called the long-distance contribution from the processes $B \to X_s + (J/\psi, J/\psi') \to X_s + \ell^+\ell^-$, whose dynamics is determined by the hadronic matrix elements of the operators O_1 and O_2. They have also been calculated via a dispersion relation[64] and data on the measured quantity $R_{\text{had}}(s) = \sigma(e^+e^- \to \text{hadrons})/\sigma(e^+e^- \to \mu^+\mu^-)$, and in some analyses are also included. As the (short-distance) contribution is expected to be a smooth function of q^2, one uses the perturbative distributions in interpolating these regions as well. The experimental distributions are in agreement with the SM, including also the zero point of $A_{FB}(q^2)$, which is a sensitive function of the ratio of the two Wilson coefficients C_9 and C_{10}.

collaboration in the ratio involving $B^\pm \to K^\pm\mu^+\mu^-$ and $B^\pm \to K^\pm e^+e^-$ measured in the low-q^2 region, which can be calculated rather accurately. In the interval $1 \le q^2 \le 6$ GeV^2, LHCb finds[75]

$$R_K \equiv \frac{\Gamma(B^\pm \to K^\pm\mu^+\mu^-)}{\Gamma(B^\pm \to K^\pm e^+e^-)} = 0.745^{+0.090}_{-0.074}(\text{stat}) \pm 0.035(\text{syst}). \quad (10)$$

This ratio in the SM is close to 1 to a very high accuracy[76] over the entire q^2 region measured by the LHCb. Thus, the measurement in (10) amounts to about 2.6σ deviation from the SM. Several BSM scenarios have been proposed to account for the R_K anomaly, discussed below, including a Z'-extension of the SM.[77] It should, however, be noted that the currently measured branching ratios $\mathcal{B}(B^\pm \to K^\pm e^+e^-) = (1.56^{+0.19+0.06}_{-0.15-0.4}) \times 10^{-7}$ and $\mathcal{B}(B^\pm \to K^\pm\mu^+\mu^-) = (1.20 \pm 0.09 \pm 0.07) \times 10^{-7}$ are also lower than the SM estimates $\mathcal{B}^{\text{SM}}(B^\pm \to K^\pm e^+e^-) = \mathcal{B}^{\text{SM}}(B^\pm \to K^\pm\mu^+\mu^-) = (1.75^{+0.60}_{-0.29}) \times 10^{-7}$, and the experimental error on the $\mathcal{B}(B^\pm \to K^\pm e^+e^-)$ is twice as large. One has to also factor in that the electrons radiate very profusely (compared to the muons) and implementing the radiative corrections in hadronic machines is anything but straight forward. In coming years, this and similar ratios, which can also be calculated to high accuracy, will be measured with greatly improved precision at the LHC and Belle II.

The other place where lepton non-universality is reported is in the ratios of the decays $B \to D^{(*)}\tau\nu_\tau$ and $B \to D^{(*)}\ell\nu_\ell$. Defining

$$R^{\tau/\ell}_{D^{(*)}} \equiv \frac{\mathcal{B}(B \to D^{(*)}\tau\nu_\tau)/\mathcal{B}^{\text{SM}}(B \to D^{(*)}\tau\nu_\tau)}{\mathcal{B}(B \to D^{(*)}\ell\nu_\ell)/\mathcal{B}^{\text{SM}}(B \to D^{(*)}\ell\nu_\ell)}, \quad (11)$$

the current averages of the BaBar, Belle, and the LHCb data are:[39]

$$R^{\tau/\ell}_D = 1.37 \pm 0.17; \quad R^{\tau/\ell}_{D^*} = 1.28 \pm 0.08. \quad (12)$$

This amounts to about 3.9σ deviation from the τ/ℓ ($\ell = e, \mu$) universality. Interestingly, this happens in a tree-level charged current process. If confirmed experimentally, this would call for a drastic contribution to an effective four-Fermi LL operator $(\bar{c}\gamma_\mu b_L)(\tau_L\gamma_\mu\nu_L)$. It is then conceivable that the non-universality in R_K (which is a loop-induced $b \to s$ process) is also due to an LL operator $(\bar{s}\gamma_\mu b_L)(\bar{\mu}_L\gamma_\mu\mu_L)$. Several suggestions along these lines involving a leptoquark have been made.[78–80] It is worth recalling that leptoquarks were introduced by Pati and Salam in 1973 in an attempt to unify leptons and quarks in $SU(4)$.[6,7] The lepton non-universality in B decays has revived the interest in theories with low-mass leptoquarks, discussed recently in a comprehensive work on this topic.[81]

3.4. *Angular analysis of the decay* $B^0 \to K^{*0}(\to K^+\pi^-)\mu^+\mu^-$

For the inclusive decays $B \to X_s\ell^+\ell^-$, the observables which have been measured are the integrated rates, the dilepton invariant mass $d\Gamma/dq^2$ and the FB asymmetry $A_{\text{FB}}(q^2)$. They are all found to be in agreement with the SM. In the exclusive

Table 2. Measurements [PDG 2014] and SM-based estimates[60] of the branching ratios $\mathcal{B}(B \to (X_s, K, K^*)\ell^+\ell^-)$ in units of 10^{-6}

Decay Mode	Expt. (BELLE & BABAR)	Theory (SM)
$B \to K\ell^+\ell^-$	0.48 ± 0.04	0.35 ± 0.12
$B \to K^*e^+e^-$	1.19 ± 0.20	1.58 ± 0.49
$B \to K^*\mu^+\mu^-$	1.06 ± 0.09	1.19 ± 0.39
$B \to X_s\mu^+\mu^-$	4.3 ± 1.2	4.2 ± 0.7
$B \to X_s e^+e^-$	4.7 ± 1.3	4.2 ± 0.7
$B \to X_s\ell^+\ell^-$	4.5 ± 1.0	4.2 ± 0.7

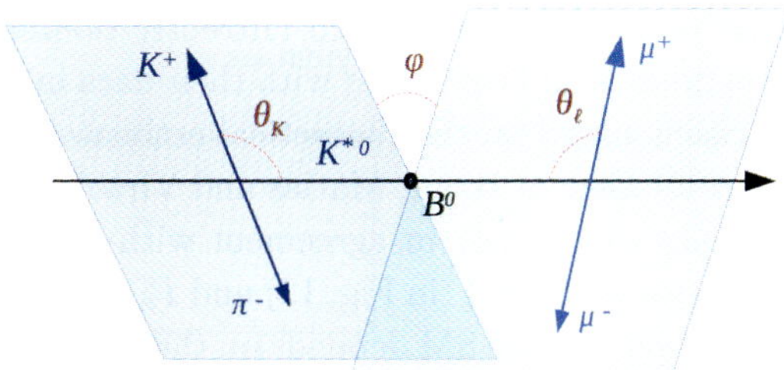

Fig. 11. Definitions of the angles in $B^0 \to K^{*0}(\to K^+\pi^-)\mu^+\mu^-$.

decays such as $B \to K^*\ell^+\ell^-$ and $B_s \to \phi\ell^+\ell^-$, a complete angular analysis of the decay is experimentally feasible. This allows one to measure a number of additional observables, defined below.

$$\begin{aligned}
\frac{1}{d(\Gamma+\bar{\Gamma})}\frac{d^4(\Gamma+\bar{\Gamma})}{dq^2 d\Omega} = \frac{9}{32\pi}\Bigg[&\frac{3}{4}(1-F_L)\sin^2\theta_K + F_L\cos^2\theta_K \\
&+\frac{1}{4}(1-F_L)\sin^2\theta_K\cos 2\theta_\ell - F_L\cos^2\theta_K\cos 2\theta_\ell \\
&+S_3\sin^2\theta_K\sin^2\theta_\ell\cos 2\phi + S_4\sin 2\theta_K\sin 2\theta_\ell\cos\phi \\
&+S_5\sin 2\theta_K\sin\theta_\ell\cos\phi + \frac{4}{3}A_{\rm FB}\sin^2\theta_K\cos\theta_\ell \\
&+S_7\sin 2\theta_K\sin\theta_\ell\sin\phi + S_8\sin 2\theta_K\sin 2\theta_\ell\sin\phi \\
&+S_9\sin^2\theta_K\sin^2\theta_\ell\sin 2\phi\Bigg]. \qquad (13)
\end{aligned}$$

The three angles θ_K, θ_ℓ and ϕ for the decay $B^0 \to K^{*0}(\to K^+\pi^-)\mu^+\mu^-$ are defined in Fig. 11. An angular analysis of the decay chains $B^0 \to K^{*0}(\to K^+\pi^-)\mu^+\mu^-$[82] and $B_s^0 \to \phi(\to K^+K^-)\mu^+\mu^-$[83] has been carried out by LHCb.

The observables in (13) are q^2-dependent coefficients of the Wilson coefficients and hence they probe the underlying dynamics. Since these coefficients have been calculated to a high accuracy, the remaining theoretical uncertainty lies in the form factors and also from the charm-quark loops. The form factors have been calculated

These form factors are related to the ones in the decay $B \to K\ell^+\ell^-$, called $f_i^K(q^2)$, discussed earlier, by $SU(3)_F$ symmetry. Of these, the form factors $f_+^\pi(q^2)$ and $f_0^\pi(q^2)$ are related by isospin symmetry to the corresponding ones measured in the charged current process $B^0 \to \pi^-\ell^+\nu_\ell$ by Babar and Belle, and they can be extracted from the data. This has been done using several parameterisations of the form factors with all of them giving an adequate description of the data.[92] Due to their analytic properties, the so-called z-expansion methods, in which the form factors are expanded in a Taylor series in z, employed in the Boyd–Grinstein–Lebed (BGL) parametrisation[93] and the Bourrely–Caprini–Lellouch (BCL)[94] parametrisation, are preferable.

The BGL parametrisation is used in working out the decay rate and the invariant dilepton mass distribution[92] for $B^+ \to \pi^+\ell^+\ell^-$, which is discussed below. The BCL-parametrisation is used by the lattice-QCD groups, the HPQCD[70,71] and Fermilab/MILC[72] collaborations, to determine the form factors $f_i^\pi(q^2)$ and $f_i^K(q^2)$. In particular, the Fermilab/MILC collaboration has worked out the dilepton invariant mass distribution in the decay of interest $B^+ \to \pi^+\ell^+\ell^-$, making use of their simulation in the large-q^2 region and extrapolating with the BCL parametrisation.

We first discuss the low-q^2 region ($q^2 \ll m_b^2$). In this case,, heavy quark symmetry (HQS) relates all three form factors of interest $f_i^\pi(q^2)$ and this can be used advantageously to have a reliable estimate of the dilepton invariant mass spectrum in this region. Including lowest order HQS-breaking, the resulting expressions for the form factors (for $q^2/m_b^2 \ll 1$) are worked out by Beneke and Feldmann.[73] Thus, fitting the form factor $f_+(q^2)$ from the charged current data on $B \to \pi\ell^+\nu_\ell$ decay, and taking into account the HQS and its breaking, lead to a model-independent predictions of the differential branching ratio (dimuon mass spectrum) in the neutral current process $B^+ \to \pi^+\ell^+\ell^-$ for low-q^2 values. However, the long-distance contribution, which arises from the processes $B^+ \to \pi^+(\rho^0,\omega) \to \pi^+\mu^+\mu^-$ are not included here. The SM invariant dilepton mass distribution in $B^+ \to \pi^+\ell^+\ell^-$ integrated over the range $1\text{GeV}^2 \leq q^2 \leq 8\text{GeV}^2$ yields a partial branching ratio

$$\mathcal{B}(B^+ \to \pi^+\mu^+\mu^-) = (0.57^{+0.07}_{-0.05}) \times 10^{-8}. \tag{22}$$

Thanks to the available data on the charged current process and heavy quark symmetry, this yields an accuracy of about 10% for an exclusive branching ratio, comparable to the theoretical accuracy in the inclusive decay $B \to X_s\gamma$, discussed earlier. Thus, the decay $B^+ \to \pi^+\mu^+\mu^-$ offers a key advantage compared to the decay $B^+ \to K^+\ell^+\ell^-$, in which case the charged current process is not available.

The differential branching ratio in the entire q^2 region is given by

$$\frac{d\mathcal{B}(B^+ \to \pi^+\ell^+\ell^-)}{dq^2} = C_B|V_{tb}V_{td}^*|^2\sqrt{\lambda(q^2)}\sqrt{1 - \frac{4m_\ell^2}{q^2}}F(q^2), \tag{23}$$

where the constant $C_B = G_F^2\alpha_{\text{em}}^2\tau_B/1024\pi^5 m_B^3$ and $\lambda(q^2)$ is the usual kinematic function $\lambda(q^2) = (m_B^2 + m_\pi^2 - q^2)^2 - 4m_B^2m_\pi^2$. The function $F(q^2)$ depends on the effective Wilson coefficients, $C_7^{\rm eff}$, $C_9^{\rm eff}$, and $C_{10}^{\rm eff}$, and the three form factors $f_+^\pi(q^2)$, $f_0^\pi(q^2)$ and $f_T^\pi(q^2)$. A detailed discussion of the determination of the form factors, of which only $f_+^\pi(q^2)$ and $f_T^\pi(q^2)$ are numerically important for $\ell^\pm = e^\pm, \mu^\pm$ is given elsewhere.[92] We recall that $f_+^\pi(q^2)$ is constrained by the data on the charged current process in the entire q^2 domain. In addition, the lattice-QCD results on the form factors in the large-q^2 domain and the HQS-based relations in the low-q^2 region provide sufficient constraints on the form factor. This has enabled a rather precise determination of the invariant dilepton mass distribution in $B^+ \to \pi^+\ell^+\ell^-$.

Taking into account the various parametric and form-factor dependent uncertainties, this yields the following estimate for the branching ratio for $B^+ \to \pi^+\mu^+\mu^-$[96]

$$\mathcal{B}_{\rm SM}(B^+ \to \pi^+\mu^+\mu^-) = (1.88^{+0.32}_{-0.21}) \times 10^{-8}, \tag{24}$$

to be compared with the measured branching ratio by the LHCb collaboration[96] (based on 3fb^{-1} data):

$$\mathcal{B}_{\rm LHCb}(B^+ \to \pi^+\mu^+\mu^-) = (1.83 \pm 0.24 \pm 0.05) \times 10^{-8}, \tag{25}$$

where the first error is statistical and the second systematic, resulting in excellent agreement. The dimuon invariant mass distribution measured by the LHCb collaboration[96] is shown in Fig. 16, and compared with the SM-based theoretical prediction, called APR13,[92] and the lattice-based calculation, called FNAL/MILC 15.[72] Also shown is a comparison with a calculation, called HKR,[95] which has essentially the same short-distance contribution in the low-q^2 region, as discussed earlier, but additionally takes into account the contributions from the lower resonances ρ, ω and ϕ. This adequately describes the distribution in the q^2 bin, around 1 GeV^2.

With the steadily improving lattice calculations for the various input hadronic quantities and the form factors, theoretical error indicated in Eq. (24) will go down

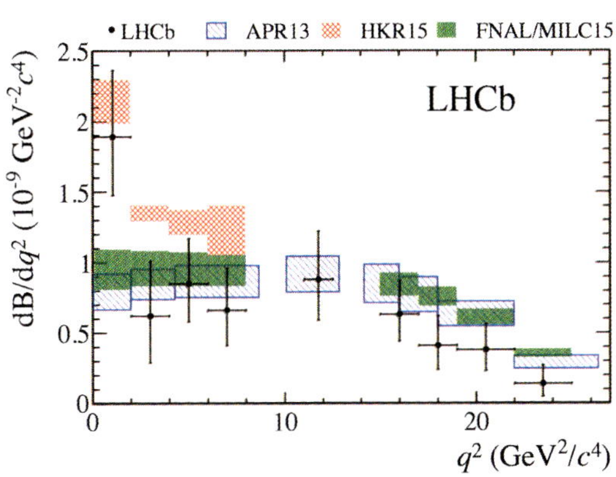

Fig. 16. Comparison of the dimuon invariant mass distribution in $B^+ \to \pi^+\mu^+\mu^-$ in the SM with the LHCb data.[96] Theoretical distributions shown are: APR13,[92] HKR15,[95] and FNAL/MILC.[72]

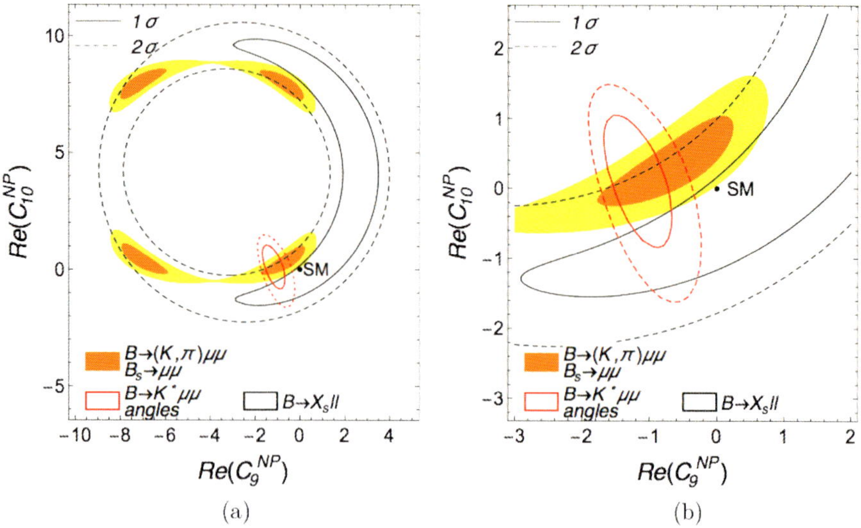

Fig. 18. Present constraints on the Wilson coefficients $\mathrm{Re}C_{10}^{\mathrm{NP}}$ vs. $\mathrm{Re}C_9^{\mathrm{NP}}$ from the semileptonic rare B-decays and $B_s \to \mu^+\mu^-$. The SM-point is indicated. (From Fermilab/MILC Lattice Collaboration.[72])

observables $F_{\mathrm{L}}, A_{\mathrm{FB}}, S_3, \ldots, S_9$ in $B^0 \to K^{*0}(\to K^+\pi^-)\mu^+\mu^-$ with the SM-based estimates was shown in Fig. 15, yielding a value of $\mathrm{Re}(C_9)$ which deviates from the SM by about 3σ. A number of groups has undertaken similar fits of the data and the outcome depends on the assumed correlations. However, it should be stressed that there are still non-perturbative contributions present in the current theoretical estimates which are not yet under complete quantitative control. The contributions from the charm quarks in the loops is a case in point. Also, form factor uncertainties are probably larger than assumed in some of these global fits.

A representative example of the kind of constraints on the Wilson coefficients C_9 and C_{10} that follow from the data on semileptonic and leptonic decays of the B mesons is shown in Fig. 18 from the Fermilab/MILC collaboration.[72] This shows that the SM point indicated by $(0,0)$ in the $\mathrm{Re}(C_9^{NP}, \mathrm{Re}(C_{10}^{NP})$-plane lies a little beyond 2σ. In some other fits, the deviations are larger but still far short for a discovery of BSM effects. As a lot of the experimental input in this and similar analysis is due to the LHCb data, this has to be confirmed by an independent experiment. This, hopefully, will be done by Belle II. We are better advised to wait and see if these deviations become statistically significant enough to warrant new physics. Currently, the situation is tantalising but not conclusive.

7. Concluding Remarks

From the measurement by the CLEO collaboration of the rare decay $B \to X_s\gamma$ in 1995, having a branching ratio of about 3×10^{-4}, to the rarest of the measured B decays, $B^0 \to \mu^+\mu^-$, with a branching fraction of about 1×10^{-10} by the LHCb and CMS collaborations, SM has been tested over six orders of magnitude. This is an impressive feat, made possible by dedicated experimental programmes carried out with diverse beams and detection techniques over a period of more than 20 years. A sustained theoretical effort has accompanied the experiments all along, underscoring both the continued theoretical interest in b physics and an intense exchange between the two communities. With the exception of a few anomalies, showing deviations from the SM ranging between 2 to 4σ in statistical significance, a vast majority of the measurements is in quantitative agreement with the SM. In particular, all quark flavour transitions are described by the CKM matrix whose elements are now determined. The CP asymmetry measured so far in laboratory experiments is explained by the Kobayashi–Maskawa phase. FCNC processes, of which rare B-decays discussed here is a class, are governed by the GIM mechanism, with the particles in the SM (three families of quarks and leptons, electroweak gauge bosons, gluons, and the Higgs) accounting for all the observed phenomena — so far. Whether this astounding consistence will continue will be tested in the coming years, as the LHC experiments analyse more data, enabling vastly improved precision in some of the key measurements discussed here. In a couple of years from now, Belle II will start taking data, providing independent and new measurements. For now, one has to give the SM the benefit of doubt.

Acknowledgments

I thank the organisers of the Salam memorial conference for inviting me and Prof. K.K. Phua for the warm hospitality in Singapore.

References

1. A. Salam and J. C. Ward, *Phys. Lett.* **13**, 168 (1964).
2. D. Eisenhower, *Draft of the Presidential Speech before the General Assembly of the United Nations, Nov. 28, 1953*, Dwight D. Eisenhower Presidential Library, Museum and Boyhood home. https://eisenhower.archives.gov.
3. S. W. Leslie, Pakistan's Nuclear Taj Mahal, in *Physics Today, February 2015* (American Institute of Physics, Madison, 2015).
4. See, for example, Shahid-ur-Rehman, "Z.A. Bhutto", *Long Road to Chagai* (Print Wise Publication, Islamabad, 1999).
5. F. J. Hasert *et al.* [Gargamelle Neutrino Collaboration], *Phys. Lett. B* **46**, 138 (1973).
6. J. C. Pati and A. Salam, *Phys. Rev. D* **8**, 1240 (1973).
7. J. C. Pati and A. Salam, *Phys. Rev. D* **10**, 275 (1974) [Erratum: *ibid.* **11**, 703 (1975)].
8. J. C. Pati and A. Salam, *Phys. Rev. Lett.* **31**, 661 (1973).
9. A. Salam and J. A. Strathdee, *Nucl. Phys. B* **76**, 477 (1974).
10. H. Georgi and S. L. Glashow, *Phys. Rev. Lett.* **32**, 438 (1974).
11. H. Fritzsch and P. Minkowski, *Annals Phys.* **93**, 193 (1975).
12. A. Ali, C. Isham, T. Kibble and Riazuddin (eds.), Singapore, Singapore: World Scientific (1994) 679 p. (World Scientific series in 20th century physics, 5).
13. S. L. Glashow, *Nucl. Phys.* **22**, 579 (1961).
14. S. Weinberg, *Phys. Rev. Lett.* **19**, 1264 (1967).

15. A. Salam, *Conf. Proc. C* **680519**, 367 (1968).
16. S. L. Glashow, J. Iliopoulos and L. Maiani, *Phys. Rev. D* **2**, 1285 (1970).
17. N. Cabibbo, *Phys. Rev. Lett.* **10**, 531 (1963). M. Kobayashi and T. Maskawa, *Prog. Theor. Phys.* **49**, 652 (1973).
18. K. A. Olive *et al.* [Particle Data Group Collaboration], *Chin. Phys. C* **38**, 090001 (2014).
19. A. V. Manohar and M. B. Wise, *Camb. Monogr. Part. Phys. Nucl. Phys. Cosmol.* **10**, 1 (2000).
20. C. W. Bauer, S. Fleming, D. Pirjol and I. W. Stewart, *Phys. Rev. D* **63**, 114020 (2001) [hep-ph/0011336].
21. C. W. Bauer, S. Fleming and M. E. Luke, *Phys. Rev. D* **63**, 014006 (2000) [hep-ph/0005275].
22. C. W. Bauer, D. Pirjol and I. W. Stewart, *Phys. Rev. D* **65**, 054022 (2002) [hep-ph/0109045].
23. M. Beneke, A. P. Chapovsky, M. Diehl and T. Feldmann, *Nucl. Phys. B* **643**, 431 (2002) [hep-ph/0206152].
24. For an excellent review, see T. Becher, A. Broggio and A. Ferroglia, *Lect. Notes Phys.* **896** (2015) [arXiv:1410.1892 [hep-ph]].
25. S. Aoki *et al.*, arXiv:1607.00299 [hep-lat].
26. A. Ali, V. M. Braun and H. Simma, *Z. Phys. C* **63**, 437 (1994) [hep-ph/9401277].
27. P. Colangelo and A. Khodjamirian, in *At the Frontier of Particle Physics*, Vol. 3: ed. M. Shifman, pp. 1495–1576 [hep-ph/0010175].
28. P. Ball and R. Zwicky, *Phys. Rev. D* **71**, 014015 (2005) [hep-ph/0406232].
29. A. Bharucha, D. M. Straub and R. Zwicky, arXiv:1503.05534 [hep-ph].
30. R. Ammar *et al.* [CLEO Collaboration], *Phys. Rev. Lett.* **71**, 674 (1993).
31. M. S. Alam *et al.* [CLEO Collaboration], *Phys. Rev. Lett.* **74**, 2885 (1995).
32. A. Ali and C. Greub, *Phys. Lett. B* **259**, 182 (1991); *Z. Phys. C* **49**, 431 (1991).
33. S. Chen *et al.* [CLEO Collaboration], *Phys. Rev. Lett.* **87**, 251807 (2001) [hep-ex/0108032].
34. P. Koppenburg *et al.* [Belle Collaboration], *Phys. Rev. Lett.* **93**, 061803 (2004) [hep-ex/0403004].
35. T. Blake, G. Lanfranchi and D. M. Straub, arXiv:1606.00916 [hep-ph].
36. S. Descotes-Genon, L. Hofer, J. Matias and J. Virto, *JHEP* **1606**, 092 (2016) [arXiv:1510.04239 [hep-ph]].
37. P. Koppenburg, Z. Dolezal and M. Smizanska, *Scholarpedia* **11**, 32643 (2016) [arXiv:1606.00999 [hep-ex]].
38. The physics of the B factories is reviewed excellently in A. J. Bevan *et al.* [BaBar and Belle Collaborations], *Eur. Phys. J. C* **74**, 3026 (2014) [arXiv:1406.6311 [hep-ex]].
39. Y. Amhis *et al.* [Heavy Flavor Averaging Group (HFAG) Collaboration], arXiv:1412.7515 [hep-ex].
40. M. Misiak *et al.*, *Phys. Rev. Lett.* **114**, 221801 (2015) [arXiv:1503.01789 [hep-ph]].
41. M. Czakon, P. Fiedler, T. Huber, M. Misiak, T. Schutzmeier and M. Steinhauser, *JHEP* **1504**, 168 (2015) [arXiv:1503.01791 [hep-ph]].
42. M. Benzke, S. J. Lee, M. Neubert and G. Paz, *JHEP* **1008**, 099 (2010) [arXiv:1003.5012 [hep-ph]].
43. M. Misiak *et al.*, *Phys. Rev. Lett.* **98**, 022002 (2007) [hep-ph/0609232].
44. T. Hermann, M. Misiak and M. Steinhauser, *JHEP* **1211**, 036 (2012) [arXiv:1208.2788 [hep-ph]].
45. A. Khodjamirian, T. Mannel, A. A. Pivovarov and Y.-M. Wang, *JHEP* **1009**, 089 (2010) [arXiv:1006.4945 [hep-ph]].

46. M. Beneke, G. Buchalla, M. Neubert and C. T. Sachrajda, *Phys. Rev. Lett.* **83** (1999) 1914 [hep-ph/9905312].
47. M. Beneke, T. Feldmann and D. Seidel, *Nucl. Phys. B* **612**, 25 (2001) [hep-ph/0106067].
48. A. Ali and A. Y. Parkhomenko, *Eur. Phys. J. C* **23**, 89 (2002) [hep-ph/0105302].
49. S. W. Bosch and G. Buchalla, *Nucl. Phys. B* **621**, 459 (2002) [hep-ph/0106081].
50. M. Beneke, T. Feldmann and D. Seidel, *Eur. Phys. J. C* **41**, 173 (2005) [hep-ph/0412400].
51. H. n. Li and H. L. Yu, *Phys. Rev. D* **53**, 2480 (1996) [hep-ph/ 9411308].
52. Y. Y. Keum, H. N. Li and A. I. Sanda, *Phys. Rev. D* **63**, 054008 (2001) [hep-ph/0004173].
53. Y. Y. Keum, M. Matsumori and A. I. Sanda, *Phys. Rev. D* **72**, 014013 (2005) [hep-ph/0406055].
54. C. D. Lu, M. Matsumori, A. I. Sanda and M. Z. Yang, *Phys. Rev. D* **72**, 094005 (2005) [Erratum: *ibid.* **73**, 039902 (2006)] [hep-ph/0508300].
55. J. g. Chay and C. Kim, *Phys. Rev. D* **68**, 034013 (2003) [hep-ph/0305033].
56. T. Becher, R. J. Hill and M. Neubert, *Phys. Rev. D* **72**, 094017 (2005) [hep-ph/0503263].
57. A. Ali, B. D. Pecjak and C. Greub, *Eur. Phys. J. C* **55**, 577 (2008) [arXiv:0709.4422 [hep-ph]].
58. A. Ghinculov, T. Hurth, G. Isidori and Y. P. Yao, *Nucl. Phys. B* **685**, 351 (2004) [hep-ph/0312128].
59. H. H. Asatryan, H. M. Asatrian, C. Greub and M. Walker, *Phys. Rev. D* **65**, 074004 (2002) [hep-ph/0109140].
60. A. Ali, E. Lunghi, C. Greub and G. Hiller, *Phys. Rev. D* **66**, 034002 (2002) [hep-ph/0112300].
61. T. Huber, E. Lunghi, M. Misiak and D. Wyler, *Nucl. Phys. B* **740**, 105 (2006) [hep-ph/0512066].
62. T. Huber, T. Hurth and E. Lunghi, *Nucl. Phys. B* **802**, 40 (2008) [arXiv:0712.3009 [hep-ph]].
63. A. Ali, T. Mannel and T. Morozumi, *Phys. Lett. B* **273**, 505 (1991).
64. F. Kruger and L. M. Sehgal, *Phys. Lett. B* **380**, 199 (1996) [hep-ph/9603237].
65. T. Huber, T. Hurth and E. Lunghi, *JHEP* **1506**, 176 (2015) [arXiv:1503.04849 [hep-ph]].
66. J. P. Lees *et al.* [BaBar Collaboration], *Phys. Rev. Lett.* **112**, 211802 (2014) [arXiv:1312.5364 [hep-ex]].
67. Y. Sato *et al.* [Belle Collaboration], *Phys. Rev. D* **93**, 032008 (2016) [Addendum: *ibid.* **93**, 059901 (2016)] [arXiv:1402.7134 [hep-ex]].
68. A. Bharucha, T. Feldmann and M. Wick, *JHEP* **1009**, 090 (2010) [arXiv:1004.3249 [hep-ph]].
69. E. Dalgic, A. Gray, M. Wingate, C. T. H. Davies, G. P. Lepage and J. Shigemitsu, *Phys. Rev. D* **73**, 074502 (2006) [Erratum: *ibid.* **75**, 119906 (2007)] [hep-lat/0601021].
70. C. M. Bouchard, G. P. Lepage, C. J. Monahan, H. Na and J. Shigemitsu, *PoS LATTICE* **2013**, 387 (2014) [arXiv:1310.3207 [hep-lat]].
71. C. Bouchard *et al.* [HPQCD Collaboration], *Phys. Rev. D* **88**, no. 5, 054509 (2013) [Erratum: *ibid.* **88**, 079901 (2013)] [arXiv:1306.2384 [hep-lat]].
72. J. A. Bailey *et al.* [Fermilab Lattice and MILC Collaborations], *Phys. Rev. Lett.* **115**, 152002 (2015) [arXiv:1507.01618 [hep-ph]].
73. M. Beneke and T. Feldmann, *Nucl. Phys. B* **592**, 3 (2001) [hep-ph/0008255].
74. A. Ali, G. Kramer and G. h. Zhu, *Eur. Phys. J. C* **47**, 625 (2006) [hep-ph/0601034].

75. R. Aaij *et al.* [LHCb Collaboration], *Phys. Rev. Lett.* **113**, 151601 (2014) [arXiv:1406.6482 [hep-ex]].
76. C. Bobeth, G. Hiller and G. Piranishvili, *JHEP* **0712**, 040 (2007) [arXiv:0709.4174 [hep-ph]].
77. C. W. Chiang, X. G. He and G. Valencia, *Phys. Rev. D* **93**, 074003 (2016) [arXiv:1601.07328 [hep-ph]].
78. G. Hiller and M. Schmaltz, *Phys. Rev. D* **90**, 054014 (2014) [arXiv:1408.1627 [hep-ph]].
79. M. Bauer and M. Neubert, *Phys. Rev. Lett.* **116**, 141802 (2016) [arXiv:1511.01900 [hep-ph]].
80. R. Barbieri, G. Isidori, A. Pattori and F. Senia, *Eur. Phys. J. C* **76**, 67 (2016) [arXiv:1512.01560 [hep-ph]].
81. I. Dorsner, S. Fajfer, A. Greljo, J. F. Kamenik and N. Kosnik, arXiv:1603.04993 [hep-ph].
82. R. Aaij *et al.* [LHCb Collaboration], *JHEP* **1602**, 104 (2016) [arXiv:1512.04442 [hep-ex]].
83. R. Aaij *et al.* [LHCb Collaboration], *JHEP* **1307**, 084 (2013) [arXiv:1305.2168 [hep-ex]].
84. S. Descotes-Genon, T. Hurth, J. Matias and J. Virto, *JHEP* **1305**, 137 (2013) [arXiv:1303.5794 [hep-ph]].
85. S. Jäger and J. Martin Camalich, *JHEP* **1305**, 043 (2013) [arXiv:1212.2263 [hep-ph]].
86. S. Jäger and J. Martin Camalich, *Phys. Rev. D* **93**, 014028 (2016) [arXiv:1412.3183 [hep-ph]].
87. W. Altmannshofer and D. M. Straub, *Eur. Phys. J. C* **75**, 382 (2015) [arXiv:1411.3161 [hep-ph]].
88. T. Hurth and F. Mahmoudi, *JHEP* **1404**, 097 (2014) [arXiv:1312.5267 [hep-ph]].
89. S. Descotes-Genon, L. Hofer, J. Matias and J. Virto, *JHEP* **1412**, 125 (2014) [arXiv:1407.8526 [hep-ph]].
90. A. Abdesselam *et al.* [Belle Collaboration], arXiv:1604.04042 [hep-ex].
91. M. Ciuchini, M. Fedele, E. Franco, S. Mishima, A. Paul, L. Silvestrini and M. Valli, *JHEP* **1606**, 116 (2016) [arXiv:1512.07157 [hep-ph]].
92. A. Ali, A. Y. Parkhomenko and A. V. Rusov, *Phys. Rev. D* **89**, 094021 (2014) [arXiv:1312.2523 [hep-ph]].
93. C. G. Boyd, B. Grinstein and R. F. Lebed, *Phys. Rev. Lett.* **74**, 4603 (1995) [hep-ph/9412324].
94. C. Bourrely, I. Caprini and L. Lellouch, *Phys. Rev. D* **79**, 013008 (2009) [Erratum: *ibid.* **82**, 099902 (2010)] [arXiv:0807.2722 [hep-ph]].
95. C. Hambrock, A. Khodjamirian and A. Rusov, *Phys. Rev. D* **92**, 074020 (2015) [arXiv:1506.07760 [hep-ph]].
96. R. Aaij *et al.* [LHCb Collaboration], *JHEP* **1510** (2015) 034 [arXiv:1509.00414 [hep-ex]].
97. V. Khachatryan *et al.* [CMS and LHCb Collaborations], *Nature* **522**, 68 (2015) [arXiv:1411.4413 [hep-ex]].
98. C. Bobeth, M. Gorbahn, T. Hermann, M. Misiak, E. Stamou and M. Steinhauser, *Phys. Rev. Lett.* **112**, 101801 (2014) [arXiv:1311.0903 [hep-ph]].
99. T. Hermann, M. Misiak and M. Steinhauser, *JHEP* **1312**, 097 (2013) [arXiv:1311.1347 [hep-ph]].
100. C. Bobeth, M. Gorbahn and E. Stamou, *Phys. Rev. D* **89**, 034023 (2014) [arXiv:1311.1348 [hep-ph]].

101. K. De Bruyn, R. Fleischer, R. Knegjens, P. Koppenburg, M. Merk, A. Pellegrino and N. Tuning, *Phys. Rev. Lett.* **109**, 041801 (2012) [arXiv:1204.1737 [hep-ph]].
102. S. Aoki *et al.*, *Eur. Phys. J. C* **74**, 2890 (2014) [arXiv:1310.8555 [hep-lat]].
103. A. J. Buras, F. De Fazio and J. Girrbach, *JHEP* **1402**, 112 (2014) [arXiv:1311.6729 [hep-ph]].

The World of Long-Range Interactions: A Bird's Eye View

Shamik Gupta

Department of Physics, Ramakrishna Mission Vivekananda University, Belur Math, Howrah 711 202, West Bengal, India
shamikg1@gmail.com

Stefano Ruffo

SISSA, INFN and ISC-CNR, Via Bonomea 265, I-34136 Trieste, Italy
ruffo@sissa.it

In recent years, studies of long-range interacting (LRI) systems have taken center stage in the arena of statistical mechanics and dynamical system studies, due to new theoretical developments involving tools from as diverse a field as kinetic theory, non-equilibrium statistical mechanics, and large deviation theory, but also due to new and exciting experimental realizations of LRI systems. In the first, introductory, Section 1, we discuss the general features of long-range interactions, emphasizing in particular the main physical phenomenon of non-additivity, which leads to a plethora of distinct effects, both thermodynamic and dynamic, that are not observed with short-range interactions: Ensemble inequivalence, slow relaxation, broken ergodicity. In Section 2, we discuss several physical systems with long-range interactions: mean-field spin systems, self-gravitating systems, Euler equations in two dimensions, Coulomb systems, one-component electron plasma, dipolar systems, free-electron lasers. In Section 3, we discuss the general scenario of dynamical evolution of generic LRI systems. In Section 4, we discuss an illustrative example of LRI systems, the Kardar–Nagel spin system, which involves discrete degrees of freedom, while in Section 5, we discuss a paradigmatic example involving continuous degrees of freedom, the so-called Hamiltonian mean-field (HMF) model. For the former, we demonstrate the effects of ensemble inequivalence and slow relaxation, while for the HMF model, we emphasize in particular the occurrence of the so-called quasistationary states (QSSs) during relaxation towards the Boltzmann–Gibbs equilibrium state. The QSSs are non-equilibrium states with lifetimes that diverge with the system size, so that in the thermodynamic limit, the systems remain trapped in the QSSs, thereby making the latter the effective stationary states. In Section 5, we also discuss an experimental system involving atoms trapped in optical cavities, which may be modelled by the HMF system. In Section 6, we address the issue of ubiquity of the quasistationary behavior by considering a variety of models and dynamics, discussing in each case the conditions to observe QSSs. In Section 7, we investigate the issue of what happens when a long-range system is driven out of thermal equilibrium. Conclusions are drawn in Section 8.

Keywords: Long-range interactions; non-additivity; ensemble inequivalence; slow relaxation; quasi-stationary states.

1. Introduction: General Considerations

In this Section, we discuss the generalities of long-range interacting systems. A detailed discussion, with extensive lists of references, may be found in several recent articles and books, see Refs. 1–8. More recent works discussed in the later parts of this article are covered in Refs. 20–31.

Long-range interacting (LRI) systems are those in which the two-body interparticle potential decays at large separation r as

$$V(r) \sim \frac{J}{r^\alpha}; \ 0 \le \alpha \le d, \tag{1}$$

where d is the dimension of the embedding space, and J is the coupling strength.[a] The range of allowed values of the decay exponent α implies that the energy per particle, ε, scales super-linearly with the system size. This feature is easily demonstrated by considering the example of a particle placed at the center of a hypersphere of radius R in d dimensions, with the other particles homogeneously distributed with a mass density ρ. For such a system, considering the interaction potential (1), the energy per particle is given as

$$\varepsilon = \int_\delta^R \mathrm{d}^d r \ \rho \frac{J}{r^\alpha} = \frac{\rho J \Omega_d}{d-\alpha}[R^{d-\alpha} - \delta^{d-\alpha}], \tag{2}$$

where $\delta \to 0$ is a short distance cut-off introduced to exclude the contribution to the energy due to particles located in a small neighborhood of radius δ, and is motivated by the need to regularize the divergence of the potential (1) at short distances. In Eq. (2), Ω_d denotes the angular volume in d dimensions. From the equation, it follows that as R is increased, the energy ε remains finite for $\alpha > d$, implying thereby the linear scaling of the total energy E with the volume $V \sim R^d$, thus making the system extensive. These systems are called short-range interacting systems. On the other hand, for our allowed values of α, the energy ε scales with the volume as $\varepsilon \sim V^{1-\alpha/d}$ (the energy scales logarithmically with V in the marginal case $\alpha = d$), thereby implying a super-linear scaling of the total energy of LRI systems with V, as $E \propto V^{2-\alpha/d}$. The LRI systems are thus generically non-extensive. For such systems, on computing the free energy $F \equiv E - TS$, with T being the intensive temperature and S being the entropy that typically scales linearly with the volume, $S \sim V$, we find due to the super-linear scaling of E with V that the thermodynamic properties are dominated by the energy. In particular, the equilibrium state of a mechanically isolated LRI system, obtained by minimizing F, corresponds to the one with the minimum energy, that is, the ground state, allowing for no thermal fluctuations. Of course, in reality, there ought to be a competition between the energy and the entropy contributions to the free energy in order to have such phenomena as phase

[a] An alternative classification, based on dynamical considerations, namely, the conditions for the existence of the so-called quasistationary states in LRI systems, is proposed in A. Gabrielli, M. Joyce, and J. Morand, *Phys. Rev. E* **90**, 062910 (2014).

transitions that are known to occur in LRI systems. A way out from this energy dominance consists in scaling the coupling constant as

$$J \to \frac{J}{V^{\alpha/d-1}}, \tag{3}$$

thereby making the energy extensive in the volume. Note that this is just a "mathematical trick" (Kac's trick) to properly study LRI systems within the framework of equilibrium statistical mechanics that exists for short-range ones, and does not correspond to any physical effect. Indeed, no interaction whose strength changes on varying the volume is known to occur. Applying this trick, one can obtain the free energy per particle, and then revert to the actual physical description by scaling back the coupling constant. An equivalent alternative to Kac's trick, which still allows for an effective competition between the energy and the entropy contribution to the free energy, consists in rescaling the temperature as

$$T \to \frac{T}{V^{1-\alpha/d}}. \tag{4}$$

Beyond the rescaling procedures discussed above, which were implemented to obtain a meaningful large-volume limit for LRI systems and competing energy and entropy contributions to the free energy, let us illustrate how such a competition may actually occur in nature, by considering a relevant LRI system in the arena of astrophysics, namely, that of globular clusters, see Fig. 1. These clusters are gravitationally bound concentrations of $N \sim 10^4 - 10^6$ stars that are spread over a volume that has a diameter ranging from several tens to about 200 light years (1 light year $= 9.4 \times 10^{15}$ m). For a typical globular cluster (M2), one has $N = 1.5 \times 10^5$,

Fig. 1. Spherically symmetric mass distribution of stars in a globular cluster.

$R = 175$ light years, and total mass $M = 2 \times 10^{30}$ Kg. An order-of-magnitude estimate of energy and entropy may be done as follows:

$$E = \frac{GN^2M^2}{R}, \; S = k_B N \implies \frac{E}{S} \sim \frac{GNM^2}{k_B R} \sim 1.7 \times 10^{60} \text{ K}, \tag{5}$$

where G is the gravitational constant, and k_B is the Boltzmann constant. To such an extremely high temperature as $\sim 10^{60}$ K, one can associate a velocity by invoking energy equipartition (neglecting interactions), as $v = \sqrt{\frac{3k_BT}{M}} \simeq 5.9$ Km/s. Typical star velocities indeed range between a few Km/s to about 100 Km/s. Thus, for systems such as these for which the temperature is high enough, the energy, although super-linear in volume ($E \sim V^{5/3}$), can effectively compete with the entropy contribution to the free energy.

Let us now make an important remark: Although Kac's trick allows to obtain an energy that is extensive in the volume, it is not necessarily additive (additivity implies extensivity, but not the converse). A simple example will illustrate the point. Consider the well-studied Curie–Weiss model of magnetism, with the Hamiltonian given by

$$H_{\text{CW}} = -\frac{J}{2N} \sum_{1 \le i < j \le N} \sigma_i \sigma_j, \tag{6}$$

where $\sigma_i = \pm 1$ are spin variables occupying the sites i of a lattice. The model mimics a mean-field system (every spin interacting with every other with the same strength), which may be considered as the $\alpha \to 0$ limit of the potential (1). Being a mean-field system, one does not need to specify the structure of the underlying lattice, excepting to mention that every site is connected to every other. In the Hamiltonian (6), the coupling strength has been rescaled by using Kac's trick, so that the energy is extensive in the number of spins given by N. Let us consider a macrostate with zero total magnetization: $M \equiv \sum_{i=1}^{N} \sigma_i = 0$, which is composed of $N/2$ spin-$(+1)$ sites and $N/2$ spin-(-1) sites, see Fig. 2. Since the energy is proportional to the square of magnetization (see Eq. (6)), the total energy of the system is $E_{1+2} = 0$. However, the energy of the two parts, namely, $E_1 = E_2 = -J/8N$, does not vanish, and, therefore, $E_{1+2} \neq E_1 + E_2$.

As it will emerge in the rest of this article, the violation of additivity will be crucial in determining both thermodynamic and dynamic properties of LRI systems, making them quite distinct from short-range ones. This point will be demonstrated by considering several illustrative examples in the later parts of the paper. As a warm-up, we may mention that a violation of additivity implies a violation of convexity of the domain of accessible macrostates of an LRI system, for example, a magnetic one in the magnetization (M)–energy (E) plane. An example of such a violation is shown in Fig. 3, where the boundary of the region of accessible macrostates is shown to have the shape of a bean. For short-range systems for which

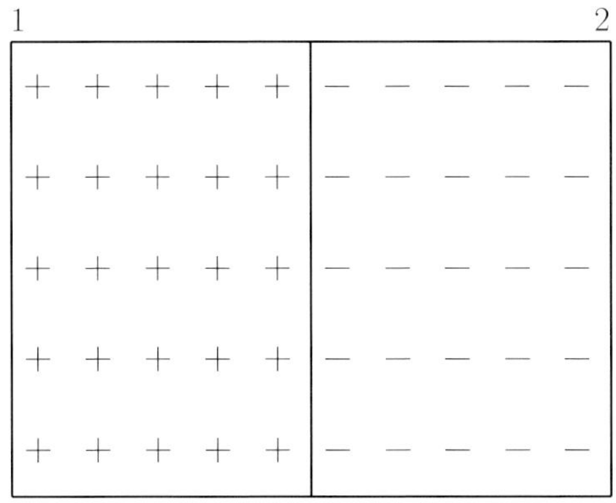

Fig. 2. Non-additivity in the extensive Curie–Weiss model: A zero magnetization macrostate of the Curie–Weiss system, Eq. (6), constituted by $N/2$ up-spins in domain 1 and $N/2$ down-spins in domain 2. Here, an up-spin is denoted by a + sign and a down-spin by a − spin.

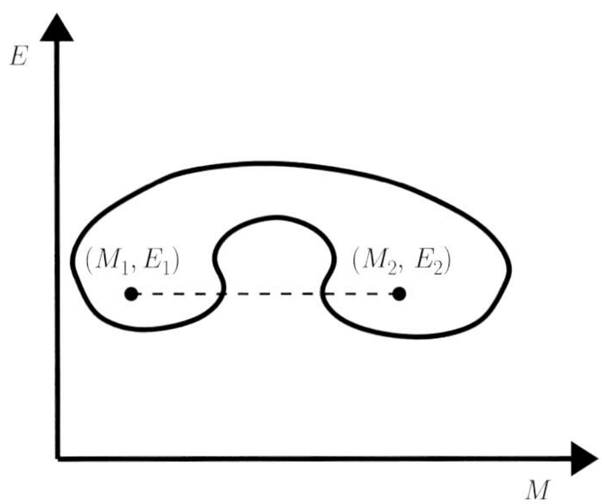

Fig. 3. Non-convex shape of the region of accessible macrostates in the magnetization-energy plane for a magnetic LRI system. The states (M_1, E_1) and (M_2, E_2) are not connected by any continuous energy-conserving dynamics.

additivity is satisfied, standard thermodynamics implies that all states satisfying

$$E = \lambda E_1 + (1 - \lambda)E_2, \ M = \lambda M_1 + (1 - \lambda)M_2, \ 0 \leq \lambda \leq 1 \tag{7}$$

must occur at the macroscopic level; this is in general not the case for LRI systems, and may imply a violation of ergodicity in the microcanonical ensemble. For example, the states (M_1, E_1) and (M_2, E_2) in Fig. 3 are not connected by any continuous energy-conserving dynamics.

On account of the violation of additivity, one should exercise caution in discussing equilibrium properties of LRI systems using the canonical ensemble whose derivation from the microcanonical ensemble relies on holding of additivity. Let us briefly recall the derivation. The microcanonical partition function for a system of

N particles contained in a volume V in $d = 3$ dimensions is given by

$$\Omega(E, V, N) \propto \int \mathrm{d}^{3N}q\mathrm{d}^{3N}p\ \delta(E - H(p, q)), \tag{8}$$

where (q, p) are the canonically conjugate variables, and H is the Hamiltonian. The entropy is defined via

$$S(E, V, N) = k_B \ln \Omega(E, V, N), \tag{9}$$

where an energy scale should be included in the logarithm to make its argument dimensionless. In deriving the canonical ensemble for a short-range system, one considers an isolated macroscopic system with energy E that is composed of a "small" part (the subsystem of interest) with energy equal to E_1, volume equal to V_1 and number of particles equal to N_1, and a "large" part that plays the role of a "bath", with energy equal to $E_2 \gg E_1$, volume equal to $V - V_1 \gg V_1$ and number of particles equal to $N - N_1 \gg N_1$. The additivity of energy implies that one has $E_2 = E - E_1$, so that the probability distribution $p(E_1)$ that the "small" system has energy E_1 is given by

$$p(E_1) \propto \Omega_2(E - E_1, V - V_1, N - N_1). \tag{10}$$

Using the definition of entropy and a Taylor expansion, one gets

$$\begin{aligned} p(E_1) &= \exp\left[S_2(E - E_1)\right] \\ &\approx \exp\left[S_2(E) - E_1 \left.\frac{\partial S_2}{\partial E}\right|_E + \cdots\right] \\ &\propto \Omega_2(E, V - V_1, N - N_1)\, e^{-\beta E_1}, \end{aligned} \tag{11}$$

where $S_2(E) \equiv S_2(E, V - V_1, N - N_1)$, and $\beta \equiv \left.\frac{\partial S_2}{\partial E}\right|_E$ is the inverse temperature: $\beta = 1/(k_B T)$. Equation (11) is the usual canonical ensemble description for the energy distribution of the system of interest. In describing LRI systems, which we have shown to be generically non-additive so that the derivation leading to (11) does not hold, we will consider both the microcanonical description, Eq. (8), and the canonical one, Eq. (11). For the latter, however, we will adopt an alternative physical interpretation, namely, that of the system of interest in interaction with an external heat bath at temperature T that induces stochastic fluctuations into the dynamics of the system.

Thermodynamic ensembles could be inequivalent for LRI systems[9–12]: a macroscopic physical state that is realizable in one ensemble is not realized in the other. We now discuss this point. Ensemble equivalence in the thermodynamic limit is mathematically based on certain properties of the partition functions. The thermodynamic limit corresponds to considering simultaneously the limits $N \to \infty$, $E \to \infty$ and $V \to \infty$, such that one has $N/V \to n$ and $E/N \to \varepsilon$, where the particle density $n \geq 0$ and the energy per particle ε are finite quantities. In this limit, the entropy per particle is given by

$$s(\varepsilon, n) \equiv \lim_{N\to\infty} \frac{1}{N} S(E, V, N). \tag{12}$$

The function $s(\varepsilon, n)$ is continuous, increasing with ε at a fixed n, so that the temperature $T = (\partial s/\partial \varepsilon)^{-1}$ is a positive quantity. For short-range systems, $s(\varepsilon, n)$ turns out to be a concave function of ε at a fixed n:

$$s\left(\lambda\varepsilon_1 + (1-\lambda)\varepsilon_2, n\right) \geq \lambda s(\varepsilon_1, n) + (1-\lambda)s(\varepsilon_2, n), \tag{13}$$

for any choice of ε_1 and ε_2, with $0 \leq \lambda \leq 1$. The partition function in the canonical ensemble is given by

$$Z(\beta, V, N) \equiv \int \mathrm{d}q^{3N} \mathrm{d}p^{3N} \ \exp\left[-\beta H(p, q)\right]. \tag{14}$$

In the thermodynamic limit, the free energy per particle is obtained as

$$f(\beta, n) \equiv -\frac{1}{\beta} \lim_{N\to\infty} \frac{1}{N} \ln Z(\beta, V, N). \tag{15}$$

Moreover, at a fixed n, the function $\phi(\beta, n) \equiv \beta f(\beta, n)$ (the rescaled free energy) is concave in β. The equivalence between the microcanonical and the canonical ensemble is a consequence of the concavity of ϕ and s and of the relation between these two functions given by the Legendre–Fenchel Transform (LFT). Indeed, one can easily prove that $\phi(\beta, n)$ is the LFT of $s(\varepsilon, n)$:

$$\phi(\beta, n) = \inf_{\varepsilon} \left[\beta\varepsilon - s(\varepsilon, n)\right], \tag{16}$$

and also the inverse LFT holds, since $s(\varepsilon, n)$ is concave in ε:

$$s(\varepsilon, n) = \inf_{\beta} \left[\beta\varepsilon - \phi(\beta, n)\right]. \tag{17}$$

These relations prove ensemble equivalence, because for each value of β, there is a value of ε that satisfies Eq. (16), and, conversely, for each value of ε, there is a value of β satisfying Eq. (17). Figure 4 provides a visual explanation of the relation between s and ϕ and of the correspondence between ε and β. Note that at a first-order phase transition, the entropy has a constant slope in the energy range $[\varepsilon_1, \varepsilon_2]$ (the phase coexistence region), resulting in a free energy with a cusp at the transition inverse temperature β_t, see Fig. 5.

Now, for LRI systems, the entropy may be a non-concave function of the energy. In this case, the Legendre–Fenchel transform is no more involutive: if applied to the entropy, it returns the correct free energy. However, the Legendre–Fenchel transform of the free energy does not coincide with the entropy, but rather with its concave envelope; this is the basic feature causing ensemble inequivalence.

violation of additivity due to the long-range nature of the interaction persists in all cases, as is represented by the occurrence of a negative specific heat. The latter phenomenon may be heuristically justified by using the virial theorem, which for the gravitational potential reads

$$\langle K\rangle = -\frac{1}{2}\langle U\rangle,\ \langle K\rangle = -E, \tag{24}$$

where $\langle\cdot\rangle$ denotes a temporal (i.e., dynamical) average. Since the kinetic energy K is always positive, it is clear that the virial theorem can only be valid for bound states for which E is negative. Using the equipartition theorem, we obtain the average kinetic energy as proportional to the temperature, and, hence, Eq. (24) implies that the specific heat c_V, which is proportional to $\mathrm{d}E/\mathrm{d}T$, is negative. More rigorously, it may be shown that regularized self-gravitating systems confined to a box have an entropy that is a non-concave function of the energy, see Fig. 6. Since the specific heat is related to the second derivative of the entropy with respect to energy, i.e.,

$$\partial^2 s/\partial\varepsilon^2 = -(c_V T^2)^{-1}, \tag{25}$$

it follows that in the energy range $[\varepsilon_a, \varepsilon_b]$, where the entropy is convex, the specific heat becomes negative. For short-range additive interactions, all states within the wider range $[\varepsilon_1, \varepsilon_2]$ would have an entropy that is represented by the thick dashed line in Fig. 6. In the figure, the inverse temperature β is also plotted as a function of ε, where note that in the region of negative specific heat, the temperature decreases as the energy increases.

Another important example of LRI systems is that of the Euler equations in two dimensions, governing incompressible, inviscid fluid flow:

$$\frac{\partial\vec{v}}{\partial t} + (\vec{v}\cdot\vec{\nabla})\vec{v} = 0,\ \vec{\nabla}\cdot\vec{v} = 0,\ \vec{v} = (v_x, v_y). \tag{26}$$

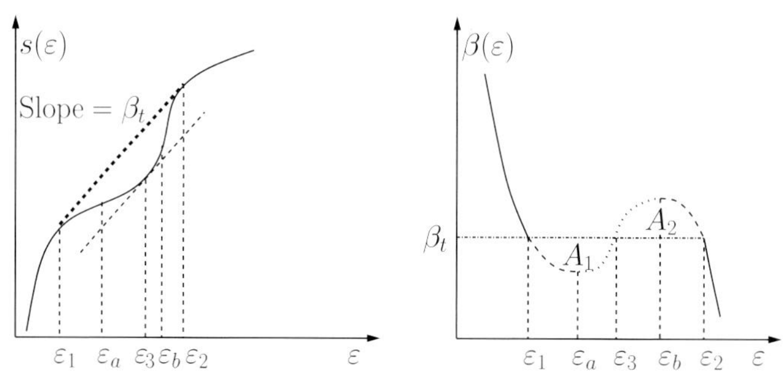

Fig. 6. (Left panel) Schematic shape of the microcanonical entropy per particle as a function of the energy per particle is denoted by the solid line, which shows a "globally" convex region in the range $[\varepsilon_1, \varepsilon_2]$; the thick dashed line realizes the "concave envelope". (Right panel) Inverse temperature β as a function of ε. According to the Maxwell's construction, the areas A_1 and A_2 must be equal: $A_1 = A_2$. The curve $\beta(\varepsilon)$ represents states that are stable (solid line), unstable (dotted line) and metastable (dashed lines).

Using the vorticity

$$\omega(x,y) \equiv \frac{\partial v_y}{\partial x} - \frac{\partial v_x}{\partial y}, \tag{27}$$

the Euler equations may be rewritten as

$$\frac{\partial \omega}{\partial t} + \vec{v} \cdot \vec{\nabla}\omega = 0. \tag{28}$$

The long-range features of this equation may be made explicit by introducing the stream function $\psi(x,y)$, as

$$v_x = +\frac{\partial \psi}{\partial y}, \ v_y = -\frac{\partial \psi}{\partial x}, \tag{29}$$

which is related to the vorticity by the Poisson equation:

$$\omega = -\Delta\psi. \tag{30}$$

Using the Green's function $G(\vec{r},\vec{r'})$, one may find the solution of the Poisson equation in a given domain D as

$$\psi(\vec{r}) = \int_D \mathrm{d}\vec{r'} \ \omega(\vec{r'}) \ G(\vec{r},\vec{r'}), \tag{31}$$

plus surface terms. In an infinite domain, one has

$$G\left(\vec{r},\vec{r'}\right) \equiv -\frac{1}{2\pi} \ln|\vec{r}-\vec{r'}|. \tag{32}$$

The energy is conserved for the Euler equation, and is given by

$$E = \int_D \mathrm{d}\vec{r} \ \frac{1}{2}(v_x^2 + v_y^2) = \int_D \mathrm{d}\vec{r} \ \frac{1}{2}\left(\nabla\psi\right)^2 = \frac{1}{2}\int_D \mathrm{d}\vec{r} \ \omega(\vec{r})\psi(\vec{r}) \tag{33}$$

$$= -\frac{1}{4\pi}\int_D\int_D \mathrm{d}\vec{r}\mathrm{d}\vec{r'} \ \omega(\vec{r'})\omega(\vec{r}) \ln|\vec{r}-\vec{r'}|, \tag{34}$$

which implies a logarithmic interaction between vortices at distant locations, thus corresponding to a decay with an effective exponent $\alpha = 0$. For a finite domain D, the Green's function contains additional surface terms that however gives no contribution to the energy (34) if the velocity field is tangent to the boundary of the domain (no outflow or inflow). One may demonstrate the non-additive features of the energy by considering the shear flow, Fig. 7, for which one has

$$v_x = -y, \ v_y = 0, \ \omega = 1, \ \psi = -y^2/2. \tag{35}$$

The energy per unit length, given by E/L, is along the x-direction of the flow and within $-1 \le y \le 1$ larger than the energy of the separate flows: $-1 \le y \le 0$, $0 \le y \le 1$:

$$\frac{E}{L} = \frac{1}{3}, \ \frac{E_{1,2}}{L} = \frac{1}{24}, \tag{36}$$

thereby demonstrating the violation of non-additivity of the energy.

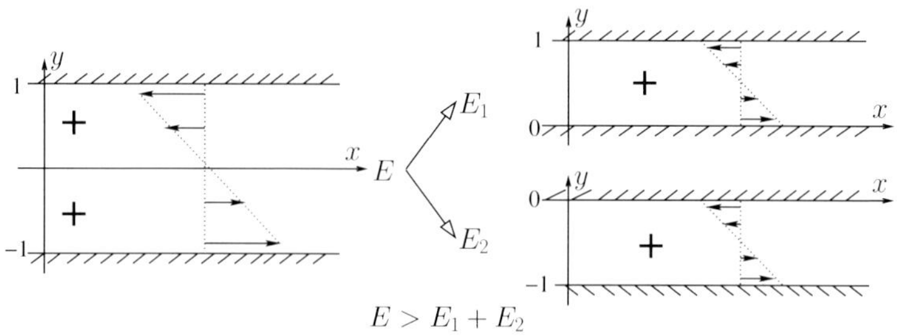

Fig. 7. An example showing the non-additivity of energy within the Euler equations for a shear flow.

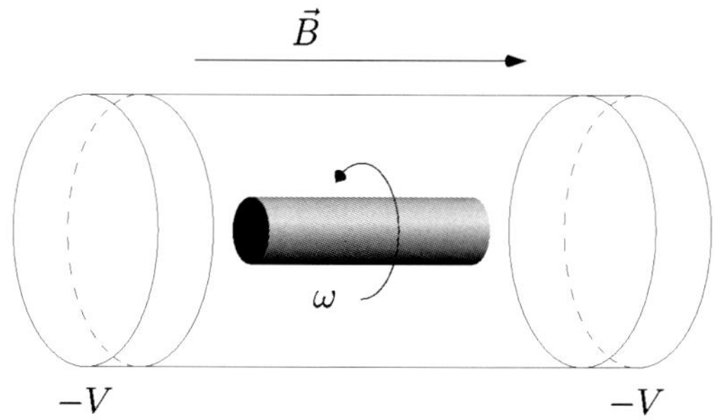

Fig. 8. Plasma of electrons confined by crossed electric and magnetic fields.

Coulomb systems constitute another relevant example of long-range interactions, which are of type (18):

$$U(\vec{r}_1, \ldots, \vec{r}_N) = \frac{1}{4\pi\varepsilon_0} \sum_{1 \le i < j \le N} e_i e_j V(|\vec{r}_i - \vec{r}_j|), \tag{37}$$

where ε_0 is the vacuum permittivity, and e_i is the charge located at position $\vec{r}_i$. For such systems, it may be shown that the excess charge is expelled to the boundary of a domain, and that the bulk is neutral. A typical configuration has a distribution of charges of equal sign surrounded by a "cloud" of particles of opposite charge, which "screens" the interactions at long range. The effective two-body potential is therefore given by

$$V_{\text{eff}} \propto \frac{\exp(-r/\lambda_D)}{r}, \tag{38}$$

where $\lambda_D \equiv (\varepsilon_0/(2ne^2\beta))^{1/2}$ is the so-called Debye length, and n is the particle density. On account of the screening, Coulomb systems are effectively short-range.

A plasma of electrons can be confined by a crossed electric field **E** and a magnetic field **B**.[13] As shown in Fig. 8, the electrons are contained axially by negative voltages and radially by a uniform axial magnetic field B_z. Under typical experimental conditions of density and temperature, electrons are collisionless; They bounce axially very rapidly and drift across the magnetic field with velocity

$$\mathbf{v} = \frac{-\nabla\phi \times \hat{z}}{B_z}. \tag{39}$$

As is the case of an effectively incompressible fluid, the electron density $n(x, y)$ obeys the evolution equations

$$\frac{\partial n}{\partial t} + \mathbf{v} \cdot \nabla n = 0, \tag{40}$$

$$\Delta\phi = \frac{en}{\varepsilon_0}, \tag{41}$$

where $-e$ is electron charge. These equations are isomorphic to the two-dimensional Euler equations with vorticity $\omega = en/\varepsilon_0$ and stream function $\psi = \phi/B_z$. The electron plasma may be regarded as the best experimental realization of two-dimensional incompressible, inviscid fluid.

Dipolar interaction is marginally long-range[14]: $\alpha = 3$ in $d = 3$. The interaction energy of two dipoles is

$$E_{ij} = \frac{\mu_0}{4\pi}\left[\frac{\vec{\mu}_i \cdot \vec{\mu}_j}{|\vec{r}_{ij}|^3} - \frac{3(\vec{\mu}_i \cdot \vec{r}_{ij})(\vec{\mu}_j \cdot \vec{r}_{ij})}{|\vec{r}_{ij}|^5}\right], \tag{42}$$

where μ_0 is vacuum permeability, and $\vec{\mu}_i$ is the dipolar moment at position site $\vec{r}_i$. Because of the anisotropy of the interaction, dipolar systems are strongly frustrated: several configurations have the same energy. For ferromagnetic samples of ellipsoidal shape, one has the total energy

$$E_{\mathrm{Dipolar}} = \frac{1}{2}\sum_{i,j} E_{ij} = E_0 V + \frac{1}{2}\mu_0 \frac{(\sum_i \vec{\mu}_i)^2}{V} D, \tag{43}$$

where E_0 is a local-energy term that depends on the crystal structure, and D is the so-called shape-dependent demagnetizing factor: $D = 1/3$ for spherical samples, $D = 0$ for needle shape samples, $D = 1$ for disk shaped samples. The free energy of a dipolar magnetic system is shape-independent, which implies that the macroscopic state cannot be ferromagnetic. However, ferromagnetism can exist in mesoscopic samples, paving the way to the possible experimental detection of long-range effects.

An experimental apparatus where long-range forces are at play is the free-electron laser.[15] In the linear free-electron laser, a relativistic electron beam propagates through a spatially periodic magnetic field, interacting with the co-propagating electromagnetic wave, see Fig. 9. Lasing occurs when the electrons bunch in a subluminar beat wave. After scaling away the time dependence of the phenomenon, and on introducing appropriate variables, e.g., the length z along the lasing direction, it is possible to capture the essence of the asymptotic state by studying the following equations of motion first introduced by Colson and Bonifacio:

$$\frac{\mathrm{d}\theta_j}{\mathrm{d}z} = p_j, \tag{44}$$

In view of the normalization of $\rho(z_1, z_2, \ldots, z_N, t)$, one has $\int \mathrm{d}z\ f_1(z,t) = N$, and in general, $\int \prod_{i=1}^{s} \mathrm{d}z_i\ f_s(z_1, z_2, \ldots, z_s, t) = N!/(N-s)!$.

Using Eq. (50), one may derive the time evolution of f_s to find that a determination of its evolution requires knowing the higher density f_{s+1}. It then follows that the time evolution of the full set $(f_1, f_2, f_3, \ldots, f_N)$ forms a coupled chain of equations, which goes by the name of the Bogoliubov–Born–Green–Kirkwood–Yvon (BBGKY) hierarchy. The first equation of the hierarchy reads

$$\left[\frac{\partial}{\partial t} + p_1 \frac{\partial}{\partial q_1}\right] f_1(z_1, t) = \int \mathrm{d}z_2\ \frac{\partial}{\partial q_1} V(|q_1 - q_2|) \frac{\partial}{\partial p_1} f_2(z_1, z_2, t). \tag{53}$$

Now, we may express the two-particle density quite generally as

$$f_2(z_1, z_2, t) = f_1(z_1, t) f_1(z_2, t) + g_2(z_1, z_2, t), \tag{54}$$

where g_2 describes the two-particle correlation. Integrating both sides with respect to (z_1, z_2), it then follows that $g_2 \sim N$. For an LRI system, let us now invoke Kac's trick, $J \to J/N$, implying $V \to V/N$. Then, using Eq. (53), and noting that $f_1 \sim N$, one obtains to leading order in $1/N$ the evolution equation[16]

$$\left[\frac{\partial}{\partial t} + p_1 \frac{\partial}{\partial q_1}\right] f_1(z_1, t) = \frac{\partial \Phi}{\partial q_1} \frac{\partial}{\partial p_1} f_1(z_1, t), \tag{55}$$

where $\Phi[f_1](q_1, t) = \int \mathrm{d}z_2\ V(|q_1 - q_2|) f_1(z_2, t)$ is the mean-field potential. Equation (55) that describes the time evolution of the single-particle phase space density is called the Vlasov equation.

In passing, we note that for a short-range system for which no Kac scaling needs to be invoked, Eq. (53) yields to leading order in $1/N$ the evolution equation

$$\left[\frac{\partial}{\partial t} + p_1 \frac{\partial}{\partial q_1}\right] f_1(z_1, t) = \int \mathrm{d}z_2\ \frac{\partial}{\partial q_1} V(|q_1 - q_2|) \frac{\partial}{\partial p_1} g_2(z_1, z_2, t). \tag{56}$$

The difference of the above equation from the Vlasov equation (55) is evident. To leading order in $1/N$, the time evolution for an LRI system is governed by the mean-field potential (besides the trivial "streaming term" on the left-hand side of both Eqs. (55) and (56) that is present even when the particles are noninteracting), with the two-particle correlation g_2 providing the next higher-order correction. Instead, for a short-range system, it is the two-particle correlation that dictates the leading time-evolution of the phase space density.

Let us remark on some relevant features of the Vlasov equation (55). The equation is evidently time-reversal invariant. One may associate with the equation a Hamilton dynamics due to a single-particle (mean-field) Hamiltonian $\mathcal{H}[f_1](p, q, t) = p^2/2 + \Phi[f_1](q, t)$. Note that the presence of the mean-field potential makes this

Hamiltonian a functional of the single-particle density f_1. One may then rewrite the Vlasov equation in terms of a Poisson bracket:

$$\frac{\partial f_1}{\partial t} + \{f_1, \mathcal{H}\} = 0, \tag{57}$$

which implies that Vlasov-stationary solutions ($\partial f_1/\partial t = 0$) are given by arbitrary normalizable functions of the single-particle Hamiltonian, as $f_1(p,q) = F(\mathcal{H})$. Another class of stationary solutions involves those that are homogeneous in the position coordinate ($\partial f_1/\partial q = 0$), that is, $f_1(p,q) \propto h(p)$, where $h(p)$ is a normalizable function of the momentum. Given the extent of arbitrariness allowed in choosing the functions F and h, one may conclude that the Vlasov equation admits an infinite number of stationary solutions. These stationary solutions define the so-called "Vlasov equilibrium" state of the system. Furthermore, the equation admits an infinite number of conserved quantities, the so-called Casimirs $C[f_1] = \int \mathrm{d}p\mathrm{d}q\ c(f_1(q,p))$, with c an arbitrary function of its argument. In particular, the single-particle entropy $S[f_1] \equiv -\int \mathrm{d}p\mathrm{d}q\ f_1 \ln f_1$ is a conserved quantity, and is thus constant in time.

The leading correction to the Vlasov equation is given by

$$\begin{aligned}&\left[\frac{\partial}{\partial t} + p_1\frac{\partial}{\partial q_1}\right] f_1(z_1,t) - \frac{\partial \Phi}{\partial q_1}\frac{\partial}{\partial p_1} f_1(z_1,t)\\ &\quad = \int \mathrm{d}z_2\ \frac{\partial}{\partial q_1} V(|q_1 - q_2|)\frac{\partial}{\partial p_1} g_2(z_1,z_2,t).\end{aligned} \tag{58}$$

Noting that $f_1 \sim N$ and $g_2 \sim N$, the above equation implies time evolution of Vlasov-stationary solutions on timescales of $O(N)$.

The second equation of the BBGKY hierarchy is

$$\begin{aligned}&\left[\frac{\partial}{\partial t} + p_1\frac{\partial}{\partial q_1} + p_2\frac{\partial}{\partial q_2} - \frac{\partial}{\partial q_1}V(|q_1-q_2|)\frac{\partial}{\partial p_1} - \frac{\partial}{\partial q_2}V(|q_1-q_2|)\frac{\partial}{\partial p_2}\right] f_2(z_1,z_2,t)\\ &= \int \mathrm{d}z_3\ \left[\frac{\partial}{\partial q_1}V(|q_1-q_3|)\frac{\partial}{\partial p_1} + \frac{\partial}{\partial q_2}V(|q_2-q_3|)\frac{\partial}{\partial p_2}\right] f_3(z_1,z_2,z_3,t).\end{aligned} \tag{59}$$

Similar to the decomposition (54), one has for the three-point correlation

$$\begin{aligned}f_3(z_1,z_2,z_3,t) &= f_1(z_1,t)f_1(z_2,t)f_1(z_3,t) + f_1(z_1,t)g_2(z_2,z_3,t)\\ &\quad + f_1(z_2,t)g_2(z_1,z_3,t) + f_1(z_3,t)g_2(z_1,z_2,t)\\ &\quad + h(z_1,z_2,z_3,t),\end{aligned} \tag{60}$$

where, arguing as for Eq. (54), one concludes that $h \sim N$. Using Eqs. (54) and (60) in Eq. (59), and using the Kac's scaling $V \to V/N$, one obtains to leading order in

5. A Model With Continuous Degrees of Freedom: The Hamiltonian Mean-Field (HMF) Model

The Hamiltonian mean-field (HMF) model is a model involving continuous degrees of freedom and evolving under Hamilton dynamics. The model has emerged over the years as a prototypical model to study and elucidate the many peculiar features resulting from long-range interactions.[18] The HMF model also mimics physical systems like gravitational sheet models and free-electron lasers. In order to derive the model, we start with the Hamiltonian (48), take the mass to be unity without loss of generality, and consider the potential to be $V(q) \propto Jq^{-\alpha}$; $0 \leq \alpha \leq 1$, for large q, so that in accordance with the Kac prescription, we scale the coupling constant J by N to make the total energy extensive in N. Next, we assume periodic coordinates so that boundary effects may be neglected. From now on, we denote the coordinates by periodic variables θ_i's, with $\theta_i \in [-\pi, \pi]$, so that $V(\theta) = V(\theta + 2\pi)$. The interparticle potential $V(\theta)$, which by definition is an even function to satisfy Newton's third law of motion, may be expanded in a cosine Fourier series: $V(\theta) = \widetilde{v}_0/2 + \sum_{k=1}^{\infty} \widetilde{v}_k \cos(k\theta)$; retaining only the first Fourier term, one obtains the HMF model. The corresponding Hamiltonian is given by

$$H = \sum_{i=1}^{N} \frac{p_i^2}{2} + \frac{1}{N} \sum_{1 \leq i < j \leq N}^{N} [1 - \cos(\theta_i - \theta_j)], \tag{79}$$

which effectively describes a system of globally interacting point particles moving on a circle, with θ_i the angular coordinate of the i-th particle on the circle, and p_i the corresponding conjugated momentum. In the Hamiltonian (79), we have without loss of generality further assumed the interparticle interaction to be attractive, by setting the coupling constant J to unity. The HMF may also be seen as a system of mean-field XY spins, only that here, the Possion bracket of the spin components is identically zero. The Hamiltonian (79) is invariant under the O(2) symmetry group. As we show below, in thermal equilibrium and for energy densities smaller than $\varepsilon_c \equiv 3/4$, the symmetry is spontaneously broken to result in a clustered state, thereby leading to a continuous phase transition at ε_c. The order parameter of clustering is the magnetization $m \equiv \sqrt{m_x^2 + m_y^2}$, with

$$(m_x, m_y) \equiv \frac{1}{N} \left(\sum_{i=1}^{N} \cos\theta_i, \sum_{i=1}^{N} \sin\theta_i \right). \tag{80}$$

We now derive the equilibrium solution of the model. After the trivial Gaussian integration over the momenta, the canonical partition function is given by

$$\begin{aligned} Z(\beta, N) &= \exp\left(-\frac{N\beta}{2}\right)\left(\frac{2\pi}{\beta}\right)^{N/2} \\ &\quad \times \int \mathrm{d}\theta_1 \ldots \mathrm{d}\theta_N \, \exp\left\{ \frac{\beta}{2N} \left[\left(\sum_{i=1}^{N} \cos\theta_i \right)^2 + \left(\sum_{i=1}^{N} \sin\theta_i \right)^2 \right] \right\}. \end{aligned} \tag{81}$$

Using the Hubbard–Stratonovich transformation, we get

$$Z(\beta, N) = \exp\left(-\frac{N\beta}{2}\right)\left(\frac{2\pi}{\beta}\right)^{N/2}$$
$$\times \frac{N\beta}{2\pi}\int \mathrm{d}x_1 \mathrm{d}x_2 \ \exp\left\{N\left[-\frac{\beta(x_1^2+x_2^2)}{2} + \ln I_0(\beta(x_1^2+x_2^2)^{\frac{1}{2}})\right]\right\}, \qquad (82)$$

where $I_0(z)$ is the modified Bessel function of order 0: $I_0(z) \equiv \int_0^{2\pi} \mathrm{d}\theta \ \exp(z_1\cos\theta + z_2\sin\theta) = \int_0^{2\pi} \mathrm{d}\theta \ \exp(z\cos\theta)$, with $z \equiv \left(z_1^2+z_2^2\right)^{1/2}$. Going to polar coordinates in the (x_1, x_2) plane yields

$$Z(\beta, N) = \exp\left(-\frac{N\beta}{2}\right)\left(\frac{2\pi}{\beta}\right)^{N/2} N\beta$$
$$\times \int_0^{\infty} \mathrm{d}x \ x \exp\left\{N\left[-\frac{\beta x^2}{2} + \ln I_0(\beta x)\right]\right\}. \qquad (83)$$

In the thermodynamic limit $N \to \infty$, the integral in (83) can be computed by using the saddle point method that involves the extremization problem of finding the particular value of x that extremizes the function $\left[-\frac{\beta x^2}{2} + \ln I_0(\beta x)\right]$, and thus involves solving the equation

$$x = \frac{I_1(\beta x)}{I_0(\beta x)}, \qquad (84)$$

where $I_1(z) = I_0'(z)$ is the modified Bessel function of order 1. In terms of the solution of this extremization problem, one finally obtains the rescaled free energy per particle as

$$\phi(\beta) \equiv \beta f(\beta) = \frac{\beta}{2} - \frac{1}{2}\ln 2\pi + \frac{1}{2}\ln\beta + \inf_{x\geq 0}\left[\frac{\beta x^2}{2} - \ln I_0(\beta x)\right], \qquad (85)$$

where note that one has to choose the particular solution of (84) that minimizes the free energy (85). For $\beta \leq 2$, the solution of Eq. (84) is given by $x = m^* = 0$, while for $\beta \geq 2$, the solution monotonically increases with β, approaching $m^* = 1$ for $\beta \to \infty$. The solution $m^* = 0$ of (84), present for all values of β, may be discarded for $\beta > 2$, since it does not minimize the free energy. One may show that the value m^* realizing the extremum in Eq. (85) is equal to the spontaneous magnetization in equilibrium. Note from the foregoing discussions that the spontaneous magnetization is defined only up to its modulus, while there is a continuous degeneracy in its direction. We have thus shown that the HMF model displays a continuous phase transition at $\beta_c = 2$ ($T_c = 0.5$). The derivative of the rescaled free energy with respect to β gives the energy per particle as

$$\varepsilon(\beta) = \frac{1}{2\beta} + \frac{1}{2} - \frac{1}{2}(m^*(\beta))^2. \qquad (86)$$

As already evident from the Hamiltonian, the lower bound of ε is 0. At the critical temperature, the energy is $\varepsilon_c = 3/4$. Since we have shown that the HMF model has a continuous phase transition in the canonical ensemble, we conclude that microcanonical and canonical ensembles are equivalent for this model.

For the system (79), one may easily write down the following Vlasov equation for the evolution of the single-particle phase space density $f(q, p, t)$ (cf. Eq. (55)):

$$\frac{\partial f}{\partial t} + p\frac{\partial f}{\partial \theta} - \frac{\partial \Phi}{\partial \theta}\frac{\partial f}{\partial p} = 0, \tag{87}$$

where one has the mean-field potential

$$\Phi[f](\theta, t) = -\int_0^{2\pi} \mathrm{d}\theta' \int_{-\infty}^{\infty} \mathrm{d}p \, \cos(\theta - \theta') f(\theta', p, t). \tag{88}$$

For distributions that are homogeneous with respect to θ, the mean-field potential evaluates to zero, implying that such distributions are stationary solutions of the Vlasov equation (87). Let us denote such homogeneous stationary solutions by $f_0(p)$. Since stationarity does not guarantee stability, one may study the stability of such homogeneous distributions with respect to small perturbations, by linearizing the Vlasov equation (87) about $f_0(p)$. One obtains the result that the homogeneous distribution $f_0(p)$ is stable if and only if the quantity

$$I \equiv 1 + \frac{1}{2}\int_{-\infty}^{\infty} \mathrm{d}p \, \frac{f_0'(p)}{p} \tag{89}$$

is positive. Such a condition reveals that there can be an infinite number of Vlasov-stable stationary distributions. Let us briefly discuss some examples of $f_0(p)$.

- The first one is the Gaussian distribution: $f_0(p) \sim \exp(-\beta p^2/2)$, which is expected at equilibrium. With the threshold condition (89), one recovers the result due to statistical mechanics reviewed above that the critical inverse temperature is $\beta_c = 2$, and its associated critical stability threshold is $\varepsilon^\star = \varepsilon_c = 3/4$.
- The second example is the so-called water-bag distribution, which has often been used in the past to numerically demonstrate the out-of-equilibrium properties of the HMF model. Such a distribution comprises momentum uniformly distributed in a given interval $[-p_0, p_0]$, where the parameter p_0 is related to the energy density as $p_0 = \sqrt{6\varepsilon - 3}$. In this case, one obtains the critical stability threshold as $\varepsilon^\star = 7/12$: the state is linearly stable for energies larger than $\varepsilon^\star$, and is unstable below.

Let us emphasize that the above examples are Vlasov-stable stationary solutions that are possible among infinitely many others, and there is no reason to emphasize one over the other.

The existence of infinitely many Vlasov-stable stationary solutions of the HMF model implies that when starting initially from one such solution in the stable regime (e.g., the water-bag distribution at energy density $\varepsilon > \varepsilon^\star = 7/12$), and

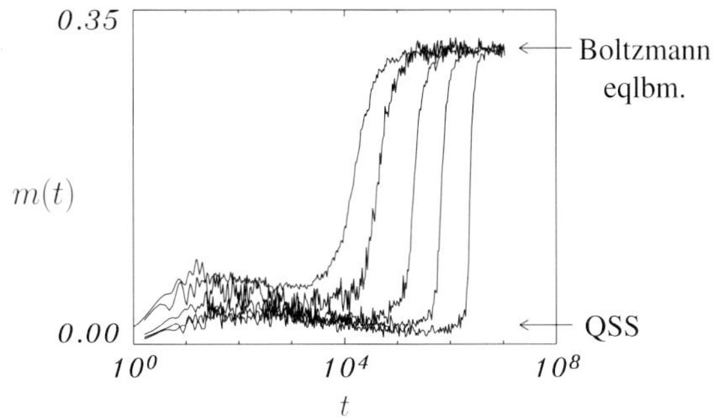

Fig. 14. Time evolution of the modulus of the magnetization $m(t)$ for different particle numbers in the HMF model (79): $N = 10^3$, 2×10^3, 5×10^3, 10^4 and 2×10^4 from left to right. The energy density is $\varepsilon = 0.69$. The values of the magnetization indicated by the horizontal arrows refer respectively to the value expected in equilibrium (labelled BG) and the one corresponding to a homogeneous QSS.

evolving under the Hamilton equations derived from the Hamiltonian (79):

$$\frac{\mathrm{d}\theta_i}{\mathrm{d}t} = p_i, \ \frac{\mathrm{d}p_i}{\mathrm{d}t} = -m_x \sin\theta_i + m_y \cos\theta_i, \tag{90}$$

the magnetization in an infinite system should remain zero at all times. For finite N, however, finite-N effects drive the system away from the water-bag state, and through other stable stationary states. Such a slow quasi-stationary evolution across an infinite number of Vlasov-stable stationary states (note: stationary only in the limit $N \to \infty$) ends with the system in the Boltzmann–Gibbs (BG) equilibrium state, see Fig. 14. For the HMF model, it has been rigorously proven that Vlasov-stable homogeneous distributions do not evolve on timescales of order smaller or equal to N. Indeed, a scaling $\sim N^\gamma$; $\gamma > 0$, for the timescale of relaxation towards the BG equilibrium state has been observed in simulations.[19] At long times, one has $m(t) \sim (1/\sqrt{N})e^{t/N^\gamma}$ for $t \gg N^\gamma$, where the prefactor accounts for finite-N fluctuations. For $\varepsilon < \varepsilon^*$, linear instability results in a faster relaxation towards equilibrium as $m(t) \sim (1/\sqrt{N})e^{\gamma t}$ for $t \gg 1/\gamma$. Here, $\gamma^2 = 6\,(7/12 - \varepsilon)$ is independent of N. Thus, there are no QSSs for energies below ε^*. Note that the slow relaxation to BG equilibrium depicted in Fig. 14 is consistent with the general scenario shown in Fig. 10.

5.1. *An experimental realization of the HMF model: Atoms in optical cavities*

Atoms interacting with a single-mode standing electromagnetic wave due to light trapped in a high-finesse optical cavity are subject to an inter-particle interaction that is long-ranged owing to multiple coherent scattering of photons by the atoms into the wave mode.[20–22] The set-up is shown in Fig. 15, which also shows optical pumping by a transverse laser of intensity Ω^2 to counter the inevitable cavity losses quantified by the cavity linewidth κ. As regards the interaction with the

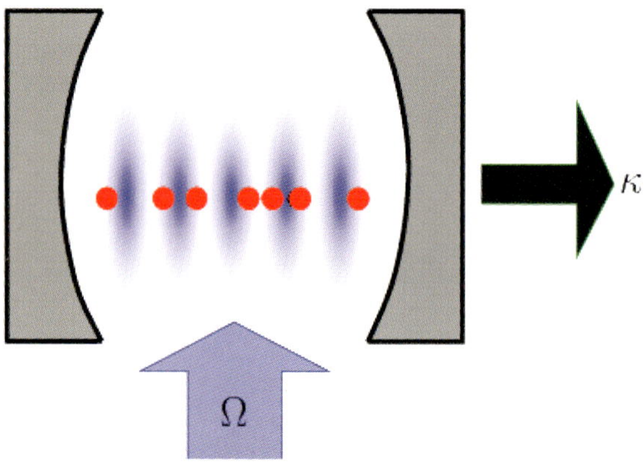

Fig. 15. Atoms interacting with a single-mode standing electromagnetic wave in a cavity of linewidth κ, and being driven by a transverse laser with intensity Ω^2.

electromagnetic field, each atom may be regarded as a two-level system, where the transition frequency between the two levels is ω_0. Considering N identical atoms of mass m confined in one dimension along the cavity axis (taken to be the x-axis), and denoting the standing wave with wave number k by $\cos(kx)$, the sum of the electric-field amplitudes coherently scattered by the atoms at time t depends on their instantaneous positions $x_1, \ldots, x_N$, and is proportional to the quantity $\Theta \equiv \sum_{j=1}^{N} \cos(kx_j)/N$, so that the cavity electric field at time t is $E(t) \propto \sqrt{N\overline{n}}\Theta$. Here, $\overline{n}$ is the maximum intra-cavity photon number per atom, given by $\overline{n} \equiv N\Omega^2\alpha^2/(\kappa^2 + \Delta_c^2)$, with $\alpha \equiv g/\Delta_a$ being the ratio between the cavity vacuum Rabi frequency and the detuning $\Delta_a \equiv \omega_L - \omega_0$ between the laser and the atomic transition frequency, and $\Delta_c \equiv \omega_L - \omega_c$ being the detuning between the laser and the cavity-mode frequency. The quantity Θ characterizes the amount of spatial ordering of atoms within the cavity mode, with $\Theta = 0$ corresponding to atoms being uniformly distributed and the resulting vanishing of the cavity field, and $|\Theta| \neq 0$ implying spatial ordering. The wave number k is related to the linear dimension L of the cavity through $k = 2\pi/\lambda$ and $L = q\lambda$, where λ is the wavelength of the standing wave, and $q \in \mathbb{N}$.

The dynamics of the system is studied by analyzing the time evolution of the N-atom phase space distribution $f_N(x_1, \ldots, x_N, p_1, \ldots, p_N, t)$ at time t, with p_j's denoting the momenta conjugate to the positions x_j. Treating the cavity field quantum mechanically, and regarding the atoms as classically polarizable particles with semi-classical center-of-mass dynamics, it may be shown that the distribution f_N evolves in time according to the Fokker–Planck equation (FPE)[20,21]

$$\partial_t f_N + \{f_N, H\} = -\overline{n}\Gamma \sum_{i=1}^{N} \sin(kx_i) \sum_{j=1}^{N} \partial_{p_i} \sin(kx_j) \left(p_j + \frac{m}{\beta}\partial_{p_j} \right) f_N. \tag{91}$$

Here, $\Gamma \equiv 8\omega_r\kappa\Delta_c/(\Delta_c^2+\kappa^2)$, $\hbar\beta \equiv -4\Delta_c/(\Delta_c^2+\kappa^2)$, $\hbar$ is the reduced Planck constant, $\omega_r \equiv \hbar k^2/(2m)$ is the recoil frequency due to collision between an atom and a photon, while the Hamiltonian H is given by

$$H \equiv \sum_{j=1}^{N} \frac{p_j^2}{2} - NJ\Theta^2; \quad J \equiv -\hbar\Delta_c\overline{n}. \tag{92}$$

The semi-classical limit is valid under the condition of κ being larger than ω_r, while Eq. (91) holds in a parameter regime in which processes describing a virtual scattering of cavity photons, which scale with the dynamical Stark shift of the cavity field $U = g\alpha$, are negligible. The Hamiltonian H describes the conservative dynamical evolution of f_N in the limit of vanishing cavity losses (or for times sufficiently small such that dissipative effects are negligible), and contains the photon-mediated long-ranged (mean-field) interaction between the atoms encoded in the second term on the right-hand side (rhs) of Eq. (92). Note that the interaction is attractive (respectively, repulsive) when Δ_c is negative (respectively, positive). Cavity losses lead to damping and diffusion, which is described by the rhs of Eq. (91).

Let us now consider the case of effective attractive interactions between the atoms and the cavity field ($\Delta_c < 0$), and study the dynamics of the system in the limit in which the effect of the dissipation may be neglected, that is, for sufficiently small times. In this limit, the dynamics of the N atoms is conservative and governed by the Hamiltonian (92). The positions x_j of the atoms enter the Hamiltonian only as kx_j, so that we may define the phase variables $\theta_j \equiv kx_j = 2\pi x_j/\lambda$ for $j = 1, \ldots, N$. Using $L = q\lambda$, and setting the origin of the x-axis in the center of the cavity, we have $x_j \in [-q\lambda/2, q\lambda/2]$, so that on using the periodicity of the cosine function, we can take the phase variables θ_j modulo q, yielding $\theta_j \in [-\pi, \pi]$. Then, by measuring lengths in units of the reciprocal wavenumber $k^{-1} = \lambda/(2\pi)$ of the cavity standing wave, masses in units of the mass of the atoms m, and energies in units of $\hbar\Delta_c$, the Hamiltonian may be rewritten in dimensionless form as

$$H = \sum_{j=1}^{N} \frac{(p_\theta)_j^2}{2} - \overline{n}N\Theta^2\,, \tag{93}$$

where, in terms of the θ variables, Θ is now expressed as $\Theta = \sum_{j=1}^{N} \cos\theta_j/N$. The $(p_\theta)_j$'s are the momenta canonically conjugated to the θ_j variables.

The similarity between the system with Hamiltonian (93) and the HMF model is now well apparent. Hence, the dynamics of a system of atoms interacting with light in a cavity in the dissipationless limit is equivalent to that of a model that differs from the HMF model in zero field just for the fact that particles in the former interact only with the x-component of the magnetization.

6. Ubiquity of the Quasistationary Behavior Under Different Energy-Conserving Dynamics

6.1. *HMF model in presence of three-body collisions*

In this Section, we address the question of robustness of QSSs with respect to stochastic dynamics of an isolated system within a microcanonical ensemble. To this end, we generalize the HMF model to include stochastic dynamical moves in addition to the deterministic ones, Eq. (90). The generalized HMF model follows a piecewise deterministic dynamics, whereby the Hamiltonian evolution is randomly interrupted by stochastic interparticle collisions that conserve energy and momentum.[23,24] We consider collisions in which only the momenta are updated stochastically. Since the momentum variable in the HMF model is one-dimensional, and there are two conservation laws for the momentum and the energy, one has to resort to three-particle collisions. Namely, three random particles, (i, j, k), collide and their momenta are updated stochastically, $(p_i, p_j, p_k) \to (q_i, q_j, q_k)$, while conserving energy and momentum and keeping the phases unchanged. Thus, the model evolves under the following repetitive sequence of events: deterministic evolution, Eq. (90), for a time interval whose length is exponentially distributed, followed by a single instantaneous sweep of the system for three-particle collisions, which consists of N^3 collision attempts.

In presence of collisions, in order to discuss the evolution of the single-particle phase space density in the limit $N \to \infty$, we need to consider instead of the Vlasov equation the appropriate Boltzmann equation that takes into account the collisional dynamics. The equation is given by

$$\frac{\partial f}{\partial t} + p\frac{\partial f}{\partial \theta} - \frac{\partial \Phi}{\partial \theta}\frac{\partial f}{\partial p} = \left(\frac{\partial f}{\partial t}\right)_c, \tag{94}$$

$$\begin{aligned}\left(\frac{\partial f}{\partial t}\right)_c &= \int \mathrm{d}\eta R[f(\theta, q, t)f(\theta', q', t)f(\theta'', q'', t) \\ &\quad - f(\theta, p, t)f(\theta', p', t)f(\theta'', p'', t)],\end{aligned} \tag{95}$$

$$\begin{aligned}R &= \alpha\delta(p + p' + p'' - q - q' - q'')\delta \\ &\quad \times \left(\frac{1}{2}(p^2 + p'^2 + p''^2) - \frac{1}{2}(q^2 + q'^2 + q''^2)\right),\end{aligned} \tag{96}$$

where we have $\mathrm{d}\eta \equiv \mathrm{d}p'\mathrm{d}p''\mathrm{d}q\mathrm{d}q'\mathrm{d}q''\mathrm{d}\theta'\mathrm{d}\theta''$. Equation (95) represents the three-body collision term, and R is the rate for collisions $(p, p', p'') \to (q, q', q'')$ that conserve energy and momentum. The constant α has the dimension of $1/(\text{time})$ and sets the scale for collisions: On average, there is one collision after every time interval α^{-1}. We refer to the Boltzmann equation with $\alpha = 0$ as the Vlasov-equation limit. Note that both the Boltzmann and the Vlasov equation are valid for infinite N, and have size-dependent correction terms when N is finite. In the Vlasov limit, any state that is homogeneous in angles but with an arbitrary momentum distribution is stationary; as discussed in Section 5, in this limit, the QSSs are related to the linear stability of the stationary solutions chosen as the initial state. Recall, for example, that the so-called water-bag state is linearly stable for energies in the range $\varepsilon^* \equiv 7/12 < \varepsilon < \varepsilon_c$, when it manifests as a QSS. The water-bag state may be realized by sampling independently the angles uniformly in $[-\pi, \pi]$ and the momenta uniformly in $[-p_0, p_0]$, with $p_0 = \sqrt{6\varepsilon - 3}$.

Let us now turn to a discussion of QSSs in the generalized HMF model, i.e., under noisy microcanonical evolution, in the light of the Boltzmann equation. First, we note that unlike the Vlasov equation, a homogeneous state with an arbitrary momentum distribution is not stationary under the Boltzmann equation; instead, only a Gaussian distribution is stationary. Suppose we start with an initial homogeneous state with uniformly distributed momenta. Then, under the dynamics, the momentum distribution will evolve towards the stationary Gaussian distribution. Interestingly, although the momentum distribution evolves, the initial θ distribution does not change in time, since for homogeneous θ distribution, the p and θ distributions evolve independently. In a finite system, however, there are fluctuations in the initial state. These fluctuations make the homogeneous state with Gaussian-distributed momenta linearly unstable under the Boltzmann equation at all energies $\varepsilon < \varepsilon_c$, as we demonstrate below. This results in a fast relaxation towards equilibrium.

One may study the linear instability of a homogeneous state with Gaussian-distributed momenta at energies below ε_c and under the evolution given by the Boltzmann equation. We now summarize the essential steps, for the simple case of energies just below the critical point. The stability analysis is carried out by linearizing Eq. (94) about the homogeneous state. We expand $f(\theta, p, t)$ as $f(\theta, p, t) = f^{(0)}(p)[1 + \lambda f^{(1)}(\theta, p, t)]$ with $f^{(0)}(p) = e^{-p^2/2T}/(2\pi\sqrt{2\pi T})$. Here, since the initial angles and momenta are sampled independently according to $f^{(0)}(p)$, fluctuations for finite N make the small parameter λ of $O(1/\sqrt{N})$. At long times, the dynamics is dominated by the eigenmode with the largest eigenvalue of the linearized Boltzmann equation, so that $f^{(1)}(\theta, p, t) = f_k^{(1)}(p, \omega)e^{i(k\theta+\omega t)}$. Since the mean-field potential Φ in Eq. (94) involves $e^{\pm i\theta}$, one needs to consider only $k = \pm 1$. The coefficients $f_{\pm 1}^{(1)}$ then satisfy

$$\begin{aligned} &\pm ipf_{\pm 1}^{(1)}(p,\omega) \mp \frac{2\pi}{2if^{(0)}}\frac{\partial f^{(0)}}{\partial p}\int \mathrm{d}p' f^{(0)}(p')f_{\pm 1}^{(1)}(p',\omega) \\ &+ (4\pi)^2 \int \mathrm{d}p'\mathrm{d}p''\mathrm{d}q\mathrm{d}q'\mathrm{d}q'' R f^{(0)}(p')f^{(0)}(p'') \\ &\times [f_{\pm 1}^{(1)}(p,\omega) - f_{\pm 1}^{(1)}(q,\omega)] = -i\omega f_{\pm 1}^{(1)}(p,\omega). \end{aligned} \tag{97}$$

Treating α as a small parameter, we solve the above equation perturbatively in α. In the absence of collisions ($\alpha = 0$), the above analysis reduces to that of the Vlasov equation and to the unperturbed solutions, namely, the frequencies $\omega^{(0)}$ and the coefficients $f_{\pm 1}^{(1)}(q, \omega^{(0)})$, which are obtained from the analysis. In particular, slightly below the critical point ε_c, the unperturbed real frequencies $\Omega^{(0)} = i\omega^{(0)}$ are given by $|\Omega^{(0)}| \approx (2/\sqrt{\pi})(T_c - T)$. Thus, in the Vlasov limit, the homogeneous state with Gaussian-distributed momenta is unstable below the critical energy, as already noted in Section 5. To obtain the perturbed frequencies Ω to lowest order in α, we now substitute the unperturbed solutions into Eq. (97). After a straightforward but lengthy algebra, one obtains at an energy slightly below the critical point the perturbed frequencies to be given by

$$\Omega \approx |\Omega^{(0)}|[1 + \alpha A], \text{ with } A = \frac{2\pi^{3/2}}{\sqrt{3}}\left(1 - \frac{1}{\sqrt{5}}\right). \tag{98}$$

This equation suggests that to leading order in α, the frequencies Ω are real for energies just below the critical value and vanish at the critical point. Thus, a homogeneous state with Gaussian-distributed momenta is linearly unstable under the Boltzmann equation at energies just below the critical point and neutrally stable at the critical point.

In the light of the above calculation, we may now analyze the evolution of magnetization in the generalized HMF model while starting from a water-bag initial condition. The two timescales that govern the time evolution of the magnetization are (i) the scale over which collisions occur, given by α^{-1}, and (ii) the scale $\sim N^{\gamma}$, over which finite-size effects add corrections to the Boltzmann equation. The interplay between the two timescales may be naturally analyzed by invoking a scaling approach, as we demonstrate below.

For $\alpha^{-1} \ll N^{\gamma}$, and times $\alpha^{-1} \ll t \ll N^{\gamma}$, the system size is effectively infinite and the evolution follows the Boltzmann equation. Here, frequent collisions at short times drive the momentum distribution towards a Gaussian. As noted above, until this happens, the initial magnetization does not change in time. Over the time the momentum distribution becomes Gaussian, the instability of such a state under the Boltzmann equation leads to a fast relaxation towards equilibrium, similar to the result for the Vlasov-unstable regime. The asymptotic behavior of the magnetization is thus

$$m(t) \sim \frac{1}{\sqrt{N}} e^{\alpha t}; \ N^{\gamma} \gg t \gg \alpha^{-1}. \tag{99}$$

By requiring that $m(t)$ acquires a value of $O(1)$, the above equation gives the relaxation time $\tau_{\rm S}$, determined by the stochastic process, as $\tau_{\rm S} \sim \ln N/\alpha$. In the opposite limit, $\alpha^{-1} \gg N^{\gamma}$, collisions are infrequent, and therefore, the process that drives the momentum distribution to a Gaussian is delayed. The magnetization stays close to its initial value, and relaxes only over the time $\sim N^{\gamma}$, over which finite-size effects come into play. Here, similar to the result for the Vlasov-stable regime, the magnetization at late times behaves as

$$m(t) \sim \frac{1}{\sqrt{N}} e^{t/N^{\gamma}}; \ \alpha^{-1} \gg t \gg N^{\gamma}. \tag{100}$$

This equation gives the relaxation time $\tau_{\rm D}$, determined by the deterministic process, as $\tau_{\rm D} \sim N^{\gamma} \ln N$. Interpolating between the above two limits of the timescales, one expects the relaxation time $\tau(\alpha, N)$ to obey $\tau^{-1} = \tau_{\rm S}^{-1} + \tau_{\rm D}^{-1}$, yielding $\tau(\alpha, N) \sim \ln N/(\alpha + 1/N^\gamma)$. More generally, this suggests a scaling form

$$\tau(\alpha, N) \sim \frac{\ln N}{\alpha} g(\alpha N^\gamma), \tag{101}$$

where, consistent with Eqs. (99) and (100), the scaling function $g(x)$ behaves as follows: $g(x) \sim x$ for $x \ll 1$, while $g(x) \to$ constant for $x \gg 1$. Equation (101) implies that for fixed N, the relaxation time of the water-bag initial state exhibits a crossover, from being of order $N^\gamma \ln N$ (corresponding to QSSs) for $\alpha \ll 1/N^\gamma$ to being of order $\ln N$ for $\alpha \gg 1/N^\gamma$. This brings us to the main conclusion of this Subsection: In the presence of collisions, the relaxation at long times does not occur over an algebraically growing timescale, which implies that under noisy microcanonical evolution, QSSs occur only as a crossover phenomenon and are lost in the limit of long times.

The above predictions, in particular, the scaling form in Eq. (101), may be verified by performing extensive numerical simulations of the generalized HMF model. The Hamilton equations, Eq. (90), may be integrated by using a symplectic fourth-order integrator. In realizing the stochastic process $(p, p', p'') \to (q, q', q'')$ while conserving the three-particle energy E and momentum P, we note that the updated momenta lie on a circle formed by the intersection of the plane $p + p' + p'' = P$ and the spherical surface $p^2 + p'^2 + p''^2 = 2E$. The radius of this circle is given by $r = \sqrt{2E - P^2/3}$. The new momenta may thus be parametrized in terms of an angle ϕ measured along this circle, as $q = (P/\sqrt{3}) + r\sqrt{2/3}\cos\phi, q' = (P/\sqrt{3}) - (r/\sqrt{6})\cos\phi - (r/\sqrt{2})\sin\phi, q'' = (P/\sqrt{3}) - (r/\sqrt{6})\cos\phi + (r/\sqrt{2})\sin\phi$. Stochasticity in updates is achieved through choosing the angle ϕ uniformly in $[0, 2\pi)$. Following the foregoing scheme, typical time evolution of the magnetization in the generalized HMF model for $N = 500$ and several values of α at an energy density $\varepsilon = 0.69$ are shown in Fig. 16 (Upper panel). The relaxation time $\tau(\alpha, N)$ is taken as the time for the magnetization to reach the fraction 0.8 of the final equilibrium value (the result, however, is not sensitive to this choice). At $\varepsilon = 0.69$, where the equilibrium value of the magnetization is $\simeq$0.3 and $\gamma \simeq 1.7$, we plot $\alpha\tau(\alpha, N)/\ln N$ versus αN^γ to check the scaling form in Eq. (101). Figure 16 (Lower panel) shows an excellent scaling collapse over several decades; this is consistent with our prediction for QSSs as a crossover phenomenon under noisy microcanonical dynamics.

6.2. *HMF model generalized to particles moving on a sphere*

In order to probe the ubiquity of the quasistationary behavior observed in the HMF model, various extensions of the model have been introduced and analyzed over the years. For example, the HMF model was considered with an additional term in the energy that is due to either (i) a global anisotropy in the magnetization along the x-axis, or, (ii) an onsite potential. In either case, QSSs were shown to is stationary under the Vlasov dynamics (103). The magnetization components m_x, m_y, m_z are determined self-consistently. As a specific example, consider a stationary state that is non-magnetized, that is, $m_x = m_y = m_z = 0$, and $f^{(0)}(\theta, \phi, p_\theta, p_\phi)$ is given by

$$f^{(0)}(\theta, \phi, p_\theta, p_\phi) = \begin{cases} \dfrac{1}{2\pi}\dfrac{1}{\pi}A \ \text{ if } \dfrac{1}{2}\left(p_\theta^2 + \dfrac{p_\phi^2}{\sin^2\theta}\right) < E; \\ \theta \in [0, \pi], \phi \in [0, 2\pi], A, E \geq 0, \\ 0, \qquad\quad \text{otherwise.} \end{cases} \tag{106}$$

The state (106) is a straightforward generalization of the water-bag initial condition for the HMF model. The parameters A and E are related through the normalization condition, $\int_0^\pi \mathrm{d}\theta \int_0^{2\pi} \mathrm{d}\phi \int_\Omega \mathrm{d}p_\theta \mathrm{d}p_\phi f^{(0)} = 1$, where the integration over p_θ and p_ϕ is over the domain $\Omega \equiv \Theta\Big(2E - p_\theta^2 - p_\phi^2/\sin^2\theta\Big)$, with $\Theta(x)$ denoting the unit step function. One gets $E = 1/(4A)$, while the conserved energy density $\varepsilon = 1/2 + \int_0^\pi \mathrm{d}\theta \int_0^{2\pi} \mathrm{d}\phi \int_\Omega \mathrm{d}p_\theta \mathrm{d}p_\phi (1/2)\Big(p_\theta^2 + p_\phi^2/\sin^2\theta\Big) f^{(0)}$ is related to E as $\varepsilon = (E+1)/2$.

Analyzing the linear stability of the stationary state (106) under the Vlasov dynamics (103), it may be shown that for energies $\varepsilon > \varepsilon^* = 2/3$, the non-magnetized state (106) is linearly stable, and is hence a QSS. In a finite system, the QSS eventually relaxes to BG equilibrium over a very long timescale, which, considering the magnetization as an indicator for the relaxation process for energies $\varepsilon < \varepsilon_c$, grows algebraically with the system size as $N^\gamma;\ \gamma > 0$; this is demonstrated by numerical simulation results in Fig. 17. For energies $\varepsilon < \varepsilon^*$, the state (106) is linearly unstable, and is thus not a QSS; in this case, numerical simulations show that the system exhibits a fast relaxation towards BG equilibrium over a timescale that grows with the system size as $\ln N$. These features of a linearly unstable and a linear stable regime of a non-magnetized Vlasov-stationary state, with a QSS emerging in the latter case, remain unaltered on adding a term to the Hamiltonian (102) that accounts for a global anisotropy in the magnetization.

6.3. *A long-range model of spins*

The ubiquity of QSSs may be tested in a dynamics very different from the particle dynamics of either the HMF model or any of its generalizations, including the model (102), namely, within classical spin dynamics of an anisotropic Heisenberg model with mean-field interactions.[26,27] The model comprises N globally coupled three-component Heisenberg spins of unit length, $\mathbf{S}_i = (S_{ix}, S_{iy}, S_{iz}),\ i = 1, 2, \ldots, N$. In terms of spherical polar angles $\theta_i \in [0, \pi]$ and $\phi_i \in [0, 2\pi]$ for the orientation of the i-th spin, one has $S_{ix} = \sin\theta_i \cos\phi_i, S_{iy} = \sin\theta_i \sin\phi_i, S_{iz} = \cos\theta_i$. The Hamiltonian of the model is given by

$$H = -\frac{J}{2N}\sum_{i,j=1}^{N} \mathbf{S}_i \cdot \mathbf{S}_j + D\sum_{i=1}^{N} S_{iz}^2, \tag{107}$$

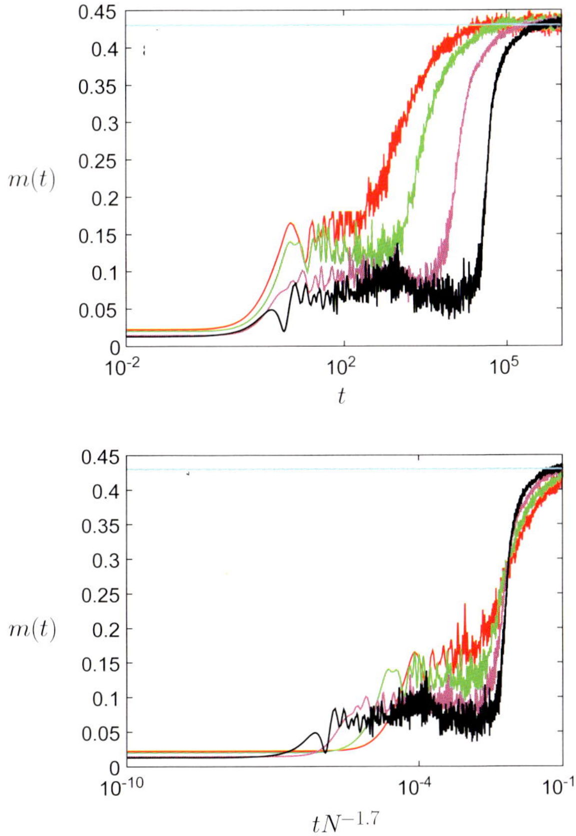

Fig. 17. For the model (102), the figures show numerical simulation results for the magnetization $m(t)$ as a function of time (Upper panel), and as a function of time scaled by $N^{1.7}$ (Lower panel) in the Vlasov-stable phase ($\varepsilon > \varepsilon^*$). The energy density is $\varepsilon = 0.7$. The blue line indicates the BG equilibrium value. The figures suggest the existence of a QSS with a lifetime scaling with the system size as $N^{1.7}$.

where the first term with $J > 0$ describes a ferromagnetic mean-field like coupling, while the last term gives the energy due to a local anisotropy. We take $D > 0$ such that at equilibrium, the energy is lowered by having the magnetization $\mathbf{m} \equiv (1/N)\sum_{i=1}^{N} \mathbf{S}_i$ pointing in the xy plane. The model (107) has an equilibrium phase diagram with a continuous transition from a low-energy magnetic phase ($m \neq 0$) to a high-energy non-magnetic phase ($m = 0$) across the critical energy density $\varepsilon_c = D(1 - 2/\beta_c)$, where β_c satisfies $2/\beta_c = 1 - 1/(2\beta_c D) + \exp(-\beta_c D)/(\sqrt{\pi\beta_c D}\mathrm{Erf}[\sqrt{\beta_c D}])$, with $\mathrm{Erf}(x)$ being the error function. The derivation of these properties is detailed in Ref. 26.

The time evolution of the model (107) is governed by the set of equations

$$\frac{\mathrm{d}\mathbf{S}_i}{\mathrm{d}t} = \{\mathbf{S}_i, H\}; \qquad i = 1, 2, \ldots, N. \tag{108}$$

Fig. 3. Photograph of the "transverse section" of the barrel part of CMS illustrating the successive layers of detection starting from the centre where the collisions occur: the inner tracker, the crystal calorimeter, the hadron calorimeter, the superconducting coil, and the iron yoke instrumented with the four muon stations. The last muon station is at a radius of 7.4 m.

3. The Experimental Challenges at the LHC

The peak instantaneous luminosity at the LHC has now passed the design value of 10^{34} cm^{-2} s^{-1}. Each of the GPDs sees an event rate of $>8 \times 10^8$ inelastic events/s. This leads to formidable experimental challenges.[13–16]

The event selection (trigger) must reduce the billion or so interactions/s to a few hundred events/s for storage and subsequent analysis. In order to avoid dead-time, *pipelined* trigger processing and readout architectures are required, where data from many bunch-crossings are processed concurrently by a chain of processing elements. The Level-1 trigger decision takes $\approx 3\,\mu$s (of which more than 50% is in signal transmission time), so the data must be stored in pipelines for these $3\,\mu$s.

An average number of 30–50 minimum bias events are superposed on the event of interest. Around 2000 charged tracks emerge from the interaction region every 25 ns. Thus highly granular detectors are required, with good time resolution, giving low *occupancy* (fraction of detector elements that contain information).

The high particle fluxes emerging from the interaction region lead to high radiation levels. Radiation hard detectors and front-end electronics are required. An integrated luminosity of $\sim$300 fb^{-1} is envisaged up till the early 2020s. Detectors placed close to the interaction vertex have to withstand doses of $\sim$1,000 kGy and fluences of 10^{15} cm^{-2} at a radius of 4 cm (these drop by a factor 100 at $\sim$75 cm) and calorimeters in the barrel region ($|\eta| < 1.5$) doses of 3 kGy (this increases to 100 to 1,000 kGy in forward region $3.5 < |\eta| < 5$). The pseudorapidity $\eta = -\ln[\tan(\theta/2)]$ where and θ is the polar angle measured from the positive z-axis (along the anti-clockwise beam direction).

Due to several extreme requirements the motto in the 1990s was that "if it doesn't exist and we need it, we will invent it". The budgets were limited so we looked around to see what existed, innovated, and then automated to drive down the costs.

The detectors are highly complex and access for maintenance is very difficult, time consuming and restricted. Hence a very high degree of long-term operational reliability, usually associated with space-bound systems, has had to be attained.

The online trigger system deals with information that is continuously generated at a rate of 40,000 Gbits/s and reduces it to hundreds of megabytes/s for storage. The tens of petabytes generated per year per experiment have to be distributed for offline analysis to scientists located across the globe. This data management required the development of the so-called 'Worldwide Computing Grid'.

As an aside, I believe Abdus Salam would have been proud of the fact that some of the structures and detectors in the CMS experiment were manufactured and tested in so many (over 40) countries of the Collaboration, including Pakistan. The experimental community in Pakistan was really formed for participation in CMS. Almost from a standing start they have produced detectors such as RPCs, and are now producing GEM detectors. Abdus Salam stated that "The creation of Physics is the shared heritage of mankind. East and West, North and South have equally participated in it." This obviously is the case for the CMS experiment, as well as for the other LHC experiments.

3.1. *Addressing the challenges: Pushing technologies*

This section is not intended to be exhaustive but discusses a few examples of detector technologies that have stretched the technical limits or are innovative.

Development of new particle detectors takes a long time and goes through many phases starting from an idea or a concept, followed by intensive R&D, prototyping, pre-series production, mass production, systems integration, installation and commissioning and finally data taking and analysis. This can be illustrated by using any one of the many detector technologies in the LHC experiments. We have chosen one from the CMS experiment, namely the lead tungstate scintillating crystals ($PbWO_4$) used in its electromagnetic calorimeter. These crystals are well suited for LHC conditions; they have a short radiation length (0.89 cm) and Moliere radius (2.0 cm), a fast response (80% of the light is emitted within 25 ns) and are radiation hard (up to 10 Mrad). Below we give some details of the major phases of their development.

- Three technologies were originally investigated for the CMS electromagnetic calorimeter: scintillating crystals ($PbWO_4$, CeF_3), scintillating glasses (HfF_3) and lead/scintillator layers pierced by wavelength shifting fibres ("shashlik"). After beam tests and weighing other considerations $PbWO_4$ crystals were chosen.

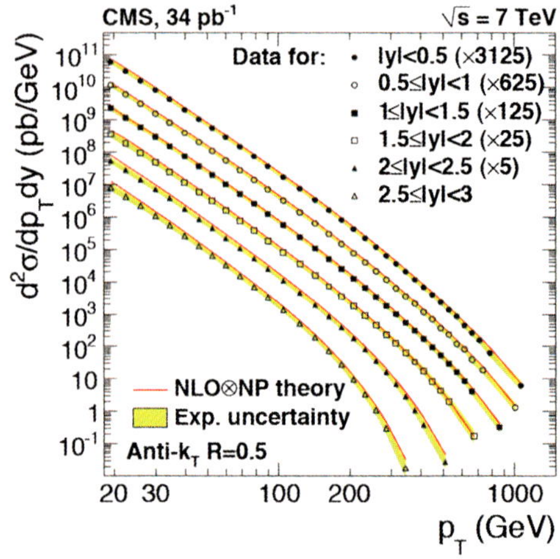

Fig. 7. The measured double differential jet cross-section in different angular ranges compared with theoretical predictions (y is the rapidity).

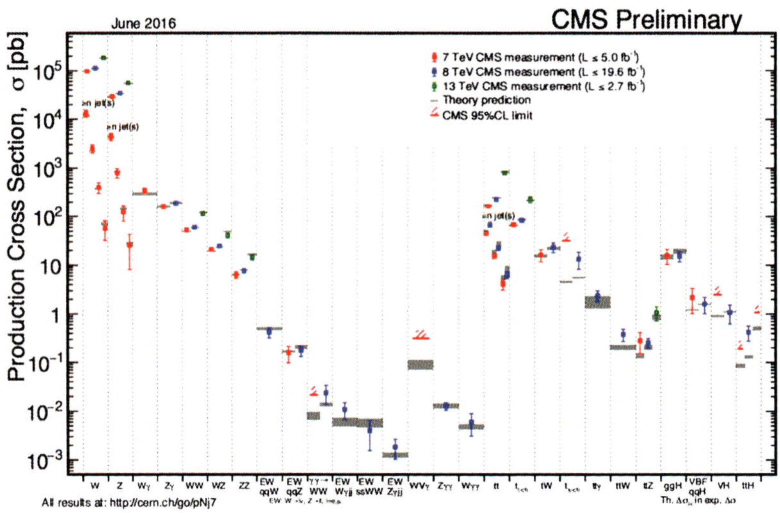

Fig. 8. A comparison of measured cross-sections for electroweak and QCD processes with theoretical predictions from the SM.

the production of bottom and top quarks and W and Z bosons, singly and in pairs, and many QCD. A summary of such measurements is shown in Fig. 8, where measurements of cross-sections for various selected electroweak and QCD processes are compared with predictions from the SM. These very diverse measurements, probing cross-sections over many orders of magnitude, established that the CMS experiment had become a “physics engine” and was ready for searching for new physics. The measurements from recent data taken at the higher centre-of-mass energy, $\sqrt{s} = 13\,\text{TeV}$, are also included. All these measurements are consistent with the predictions of the SM. Hence it can be concluded that the CMS detector performance is well understood and known SM processes are correctly observed. The volume of physics results coming out from CMS is illustrated by the large number (more than 450 so far) of physics papers published in peer-reviewed journals.

It is vital to understand SM processes correctly since they often constitute large backgrounds to signatures of new physics, such as those expected for the Higgs boson. The speed with which these measurements verified SM predictions for known physics is a tribute to the large amount of work done by many, including theorists, other collider experiments at LEP, Tevatron, HERA, and b-factories as well as the quality of the construction, and preparation of the LHC experiments.

6. Searches Beyond Known Physics

It has been shown above that CMS measures known physics as predicted in the SM. We shall now present results from the searches for physics beyond what was known at the start of LHC running:

— Search for the Higgs boson
— Search for physics beyond the SM (BSM)

- Supersymmetry,
- Extra Dimensions,
- Unexpected physics?

We shall not present results from all of these searches but concentrate on the search for the Higgs boson and a few selected searches for BSM physics that are illustrative of the capability of CMS in this area.

6.1. *Search for the Higgs boson*

The production cross-sections and the branching fractions into the various decay modes of the SM Higgs boson as a function of mass are shown in Figs. 9(a) and 9(b), respectively.[18] For m_H up to $\approx$700 GeV the dominant Higgs-boson production mechanism is gluon–gluon fusion. The W–W or Z–Z fusion mechanism, known as vector boson fusion (VBF), becomes important for the production of higher-mass Higgs bosons. The quarks that emit the W/Z bosons have transverse momenta of the order of W and Z masses. The resulting high-energy jets in the forward regions, $2.0 < |\eta| < 5.0$, can be used to tag this reaction. The tagging of forward jets from the VBF process has turned out to be very important for measurement of the properties of the newly found boson.

With an integrated luminosity of about $5\,\text{fb}^{-1}$ at $\sqrt{s} = 7\,\text{TeV}$ and about $20\,\text{fb}^{-1}$ at $\sqrt{s} = 8\,\text{TeV}$, CMS has examined some 2,000 trillion inelastic pp collisions and in which there should be about 600 k SM Higgs bosons (at a mass of $m_H = 125\,\text{GeV}$) in Run 1 (2010–2012).

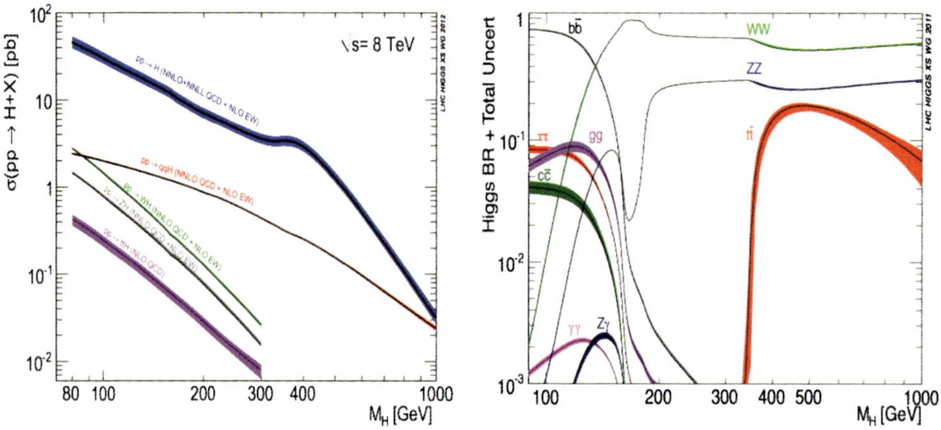

(a) (b)

Fig. 9. (a) The SM Higgs production cross-section at $\sqrt{s} = 8\,\text{TeV}$. (b) The SM Higgs branching ratios as a function of the Higgs-boson mass.

Once produced the Higgs boson disintegrates immediately in one of several ways (decay modes) into known SM particles, whose probability depends on its mass. A search had to be envisaged not only over a large range of masses but also over the many possible decay modes: into pairs of photons, Z bosons, W bosons, τ leptons, and b quarks. As mentioned earlier the natural width of the Higgs boson at $m_H \sim 125\,\text{GeV}$ is only 4 MeV and hence the observed width is entirely dominated by instrumental mass resolution. The two decay channels in which the resolution is excellent are H$\to \gamma\gamma$ and H$\to$ZZ$^{(*)} \to 4l$. However the branching fraction in the two modes are $2\cdot 10^{-3}$ and 10^{-4} respectively, leading to approximately 500 events and 20 events in the final plots, after taking account of efficiencies. Although the achievable resolution is only ~1 GeV these are nevertheless the two best decay channels for discovery.

The fact that the Higgs boson we discovered has a mass of 125 GeV means that many decay channels are observable.

6.1.1. *Bosonic decay modes of the Higgs boson*

The H$\to \gamma\gamma$ decay mode

In the H$\to \gamma\gamma$ analysis a search is made for a narrow peak in the diphoton invariant mass distribution in the mass range 110–150 GeV, on a large irreducible background of SM production of two photons. There is also a reducible background where one or more of the reconstructed photon candidates originate from misidentification of jet fragments, with the process of QCD Compton scattering dominating.

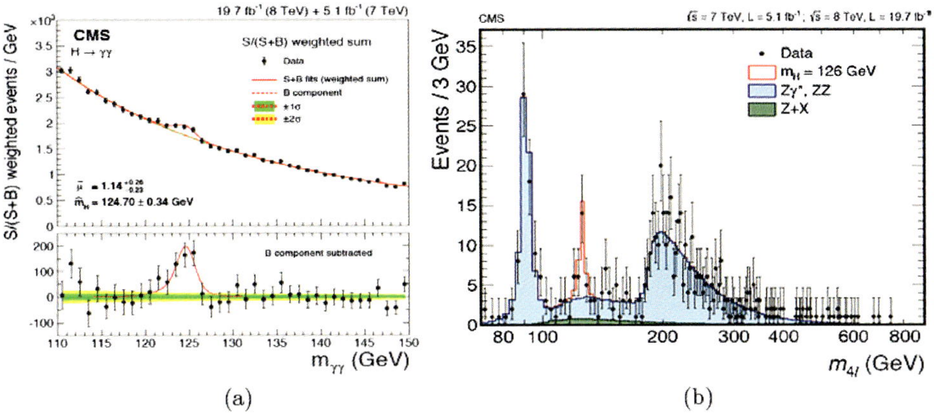

Fig. 10. (a) The two-photon invariant mass distribution of selected candidates, weighted by S/B of the category in which it falls. The lines represent the fitted background and the expected signal contribution ($m_H = 125\,\text{GeV}$). The four-lepton invariant mass distribution for selected candidates relative to the background expectation. (b) The expected signal contribution ($m_H = 125\,\text{GeV}$) is also shown.

The event selection requires two isolated[a] photon candidates satisfying p_{T} and photon identification criteria. In CMS, typically a p_{T} threshold of $m_{\gamma\gamma}/3$ ($m_{\gamma\gamma}/4$) is applied to the photon leading (sub-leading) in p_{T}, where $m_{\gamma\gamma}$ is the diphoton invariant mass. The selected events are categorised according to their estimated signal to background ratio, and whether they are tagged as likely to have originated from particular production processes. The background is estimated from data, without the use of MC simulation, by fitting the diphoton invariant mass distribution in a range ($100 < m_{\gamma\gamma} < 180\,\text{GeV}$). In the discovery analysis polynomial functions were used to describe the shape of the background.

The results from the CMS experiment are shown in Fig. 10(a). A clear peak at a diphoton mass of around 125 GeV is seen.[19] The observed (expected) significance was found to be 6.5σ (6.3σ).

The H→ZZ→4l decay mode

In the H→ZZ→4l decay mode a search is made for a narrow mass peak in the presence of a small continuum background. The background sources include an irreducible four-lepton contribution from direct ZZ production via quark–antiquark and gluon–gluon processes. Reducible background contributions arise from Z + bb

[a] Particles such as electrons, muons or photons emerging from a fundamental parton–parton interaction tend to be isolated, i.e. they are produced with no other accompanying particles in their immediate vicinity.

and tt production where the final states contain two isolated leptons and two b-quark jets producing secondary leptons.

The event selection requires two pairs of same-flavour, oppositely charged isolated leptons. Since there are differences in the reducible background rates and mass resolutions between the sub-channels 4e, 4μ, and $2e2\mu$, they are analysed separately. Electrons are typically required to have $p_T > 7$ GeV. The corresponding requirements for muons are $p_T > 5$ GeV. The ZZ background, which is dominant, is evaluated from Monte Carlo simulation studies.

The m_{4l} distribution is shown in Fig. 10(b).[20] A clear peak is observed at ~125 GeV, in addition to the one at the Z mass. The latter is due to the conversion of a photon from inner bremsstrahlung process that is emitted simultaneously with the dilepton pair. The observed (expected) significance was found to be $6.8\sigma(6.7\sigma)$.

6.1.2. *Leptonic decay modes of the Higgs boson.*

It is important to establish whether the new particle also couples to fermions, and in particular to down-type fermions, since the measurements in Section 6.1.1 mainly constrain the couplings to the up-type top quark. Determination of the couplings to down-type fermions requires direct measurement of the Higgs boson decays to bottom quarks or τ leptons.

Decays to fermions: The $H \rightarrow \tau\tau$ and the $H \rightarrow bb$ decay modes

The $H \rightarrow \tau\tau$ search is typically performed using the final-state signatures $e\mu$, $\mu\mu$, $e\tau_h$, $\mu\tau_h$, $\tau_h\tau_h$, where electrons and muons arise from leptonic τ-decays and τ_h denotes a τ lepton decaying hadronically. Each of these categories is further divided into two exclusive sub-categories based on the number and the type of the jets in the event: (i) events with one forward and one backward jet, consistent with the VBF topology, (ii) events with at least one high p_T hadronic jet but not selected in the previous category. In each of these categories, a search is made for a broad excess in the reconstructed $\tau\tau$ mass distribution. The main irreducible background, $Z \rightarrow \tau\tau$ production, and the largest reducible backgrounds (W + jets, multijet production, $Z \rightarrow ee$) are evaluated from various control samples in data.

The $H \rightarrow bb$ decay mode has by far the largest branching ratio (~54%). However since $\sigma_{bb}(\text{QCD}) \sim 10^7 \times \sigma(H \rightarrow bb)$ the search concentrates on Higgs boson production in association with a W or Z boson decaying by one of the following modes: $W \rightarrow e\nu/\mu\nu$ and $Z \rightarrow ee/\mu\mu/\nu\nu$. The $Z \rightarrow \nu\nu$ decay is identified by the requirement of a large missing transverse energy. The Higgs boson candidate is reconstructed by requiring two b-tagged jets.

Evidence for a Higgs boson decaying to a $\tau\tau$ lepton pair was first reported by CMS.[21] CMS has updated its analysis and the results reported in Table 2 come from Ref. 22 where both the $H \rightarrow \tau\tau$ and $H \rightarrow WW$ contributions are considered as

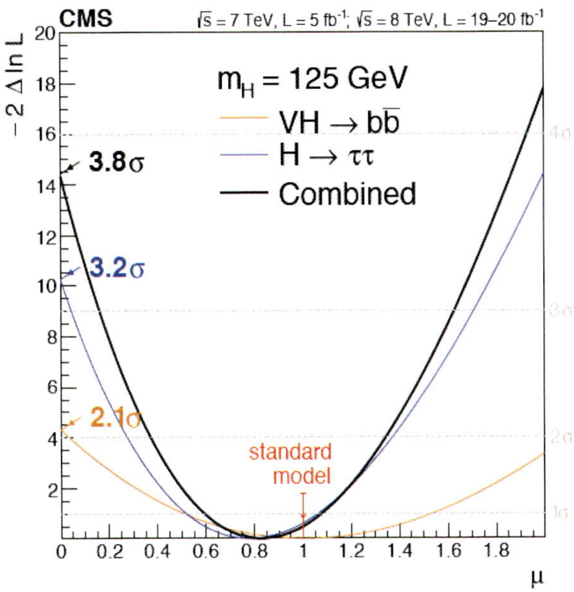

Fig. 11. Scan of the profile likelihood as a function of the signal strength relative to the expectation for the production and decay of a standard model Higgs boson, m, for $m_H = 125\,\text{GeV}$.

signal in the $\tau\tau$ decay-tag analysis. This treatment leads to an increased sensitivity to the presence of a Higgs boson that decays into both $\tau\tau$ and WW.

CMS has combined the results from the decay modes $\text{H} \rightarrow \tau\tau$ and VH with $\text{H} \rightarrow \text{bb}$.[23] Figure 11 shows the scan of the profile likelihood as a function of the signal strength relative to the expectation for the production and decay to fermions (bb and $\tau\tau$) of a SM Higgs boson with $m_H = 125\,\text{GeV}$. At this mass the excess over the background is found to have (expected) significance of $3.8\sigma(3.9\sigma)$ for the combination.

6.1.3. *Compatibility of the observed state with the SM Higgs boson hypothesis: Signal strength*

To establish whether or not the newly found state is the Higgs boson of the SM, we need to precisely measure its other properties and attributes. Several tests of compatibility of the observed excesses with those expected from a standard model Higgs boson have been made.

In one comparison labelled the signal strength μ, where $\mu = \sigma/\sigma_{SM}$, the measured production cross-section times branching fraction of the signal is compared with the SM expectation, determined for each decay mode individually and for the overall combination of all channels. A signal strength of one would be indicative of a SM Higgs boson. Table 1 summarises the measurements of μ from the ATLAS and CMS experiments.[24]

Table 1. The measured signal strength μ and their total uncertainties for different Higgs boson decay channels, from the ATLAS and CMS experiments and their data combined. The expected uncertainties in the measurements are displayed in parentheses.

Decay channel	ATLAS+CMS	ATLAS	CMS
$\mu^{\gamma\gamma}$	$1.14\ ^{+0.19}_{-0.18}$	$1.14\ ^{+0.27}_{-0.25}$	$1.11\ ^{+0.25}_{-0.23}$
	$\left(^{+0.18}_{-0.17}\right)$	$\left(^{+0.26}_{-0.24}\right)$	$\left(^{+0.23}_{-0.21}\right)$
μ^{ZZ}	$1.29\ ^{+0.26}_{-0.23}$	$1.52\ ^{+0.40}_{-0.34}$	$1.04\ ^{+0.30}_{-0.25}$
	$\left(^{+0.23}_{-0.20}\right)$	$\left(^{+0.32}_{-0.27}\right)$	$\left(^{+0.30}_{-0.25}\right)$
μ^{WW}	$1.09\ ^{+0.18}_{-0.16}$	$1.22\ ^{+0.23}_{-0.21}$	$0.90\ ^{+0.23}_{-0.21}$
	$\left(^{+0.16}_{-0.15}\right)$	$\left(^{+0.21}_{-0.20}\right)$	$\left(^{+0.23}_{-0.20}\right)$
$\mu^{\tau\tau}$	$1.11\ ^{+0.24}_{-0.22}$	$1.41\ ^{+0.40}_{-0.36}$	$0.88\ ^{+0.30}_{-0.28}$
	$\left(^{+0.24}_{-0.22}\right)$	$\left(^{+0.37}_{-0.33}\right)$	$\left(^{+0.31}_{-0.29}\right)$
μ^{bb}	$0.70\ ^{+0.29}_{-0.27}$	$0.62\ ^{+0.37}_{-0.37}$	$0.81\ ^{+0.45}_{-0.43}$
	$\left(^{+0.29}_{-0.28}\right)$	$\left(^{+0.39}_{-0.37}\right)$	$\left(^{+0.45}_{-0.43}\right)$
$\mu^{\mu\mu}$	$0.1\ ^{+2.5}_{-2.5}$	$-0.6\ ^{+3.6}_{-3.6}$	$0.9\ ^{+3.6}_{-3.5}$
	$\left(^{+2.4}_{-2.3}\right)$	$\left(^{+3.6}_{-3.6}\right)$	$\left(^{+3.3}_{-3.2}\right)$

Both the ATLAS and CMS experiments have measured μ values, by decay mode and by additional tags used to select preferentially events from a particular production mechanism. The best-fit value for the common signal strength μ, obtained in the different sub-combinations and the overall combination of all search channels in the ATLAS and CMS experiments leads to an observed μ value of $1.09 \pm 0.11(0.10)$ for CMS for a Higgs boson mass of 125.09 GeV.[24] The theoretical errors are still sizeable ($\sim$7%), so there is also work needed to reduce these as data accumulate. Some of the theoretical errors can be cancelled by taking ratios as illustrated in Fig. 12.[24]

6.2. *Moving forward: Should we really expect new physics?*

The newly found boson described in Section 6.1 has the fingerprints of the SM Higgs boson. Now the question arises whether there is new physics. The prevailing

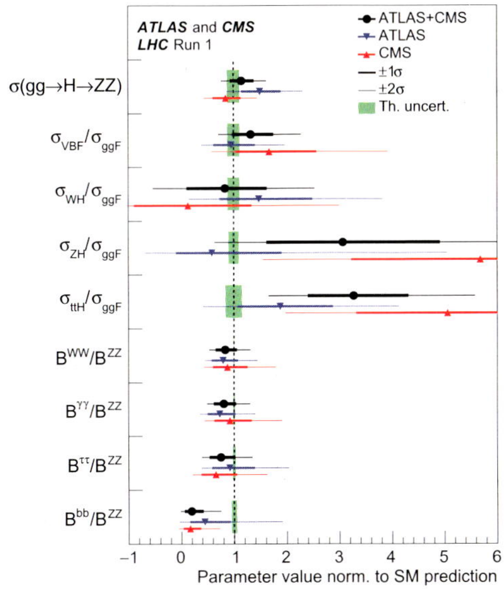

Fig. 12. Best fit values of the gg→H→WW cross section and of ratios of cross sections and branching fractions for the combination of the ATLAS and CMS measurements.

conjecture is that there must be new physics. There is ample observational evidence for physics beyond the SM. Amongst the examples are:

- neutrinos have mass and oscillate from one type to another,
- the lightness of the Higgs boson,
- the existence of dark matter,
- the existence of matter–antimatter asymmetry.

6.2.1. *Lightness of the Higgs boson and supersymmetry*

Let us first consider the lightness of the Higgs boson. What happens if we try to extend the validity of the SM to energy scales much beyond the weak scale? The mass of a fundamental scalar, such as the Higgs boson, is altered by radiative corrections ($m_{\rm H}^2 \to m_{\rm bare}^2 + \Lambda^y$) arising from quantum loops containing fermions or bosons (Fig. 13), where Λ is the cutoff momentum. The cutoff could be as high as 10^{15} GeV for all we know! Hence we would have a scalar boson with a mass around 100 GeV getting mass corrections of the order of 10^{15} GeV — clearly an unappealing feature. Inclusion of radiative corrections has been very successful in the past, e.g. correctly predicting the mass of the top quark mass (Fig. 14) and narrowing the range for a putative Higgs boson, before these particles were discovered. So the effect of radiative corrections cannot be lightly dismissed and it is not sufficient to invoke the anthropic principle and state that the Higgs boson mass is small because it is small! So the question already posed in the early 1990s remains: what protects the mass of the Higgs boson?

the above-mentioned radiative corrections meaning that the mass difference (m_{top}-m_{stop}) has to be small. This is sometimes labelled as a "natural" scenario. It is therefore particularly interesting to search for a stop at "low" masses.

Some argue that the fact that the Higgs boson was found to have a mass below 135 GeV should be considered as a prediction of SUSY. Experimentalists did pay attention to this prediction when designing the experiments as finding the Higgs boson in this mass region presented particular and difficult challenges.[11]

SUSY also has other attractive features: it provides a route to grand unification of strong, electromagnetic and weak forces at a scale of about 10^{15} GeV, and an escape from too rapid a proton decay. In some scenarios it also generates a candidate particle for dark matter. Dark matter is weakly and gravitationally interacting matter with no electromagnetic or strong interactions. These are the properties of the lightest supersymmetric particle (LSP) in models that conserve a quantum number called R-parity and thus this particle is stable ($R = (-1)^{3(B-L)+2S}$) and S, B and L are spin, baryon number and lepton number respectively. Hence the question arises: is dark matter a result of supersymmetry?

The searches for SUSY can be categorised by production process, and the characteristics of the final state. At the LHC, as the fundamental initial state interaction is a strong interaction between quarks and gluons, the cross-section for the production of squarks and gluinos is large (Fig. 16). There is also a sizeable cross-section for stop production and a smaller one for partners of electroweak particles collectively known as *ewk-inos*. The final states can be classified by those that

(i) conserve R-parity — the LSP is stable, behaves like a neutrino, and is a possible dark matter candidate
(ii) violate R-parity — the LSP decays into SM particle and the searches typically focus on large object multiplicities
(iii) lead to long-lived particles.

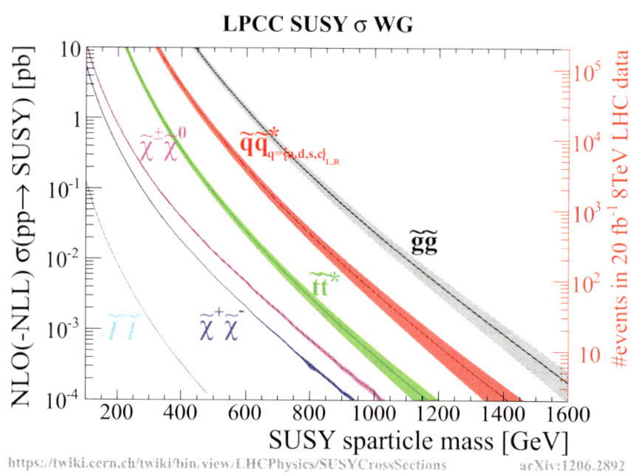

Fig. 16. The production cross-section for some pairs of supersymmetric particles.

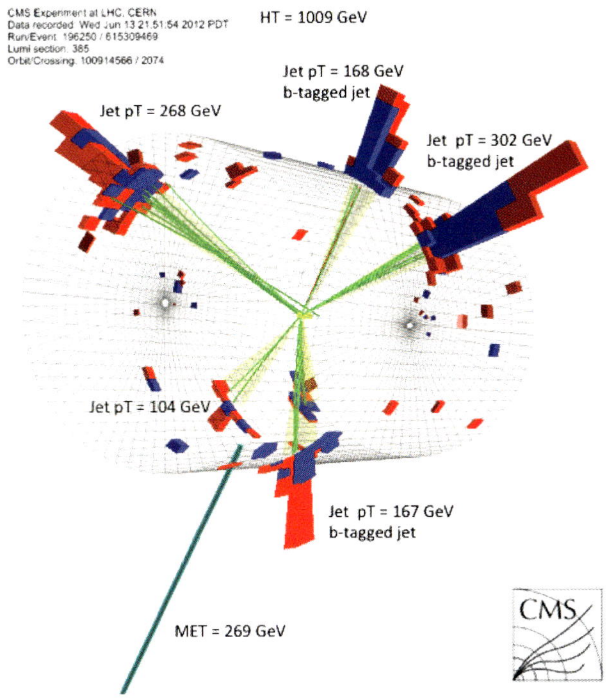

Fig. 17. Event recorded with the CMS detector with multi-jets and missing E_T with characteristics expected from production of supersymmetric particles.

In R-parity conserving scenarios the SUSY particles are produced in pairs. We look for decay chains that contain LSPs leading to significant E_T^{miss}. A candidate event is shown in Fig. 17. Such events can also be produced by known SM processes, e.g. Z+jets where the Z boson decays into neutrino pairs. A search is made for a significant excess of events over and above that precisely predicted by SM physics.

Although many searches have been made SUSY has not turned up yet at the LHC. The results of searches made by CMS are summarised in Fig. 18. Each line in the summary plot corresponds to a dedicated physics analysis by a group of 30–50 scientists, examining the predictions, the selection cuts, the efficiencies, the backgrounds, and the systematic errors. A null result is used to set limits on the mass and the production cross-sections of the hypothetical particles, that today range in mass between 0.5 and 1 TeV. Higher energy running will allow us to search for particles with higher masses. Searches make an assumption that whatever decay mode is examined its branching fraction is 100%. This assumption simplifies the presentation of the results for use by others, but is unlikely to be the case and so SUSY could still exist at lower values of $\sigma \cdot BR$.

The allowed SUSY space is large. There is a very large number of parameters (~120) that describe supersymmetry, though by making some plausible assumptions

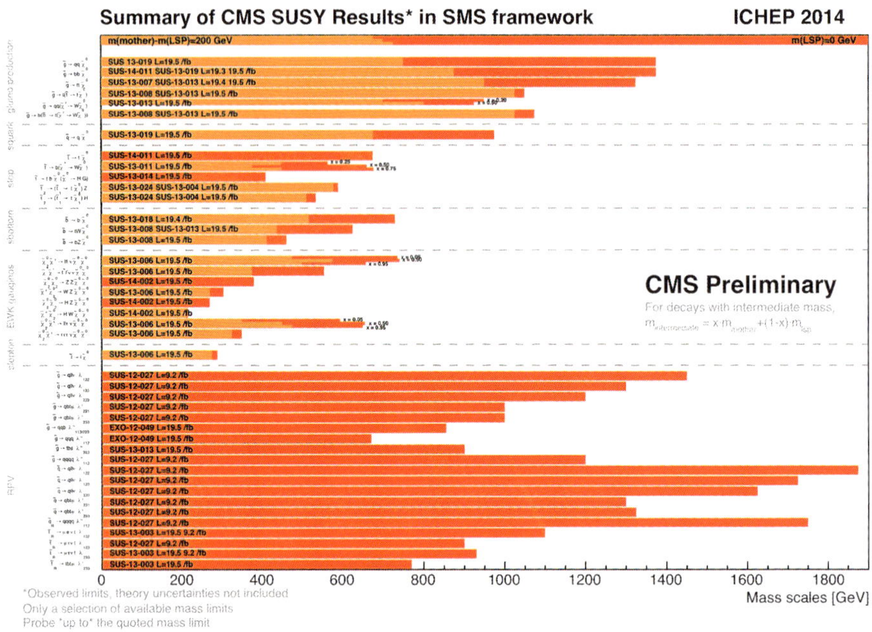

Fig. 18. A compilation of mass limits in searches for supersymmetric particles with the ATLAS experiment.

this number can be reduced to, e.g., 19 in pMSSM (phenomenological minimal supersymmetric model) — a smaller but still a large number — making it difficult to fully probe the allowed space. So it is no surprise that there are holes in the searches. It may well turn out that it is not possible to definitively exclude weak-scale SUSY at the LHC.

6.4. *Searching for new physics at the LHC: Non-SUSY BSM Physics*

These searches include those for compositeness of SM particles, extra dimensions (some of these theories predict heavy resonances that may be observable at the LHC), new heavy gauge bosons, leptoquarks, excited fermions, black holes, dark matter particles, and more.

Physics beyond the SM could involve extra space dimensions. It is known that fundamental laws of Nature, e.g. gravitation are modified by the number of space dimensions — the familiar $1/r^2$ law of gravity in 3D changes to $1/r^{n-1}$ for n dimensions. This could be a mechanism by which gravity could appear so weak at scales probed so far. Some theories of extra dimensions predict Z-like massive bosons with a mass of a few TeV. In Run 1 CMS looked for peaks in dilepton invariant mass distributions. None were found and limits for $m(Z'_{SSM})$of $\sim$2.9 TeV have been set assuming that the potential state decays into dielectron or dimuon pairs with the same branching as in the decay of the Z boson (Fig. 19). Setting limits on $\sigma{\cdot}$BR

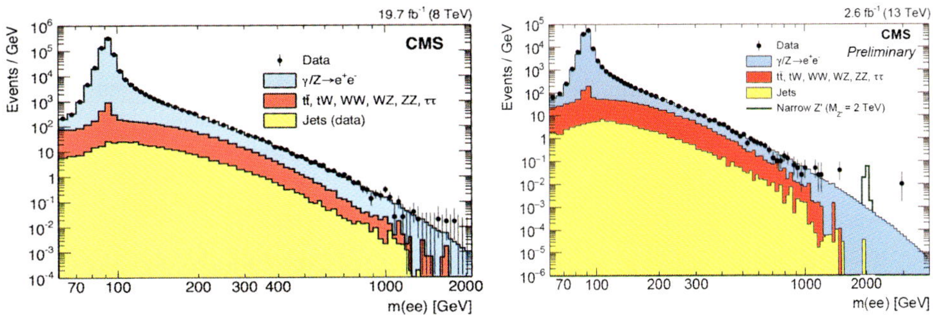

(a) (b)

Fig. 19. (a) The high mass di-muon mass spectrum measured at $\sqrt{s} = 8\,\text{TeV}$. (b) The high mass di-electron mass spectrum measured at $\sqrt{s} = 13\,\text{TeV}$.

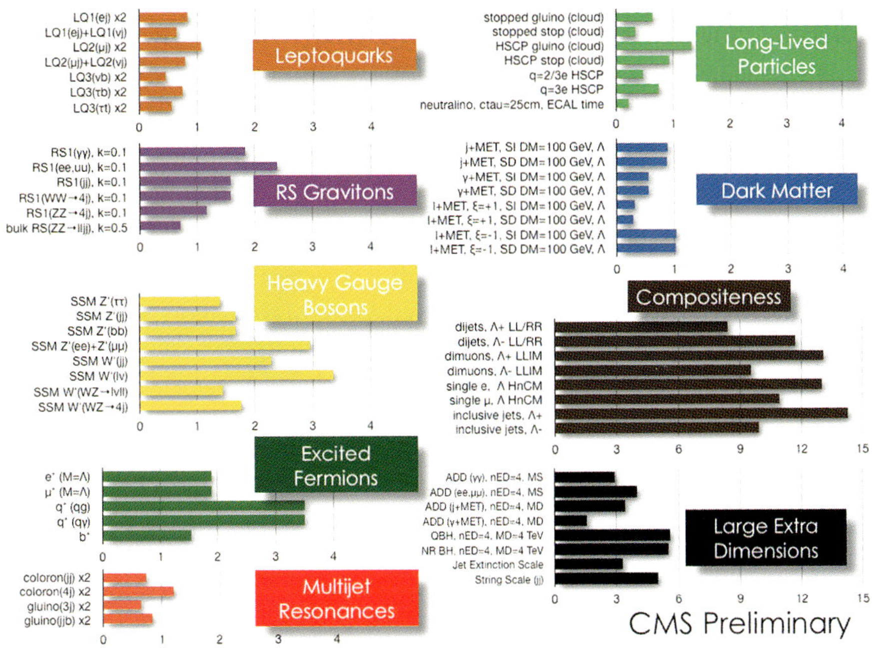

Fig. 20. A compilation of mass limits in searches for various particles predicted by theories of beyond the SM physics with the CMS experiment.

allows contact with any theory predicting the existence of such bosons — assuming they have non-zero branching to dielectrons or dimuons.

The search for physics beyond the SM is summarised in Fig. 20. Again each line represents one analysis and are grouped together according to the subject of the search, e.g. leptoquarks, gravitons, dark matter, excited fermions, etc. Some SM results have from the $\sqrt{s} = 13\,\text{TeV}$ running have been included in Fig. 8.

Table 2. Observed and expected mass limits for analyses that exclude the listed models at 95% CL for a resonance mass from 1.5 TeV up to the indicated value.

Model	Final state	Observed mass limit [TeV]	Expected mass limit [TeV]
String	qg	7.0	6.9
Scalar diquark	qq	6.0	6.1
Axigluon/coloron	$q\bar{q}$	5.1	5.1
Excited quark (q^*)	qg	5.0	4.8
Color-octet scalar	gg	3.1	3.3
Heavy W (W')	$q\bar{q}$	2.6	2.3

Even a small quantity of data collected at the higher energy opens up significant new territory in the search for heavy states. One example is the search for bumps in dijet invariant mass distribution testing the s-channel production of new heavy resonances. The highest mass dijet pair observed has an invariant mass of 6.14 TeV. The search did not reveal any structure other than a smoothly falling distribution and limits have been set for, e.g. excluding contact interaction below 12.1 TeV and the scale for ADD models of <7.7 TeV.[25] The exclusions are summarised in Table 2.

At the end of the 8 TeV run both ATLAS and CMS experiments saw some anomalies in their data beyond the predictions of the SM. These have now been cleared with the new $\sqrt{s} = 13$ TeV data.

A search in the 2015 data for a bump in diphoton invariant mass distribution at high masses recently raised particular interest. Both ATLAS and CMS experiments saw a peak at an invariant mass around 750 GeV. In CMS the local significance was found to be 3.0σ whilst the global significance was 1.7σ. Analysis of a much larger amount of data, taken in 2016, has shown that this bump is most likely a statistical fluctuation in both the experiments. Hence much care has to be taken when interpreting low significance bumps even if both experiments see it. The experiments displayed justified caution. As the data recorded and examined is currently increasing with a relatively short time constant it is sufficient just to wait a short while in order to confirm or refute any hints of new physics.

7. Looking Ahead

In the current LHC phase (2015–2023) we expect to accumulate an integrated luminosity corresponding to around 300 fb^{-1}, at a centre-of-mass energy almost twice that in Run 1 (13 TeV instead of 8 TeV). Some 5 million Higgs bosons will have been produced allowing for a much more detailed examination of its properties and to see if any deviate from the precise predictions of the SM. If they do significantly deviate, then we know that there is physics beyond the SM. A recent study[26] has given indication of the precision that can be achieved on the measured properties of the Higgs boson with an integrated luminosity of 300 fb^{-1} (Table 3).

Table 3. Precision on measurements of the signal strength for some key decay modes of a SM-like Higgs boson with integrated luminosities of 300 and 3,000 fb^{-1} at $\sqrt{s} = 14$ TeV. The coupling scale factor κ_i are defined in such a way that the cross-sections $\sigma_{i,j}$ and the partial decays widths Γ_j associate to the SM with $\kappa_i^2 . \kappa_j^2$. The numbers in brackets are % uncertainties on κ_i for [no theory uncertainty, current theory uncertainty

L (fb^{-1})	Exp.	κ_γ	κ_W	κ_Z	κ_g	κ_b	κ_t	κ_τ	$\kappa_{Z\gamma}$	$\kappa_{\mu\mu}$
300	ATLAS	[8, 13]	[6, 8]	[7, 8]	[8, 11]	N/a	[20, 22]	[13, 18]	[78, 79]	[21, 23]
	CMS	[5, 7]	[4, 6]	[4, 6]	[6, 8]	[10, 13]	[14, 15]	[6, 8]	[41, 41]	[23, 23]
3000	ATLAS	[5, 9]	[4, 6]	[4, 6]	[5, 7]	N/a	[8, 10]	[10, 15]	[29, 30]	[8, 11]
	CMS	[2, 5]	[2, 5]	[2, 4]	[3, 5]	[4, 7]	[7, 10]	[2, 5]	[10, 12]	[8, 8]

In addition we shall be looking for all the states listed in Figs. 18 and 20, e.g. new resonances, SUSY particles, "exotica". We also shall be looking for deviations from the precise predictions of numerous SM processes. It has to be remarked that in addition to the LHC now being a Higgs factory, it is also a top-quark factory, a W/Z factory, a b-quark factory, etc., producing enormous numbers of these particles.

7.1. *Looking ahead at the LHC: The high-luminosity LHC*

Thoughts have already turned to the long-term future of the LHC. The highest priority in the global strategy for particle physics is the exploitation of the full potential of the LHC, including the high-luminosity upgrade (HL-LHC) of the machine and detectors with a view to collecting ten times more data (i.e. 3,000 fb^{-1}) than in the initial design, by around 2035. The work towards this goal is proceeding well. Figure 21 illustrates the long-term evolution of the LHC programme and the arrow indicates our present position — so in a sense the LHC programme has just started. Foreseen are increases in instantaneous luminosity, from $10^{34}\ cm^{-2}\ s^{-1}$ to $5 \times 10^{34}\ cm^{-2}\ s^{-1}$, and in integrated luminosity, from 300 fb^{-1} to 3,000 fb^{-1}. This would lead to an examination of some 50 million Higgs bosons, amongst other states.

Considerable and fresh detector R&D and prototyping is being carried out to build new inner trackers and forward calorimeters, as well as extensive upgrades of electronics to allow selection of approximately one million events at the first level of triggering. In CMS we shall construct a very powerful endcap calorimeter, with fine lateral and longitudinal granularity, with each of the 6 million cells providing both the energy and precise timing information. This would allow tracking of objects such as electrons and photons, as well as jets, they be taus, merged jets or highly collimated jets arising from VBF reactions. The ability to trigger on jets from VBF reactions, without requirements in the central region, will open up a new domain of physics with respect to what is possible today. Having a more powerful Level-1 trigger is particularly important as the higher pileup conditions leads to more confusion with respect to the events of interest. However, the trigger thresholds cannot be increased much from the ones used today if one wishes to capture events

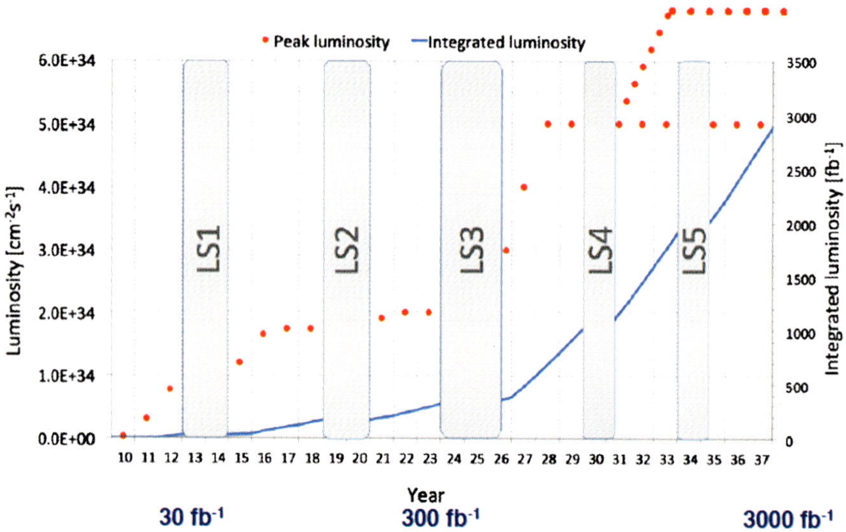

Fig. 21. The predicted evolution of the instantaneous (lhs scale) and integrated luminosity (rhs scale). The ATLAS and CMS experiments will be upgrade during the Long Shutdown 3 (LS3).

that lead to objects with low or moderate p_T, e.g. the Higgs boson events. Hence more detailed (implying a larger amount) information has to be transferred out at 40 MHz and subsequently used to form the Level-1 trigger. Fortunately, the newer generation of FPGA's are now powerful enough, and the link speeds are high enough, for this to be contemplated. Introduction of the information from the inner tracker is envisaged at this first level of triggering. This will further increase the power of the Level-1 trigger — so that it becomes radically different from what is possible today.

8. Summary

The construction of the standard model is a towering intellectual achievement of humankind and Abdus Salam was in the first rank amongst the "constructors" of the SM that include many theorists and experimentalists. The construction of the SM is the result of intense interplay between experimental discoveries and theoretical intuition over the last 50 years.

After 25 years of the design and construction of the LHC and its experiments, and following the first exploitation of physics data, we are firmly in the phase of physics exploitation. During the design and construction phases several new technologies were developed and others pushed to their limits. A "massive" discovery has been made — that of a Higgs boson. The boson appears to be just the one predicted by the SM. No evidence has yet been found for new physics.

However, there is ample experimental evidence for phenomena that the standard model does not explain. The study of the newly found Higgs boson might lead us to new physics. The "novelty" and the "lightness" of this Higgs boson calls for its in-depth study in the current LHC run, and at the HL-LHC and beyond.

The discovery of Higgs boson is just the start of the exploration of the Terascale. Lying ahead is an exciting programme of physics and innovation at the LHC and the HL-LHC. We hope new physics will appear in the $\sqrt{s} = 13\,\text{TeV}$ running of the LHC. Only experiments reveal or confirm Nature's secrets.

Acknowledgments

We would like to thank the organisers of the Memorial Meeting for Nobel laureate Professor Abdus Salam's 90th birthday at the Nanyang Technological University, Singapore, for the invitation to present the CMS experiment, from conception through construction to data taking and discovery of the Higgs boson. The discovery has only been possible through the relentless and sustained efforts of all of the collaborators in CMS, and those in the LHC accelerator and the Worldwide Computing Grid.

References

1. S. Weinberg, *Phys. Rev. Lett.* **19**, 1264 (1967).
2. A. Salam, in *Proceedings of the Eighth Nobel Symposium*, ed. N. Svartholm (Almqvist & Wiskell, 1968. C), p. 367; Jarlskog, *CP Violation* (World Scientific, Singapore, 1988).
3. S. L. Glashow, *Nucl. Phys.* **22**, 579 (1961).
4. P. Jenni, The long journey to the Higgs boson and beyond at the LHC Part II: Emphasis on ATLAS, in *Memorial Volume on Abdus Salam's 90th Birthday*, eds. L. Brink, M. Duff and K. K. Phua (World Scientific, 2017).
5. ATLAS Collaboration, *Phys. Lett.* B, **716**, 1 (2012).
6. CMS Collaboration, *Phys. Lett.* B, **716**, 30 (2012).
7. F. Englert and R. Brout, *Phys. Rev. Lett.* **13**, 321 (1964).
8. P. W. Higgs, *Phys. Lett.* **12**, 132 (1964).
9. P. W. Higgs, Phys. *Rev. Lett.* **13**, 508 (1964).
10. G. S. Guralnik, C. R. Hagen, and T. W. B. Kibble, *Phys. Rev.* **155**, 1554 (1967).
11. C. J. Seez, T. S. Virdee, L. Di Lella, R. H. Kleiss, Z. Kunszt, W. J. Stirling, in: G. Jarlskog, D. Rein (Eds.), *Proceedings of the Large Hadron Collider Workshop*, Aachen, Germany, 1990, p. 474, CERN 90-10-V-2/ECFA 90-133-Vol-2.
12. M. Della Negra, D. Froidevaux, K. Jakobs, R. Kinnunen, R. Kleiss, A. Nisati, T. Sjostrand, in: G. Jarlskog, D. Rein (Eds.), *Proceedings of the Large Hadron Collider Workshop*, Aachen, Germany, 1990, p. 509, CERN 90-10-V-2/ECFA 90- 133-V ol-2.
13. N. Ellis and T. S. Virdee, *Ann. Rev. Nucl. Part. Sci.* **44**, 609 (1994).
14. T. S. Virdee, *Physics World* **17**(9), 35 (2004).
15. T. S. Virdee, *Phys. Rep.* **403–404**, 401 (2004).
16. D. Froidevaux and P. Sphicas, *Ann. Rev. Nucl. Part. Sci.* **56**, 375 (2006).
17. CMS Collaboration, *JINST* **3**, S08004 (2008).
18. LHC Higgs Cross Section Working Group, arXiv:1101.0593 (2011), arXiv:1201.3084 (2012), arXiv:1307.1347 (2013).
19. CMS Collaboration, *Eur. Phys.* **74**, 3076 (2016).
20. CMS Collaboration, *Phys. Rev. D* **89**, 092007 (2014).

21. CMS Collaboration, *JHEP* **05**, 104 (2014).
22. CMS Collaboration, *Eur. Phys. J. C* **75**, 212 (2015).
23. CMS Collaboration, *Nat. Phys.* **10**, 557 (2014).
24. ATLAS and CMS Collaborations, *JHEP* **08**, 045 (2016).
25. CMS Collaboration, *Phys. Rev. Lett.* **116**, 071801 (2016).
26. ALICE, ATLAS, CMS and LHCb Collaborations, ECFA/13/284 (2013).

The Long Journey to the Higgs Boson and Beyond at the LHC Part II: Emphasis on ATLAS

Peter Jenni
Albert-Ludwigs-University of Freiburg, Germany, and CERN, Switzerland
peter.jenni@cern.ch

The journey in search for the Higgs boson with the ATLAS and CMS experiments at the Large Hadron Collider (LHC) at CERN started more than two decades ago. But the first discussions motivating the LHC project dream date back even further into the 1980s. This article will recall some of these early historical considerations, mention some of the LHC machine milestones and achievements, focus as an example of a technological challenge on the unique ATLAS superconducting magnet system, and then give an account of the physics results so far, leading to, and featuring particularly, the Higgs boson results, and sketching finally prospects for the future. With its emphasis on the ATLAS experiment it is complementary to the preceding article by Tejinder S. Virdee which focused on the CMS experiment.

Keywords: Large Hadron Collider; ATLAS experiment; superconducting toroid magnet system; CMS experiment; Standard Model; Higgs boson; searches for physics Beyond the Standard Model.

1. Introduction

The Large Hadron Collider (LHC) project is a global scientific adventure, which was initiated some 30 years ago, combining the accelerator, the experiments, and a worldwide computing grid, based on, and accompanied of course with lots of continuous motivation from the theory community. Having had the privilege to be involved from the beginning, it is a great pleasure to present now results from the first few years of operation.

1.1. *How the LHC came to be*

The idea that the tunnel for the future Large Electron–Positron (LEP) collider should be able to house at some time even further in the future a Large Hadron Collider (LHC) was already in the air in the late 1970s. Moreover — thankfully — those leading CERN at the time had the vision to plan for a tunnel with big enough diameter to allow the eventual installation of such an accelerator. In the broader community, however, enthusiasm for an LHC surfaced for the first time in 1984, promoted in part by members of CERN's successful proton–antiproton collider experiments UA1 and UA2 with their discovery of the W and Z bosons the previous year. A workshop in Lausanne on the "Large Hadron Collider in the LEP Tunnel"[1] organized jointly by the European Committee for Future Accelerators (ECFA) and CERN brought together working groups that comprised machine experts, theorists and many experimentalists.

With the realization of the great physics potential of an LHC, several motivating workshops and conferences followed, where the formidable experimental challenges started to appear manageable, provided that enough R&D work on detectors could be carried out. Highlights of these "LHC experiment preliminaries" were the 1987 Workshop in La Thuile initiated by the so-called "Rubbia Long-Range Planning Committee"[2] and the large Aachen LHC Workshop[3] in 1990. The late 1980s were characterized by great optimism and, seen from today, by somewhat naïve dreams to have LHC running already before 2000, eventually in alternation with LEP. Also during these early years the community became more and more convinced that sophisticated detectors would be needed to address the full physics potential of the machine. Finally, in March 1992 the famous conference "Towards the LHC Experimental Programme"[4] took place in Evian-les-Bains, where several proto-collaborations presented their designs in "Expressions of Interest". Moreover, from the early 1990s, CERN's LHC Detector R&D Committee (DRDC), which reviewed and steered R&D collaborations, greatly stimulated innovative developments in detector technology.

2. Approval and Construction of the Project

2.1. *The approval and construction of the collider*

At CERN in the second half of the 1970s ECFA recommended LEP as the flagship machine for a thorough experimental study of the Standard Model. The planned beam energy was about 100 GeV with a tunnel circumference of about 27 km. As mentioned above, already at that time there was the feeling that such a construction represented a major and perhaps unique undertaking, also in view of possible subsequent use of the tunnel for a proton machine. While the construction of the LEP tunnel was underway, in the USA the design and construction of the Superconducting Super Collider (SSC) for proton–proton collisions with a beam energy of 20 TeV was proceeding and taking up momentum. The debate was what Europe should do for a future hadron collider at the energy frontier. CERN had not only a tunnel (in construction) but also an existing, well-operating chain of injectors (linear accelerator, booster, PS and SPS), which, after some modifications, could be used for injecting protons and ions into a collider in the LEP tunnel. In this context, the feasibility study of a proton–proton collider in the LEP tunnel started at the end of 1982 and led to the Lausanne ECFA-CERN workshop in 1984 mentioned above. The 1987 workshop on physics at future accelerators in La Thuile, and the report to the CERN Council of the Long-Range Planning Committee put the LHC on the road map as the next CERN project after LEP.

The project received the first approval by the CERN Council in December 1994, albeit in a staged version with a 'missing magnet' scheme limiting its energy initially to 10 TeV. The final approval for the present configuration came at the end of 1996 after contributions from several non-CERN member states have been agreed upon.

The commissioning started in 2008, but was soon interrupted by a major technical incident. After the repair, the first collisions occurred in November 2009, and a successful three-year run started in 2010, referred to as LHC Run-1.

2.2. *The approval and construction of the ATLAS experiment*

The detection of the Standard Model Higgs boson played a particularly important role in the design of the general-purpose experiments. In the region of low mass ($114 < m_H < 150$ GeV), the two channels considered particularly suited for unambiguous discovery[5] were the decay to two photons and the decay to two Z bosons, where one or both of the Z bosons could be virtual. As the natural width of such a low-mass Higgs boson is <10 MeV, the width of any observed peak would be entirely dominated by instrumental mass-resolution. This meant that in designing the general purpose detectors, considerable care was placed on the magnetic field strength and the precision tracking systems, as well as on high-resolution electromagnetic calorimeters. The high-mass region ($m_H > 150$ GeV) and signatures from Supersymmetry (SUSY) drove the need for good hadronic resolution for jets and missing transverse energy, as well as for almost full 4π calorimetry coverage.

The choice of the field configuration determined the overall design. It was well understood that to stand the best chance of making discoveries at the new energy scale of the LHC — and in the harsh conditions generated by about a billion pairs of protons interacting every second — would require the invention of new technologies while at the same time pushing existing ones to their limits. In fact, a prevalent saying was "We think we know how to build a high-energy, high-luminosity hadron collider — but we don't have the technology to build a detector for it". But in fact both were formidable challenges. That the general purpose experiments have worked so marvellously well since the start-up of the LHC is a testament to the difficult technology-choices made by the conceivers and the critical decisions made during the construction of these experiments. It is noteworthy that the very same elements mentioned above were crucial in the 2012 discovery of a Higgs boson.

At the Aachen meeting[3] much discussion took place on which detector technologies and field configurations to deploy. At the Evian meeting four experiment designs were presented: two using toroids and two using high-field solenoids. In the aftermath of this meeting, lively discussions took place in the community on how to continue and possibly join forces. The time remaining was short as the newly formed experiment peer review committee, the LHC Committee (LHCC), had set a deadline for the submission of the letters of intent (LoI) of 1st October 1992. The designs deploying the toroidal configurations merged to form the ATLAS experiment, in the focus for this report, deploying a superconducting air-core toroid for the measurement of muons, supplemented by a superconducting 2 T solenoid to provide the magnetic field for inner tracking and by a liquid-argon/lead electromagnetic calorimeter with a novel "accordion" geometry.

It took seven rounds of intense encounters between the experiment teams and the LHCC referees before the committee decided at its 7th meeting on 8–9 June 1993 "to recommend provisionally that ATLAS and CMS should proceed to technical proposals", with agreed milestones for further reviews. So the two general-purpose experiments were launched, but the teams could not have foreseen the enormous technical, financial, industrial and human challenges that lay ahead. For the technical proposals, many difficult technology-choices now had to be made "for real" for all of the detector components, whereas the LoI had just presented options for many items. This meant, for example, that several large R&D collaborations sometimes had to give up on excellent developments in instrumentation, which had been carried out over many years, and to find their new place working on the technologies that the experimental collaborations considered best able to deliver the physics. Costs and resources were a constant struggle: they were reviewed over a period of many years by an expert costs-review committee (called CORE) of the LHCC. It was not easy for many bright physicists to accept that the chosen technology had also to remain affordable.

It cannot be stressed enough how important the many years of R&D were that preceded the final detector construction for the LHC experiments. Technologies had to be taken far beyond their state-of-the-art of the late 1980s in terms of performance criteria in the anticipated harsh LHC environment, like granularity and speed of readout, radiation resistance, reliability, but also considering buildable sizes of the detector components and number of units, and very importantly at an affordable cost. For many detector subsystems there were initially a few parallel developments pursued as options, because it was not guaranteed from the onset that a given proposed technology would finally fulfil all the necessary requirements. Increasingly more realistic prototypes were developed, in a learning process for both the detector communities and the industries involved.

The submission of the Technical Proposals followed in 1995, and they were approved in 1996. The formal approval for construction was given on 1st July 1997 by the then Director-General, Chris Llewellyn Smith, based on the recommendations of the Research Board and the LHCC (by then at meeting number 27) after the first of a long series of Technical Design Reports, imposing a material cost ceiling of 475 MCHF. These were also the years when the formal framework was set up by CERN and all of the funding agencies, first in an interim and finally via a detailed Construction Memorandum of Understanding, agreed upon in the new, biannual LHC Resources Review Boards.

Some of the major technology decisions were taken by the Collaboration before the submission of the Technical Proposal. For other choices the R&D needed more time, and they could only be made in the subsequent years from 1996 to the early

2000s, thereby defining the timing for the final Technical Design Reports of the various detector components.

Instead of giving a comprehensive account of all technological developments and constructions, well beyond the scope of this presentation, one major component is singled out as an example of the many technical challenges, namely the rather unique magnet system of the ATLAS detector. A detailed technical description of the complete detector 'as built' can be found in Ref. 6. It is worth noting that the ATLAS detector concept evolved based on the long evolution of generations of hadron collider experiments.[7]

2.2.1. *The ATLAS magnet system*

ATLAS features a unique configuration with four large superconducting magnets. A drawing of the magnet system can be seen in Fig. 1. It is an arrangement of a central solenoid (CS, not visible in the figure) providing the inner tracking detector with magnetic field, surrounded by a system of three large air-core toroids generating the magnetic field for the muon spectrometer. The overall dimensions of the magnet system are about 26 m in length and 22 m in diameter. Two end-cap toroids (ECT) are inserted in the barrel toroid (BT) at each end and line up with the CS. They have a length of 5 m, an outer diameter of 10.7 m and an inner bore of 1.65 m. The CS extends over a length of 5.8 m and has a bore of 2.4 m. The unusual configuration and large size made the magnet system a considerable challenge requiring

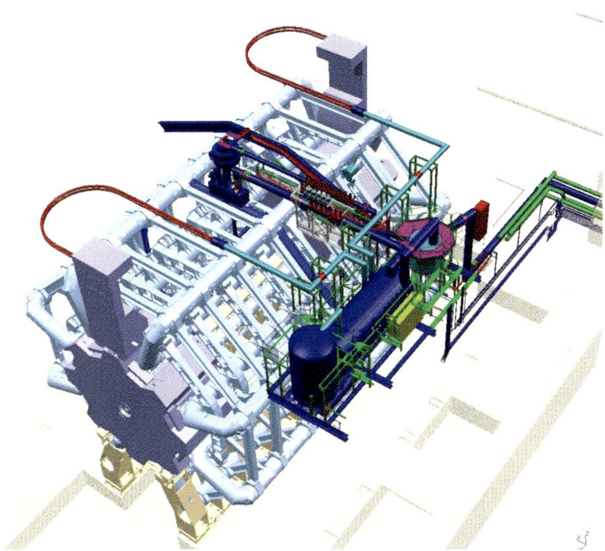

Fig. 1. Schematic view of the toroid magnet system including services connections. The eight BT coils are visible with the two castellated ECTs inserted on both ends. The central solenoid is not visible in this figure.

careful engineering. The first conceptual design of the magnet system was sketched in the early 1990s, and the technical design reports were published in 1997. The toroid cold-mass and cryostat integration work began at CERN in 2001. The first BT coil was lowered into the cavern in fall 2004, immediately followed by the CS (embedded inside the liquid Argon (LAr) electromagnetic barrel calorimeter cryostat). The remaining seven BT coils were installed in 2004 and 2005, and the ECTs in 2007.

The CS provides a central field of 2 T with a peak magnetic field of 2.6 T at the superconductor itself. The peak magnetic fields on the superconductors in the BT and ECT are 3.9 T and 4.1 T respectively. The performance in terms of bending power is characterized by the field integral $\int Bdl$, where B is the azimuthal field component and the integral is taken on a straight line trajectory between the inner and outer radius of the toroids. The BT provides 2 Tm to 6 Tm and the ECTs contribute with 1 Tm to 8 Tm in the 0.0–1.4 and 1.6–2.7 pseudorapidity (η) ranges, defined in terms of the polar angle θ as $\eta = -\ln[\tan(\theta/2)]$.

The central solenoid

The position of the CS in front of the electromagnetic calorimeter demanded a careful minimisation of the material in order to achieve the desired calorimeter performance. As a consequence, the CS and the LAr calorimeter share one common vacuum vessel, thereby eliminating two vacuum walls. The solenoid contributes a total of 0.66 radiation lengths at normal incidence. The single-layer coil is wound with high-strength Al-stabilized Nb/Ti conductor (1173 turns, 9.1 km total), specially developed to achieve a high field while optimizing thickness, inside a 12 mm thick Al support cylinder. The coil mass is 5.4 tonnes and the stored energy is 40 MJ. The flux is returned by the steel of the hadronic calorimeter and its girder structure. The CS is energized by an 8 kA power supply. The magnetic field of the CS has been mapped *in situ* with a relative precision of better than 10^{-4} over the full tracking volume. Figure 2 shows the insertion of the solenoid into the barrel cryostat.

The toroid system

Each of the three toroids consists of eight coils assembled radially and symmetrically around the beam axis. The ECT coil system is rotated by 22.5° with respect to the BT coil system in order to provide radial overlap and to optimize the bending power in the interface regions of both coil systems. The conductor and coil-winding technology is essentially the same in the BT and ECTs; it is based on winding a pure Al-stabilized Ni/Ti/Cu conductor (81.6 km overall) into pancake-shaped coils, followed by vacuum impregnation.

The BT coils are of a flat racetrack type with two double-pancake windings. The windings are housed in an aluminium alloy casing, with a total cold

Fig. 2. Insertion of the central solenoid into the barrel cryostat with the LAr calorimeter.

mass of 360 tonnes. The magnetic forces are transferred to the warm structure, counteracting the net Lorentz forces of about 1400 tonnes per coil directed inwards. The coils are housed in individual cryostats. The toroidal structure consists of eight cryostats and the linking elements between them, called voussoirs and struts that provide mechanical stability. Services are brought to the coils through a cryogenic ring linking the eight cryostats to a separate service cryostat (visible in Fig. 1), which provides connections to the power supply, the helium refrigerator, the vacuum systems and the control system. Figure 3 shows the BT after its installation in the underground cavern.

Each ECT also consists of eight coils in an aluminium alloy housing. They are cold-linked and assembled as a single cold mass of 140 tonnes, housed in one large cryostat. Therefore the internal forces are taken by the cold supporting structure between the coils, a different design solution than in the BT. Due to the magnetic forces, the ECT magnets are pulled into the BT and the corresponding axial forces are transferred to the BT cryostats via axial transfer points linking both magnet systems. The cryostats rest on a rail system facilitating the movement and parking of the ECT magnets for access to the detector centre.

The magnets are indirectly cooled by forced flow of helium at 4.5 K through tubes welded on the casing of the windings. The cooling power is supplied by a central refrigeration plant located in the side cavern and the services are distributed among the four magnets. Electrically the eight coils of the BT are connected in series, as are the 16 coils in the two ECTs. The toroid coil systems have a 21 kA power supply and are equipped with control systems for fast and slow energy dumps. The stored

Fig. 3. Barrel Toroid installed in the underground cavern, with the barrel calorimeters including the central solenoid in the common cryostat with the LAr in the background before its insertion into the centre of the detector.

energy is 1.1 GJ in the BT and 0.5 GJ together in the two ECTs. An adequate and proven quench protection system has been designed to safely dissipate the total stored energy of 1.6 GJ without overheating the coil windings.

The full magnet system, including its fast safety discharge procedures, has been commissioned in its final configuration in the cavern in 2008, and worked since then very reliably during all data taking periods of the experiment. It is arguably the largest and most complex superconducting magnet system ever built for a particle physics experiment.

2.2.2. *Comments on installation, testing and calibration*

The Individual detector components (e.g. chambers) were built and assembled in a distributed way all around the globe in the numerous participating institutes and were typically first tested at their production sites, then after delivered to CERN, and finally again after their installation in the underground caverns. The collaboration also invested enormous efforts in testing representative samples of the detectors in test beams at CERN and other accelerator laboratories around the world. These test beam campaigns not only verified that performance criteria were met over the several years of production of detector components, but were also used to prepare the initial calibration and alignment data for the future LHC operation. Very important were the so-called large combined test beam set-ups, which represented whole 'slices' of the different layers of the final detector.

Fig. 4. Photograph of one end of the ATLAS detector barrel with the calorimeter end-cap still retracted before its insertion into the barrel toroid magnet structure (February 2007 during the installation phase).

Given its size and structure the ATLAS detector had to be assembled directly in the underground cavern. Figure 4 shows one end of the cylindrical barrel detector after about four years of installation work, 1.5 years before completion. The ends of four of the barrel toroid coils are visible, illustrating the eightfold symmetry of the structure.

During the progressing installation the experiments made extensive use of the constant flow of cosmic rays impinging on Earth providing a reasonable flux of muons even at a depth of 100 m underground, typically a few hundred per second traversing the detectors. These muons were used to check the whole chain from hardware to analysis programs of the experiments, and to align the detector elements and calibrate their response prior to the pp collisions. In particular, after the LHC incident[8] on 19th September 2008 the experiments used the 15 months LHC down time, before the first collisions on 23rd November 2009, to run the full detectors in very extensive cosmic ray campaigns, collecting many hundreds of millions of muon events. These runs allowed the experiments to be ready for physics operation, with already accurately pre-calibrated and pre-aligned detectors, by the time of the first pp collisions.

2.3. *The worldwide LHC computing grid*

A particular challenge for ATLAS and CMS are the very high collision rates in the LHC, necessary for the Higgs search and studies, given the small production cross section combined with the need to investigate final states with very small branching fractions. In the first three years of operation the LHC reached a peak instantaneous luminosity of $7 \times 10^{33}\,\mathrm{cm}^{-2}\mathrm{s}^{-1}$ with a 50 ns bunch spacing, which meant that the detectors had to simultaneously cope with up to ~50 overlapping (pile-up) events in a given bunch crossing. In the years ahead, the instantaneous luminosity is still expected to rise two- to three-fold. It is technically not possible to store all data for all events, therefore a trigger system is used to reject large numbers of events and retain only the interesting ones from crossings with potential physics processes of interest. This is done in real time by sophisticated integrated trigger and data acquisition systems, involving custom-made fast electronics in a first stage and large computing farms in subsequent stages before the data, a few hundred events per second, are transferred to mass storage for further analyses.

The LHC experiments generate huge amounts of data (tens of petabytes of data per year) requiring a fully distributed computing model. The worldwide LHC Computing Grid (WLCG) was developed to deal with this task.[8,9] With its hierarchical architecture of tiered centres it allows any user anywhere access to any data recorded or produced in the analyses steps during the lifetime of the experiments. The centre at CERN receives the raw data, carries out prompt reconstruction, almost in real time, and exports the raw and reconstructed data to the 12 Tier-1 centres and also to more than 100 Tier-2 centres for physics analyses. These are located all over the globe and give physicists access to the data from their home institutions in the worldwide collaborations. Significant parts of the infrastructure are shared between the LHC communities in a given region.

3. Standard Model Physics

3.1. *Data samples*

The first stable collisions were delivered by the LHC, and recorded by the experiments, on 23rd November 2009, at a collision energy of 900 GeV, corresponding to the injected beams from the SPS. These low-energy, low-luminosity collisions were very useful for the commissioning of the detectors, and resulted also in some valuable physics results about minimum bias events. The first high-energy collisions were recorded on 31st March 2010 at 7 TeV, resulting in a wealth of Standard Model measurements, and first exploratory searches.

But the bulk of the high-energy data, leading to the Higgs boson discovery, followed in the two years after. By the end of 2012 (LHC Run-1) the total amount of data examined corresponded to about 5 fb^{-1} at 7 TeV and about 20 fb^{-1} at 8 TeV collision energy. This corresponds to some 2000 trillion pp collisions in each of the ATLAS and CMS detectors. Remarkably, almost 95% of the stably delivered luminosity by the LHC was recorded, and typically 90% used for high-quality physics results (Fig. 5). After the first long shut-down (LS1) used to maintain and consolidate the LHC machine and experiments, the collider restarted in summer 2015

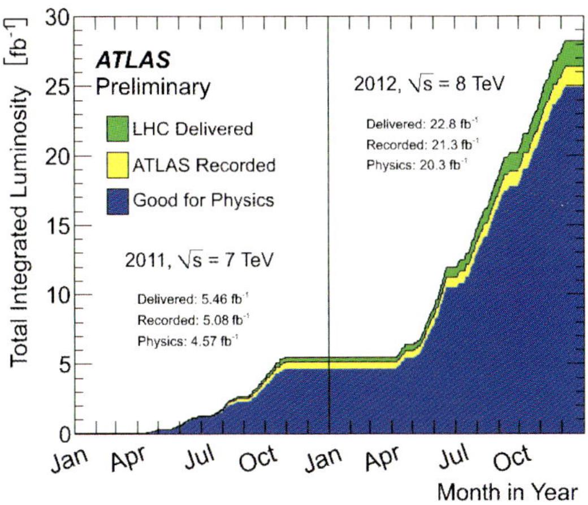

Fig. 5. Evolution of the accumulated luminosity for LHC Run-1 over the years 2011 and 2012, showing separately the stably delivered, recorded, and used for physics portions.

the LHC Run-2 at 13 TeV collision energy, delivering about $4\,\text{fb}^{-1}$ by the end of the year.

3.2. *Some examples of Standard Model measurements*

Observing, and measuring accurately, at the LHC collision energies, the production of known particles of the SM, was always considered to be a necessary stepping stone towards exploring the full potential of the LHC with its promise of new physics, firstly of the discovery of the Higgs boson.[10–18] The SM processes, such as W and Z production, are often referred to as 'standard candles' for the experiments. However, there is much more value to measuring SM processes than this: never before could the SM physics be studied at a hadron collider with such sophisticated and highly accurate detectors, ultimately allowing tests of detailed predictions of the SM with unprecedented precision and minimal instrumental systematic errors. All ATLAS physics results can be found on the web,[19] only few of them will be referenced individually in the following.

A very basic measurement is the evaluation of the total pp cross-section which has been published at the LHC by the TOTEM and the ATLAS experiments. The result is shown in Fig. 6(a) and exhibits the expected rise at this energy. As a new result from Run-2 Fig. 6(b) reports the first inelastic cross-section measurement at 13 TeV.[20]

Studies of charged and neutral Intermediate Vector Bosons (IVB) W and Z are the major benchmarks at the LHC for demonstrating the excellent detector performance, as well as for testing SM predictions. The W decays into a charged lepton (electron or muon) and its neutrino (experimentally giving rise to missing transverse energy E_T^{miss}), and Z decays into charged lepton pairs, can be extracted almost free of backgrounds, as shown in Fig. 7.

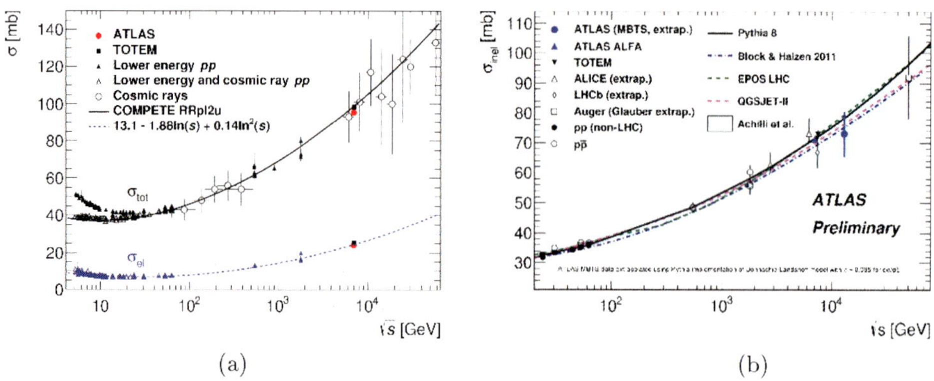

Fig. 6. (a) Total pp cross-section and (b) inelastic cross-section measurements at LHC.

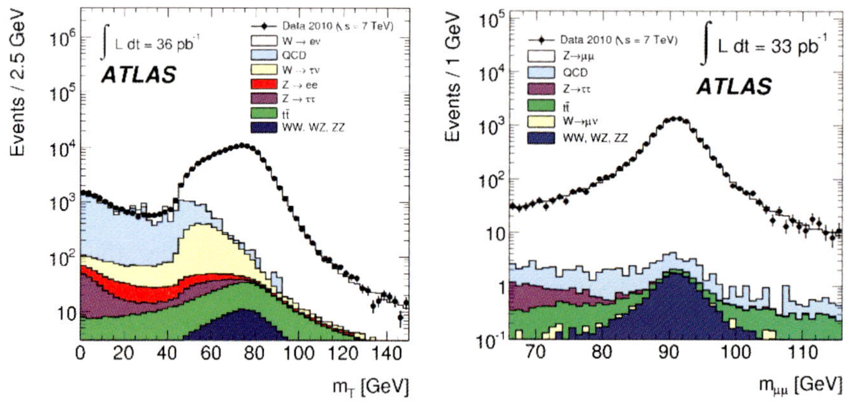

Fig. 7. The transverse mass distribution for electron events with missing transverse energy (left) and the di-muon invariant mass distribution (right) on logarithmic vertical scales. The estimated small background contributions are indicated, as well as the expected signal shape from MC simulations.

As an example, the good agreement between the measured and expected W cross-sections times the leptonic decay branching ratios (which is the expected rate for W bosons to be produced and then decay to leptons) is illustrated in Fig. 8, including some preliminary measurements[21] at the 13 TeV collision energy of LHC Run-2. With the presently available data samples the measurements are expected to already strongly constrain the theoretical model parameters. Detailed measurements of properties for IVB production and decay at the LHC have been published from the full Run-1 data samples. They include, for example, the lepton charge asymmetry measurements for W decays, which were an important signature of the electroweak nature of the W at the time of their discovery in the early 1980s.

Hard collisions (characterized by having final state particles with large transverse energy, E_T) at the LHC are dominated by the production of high transverse

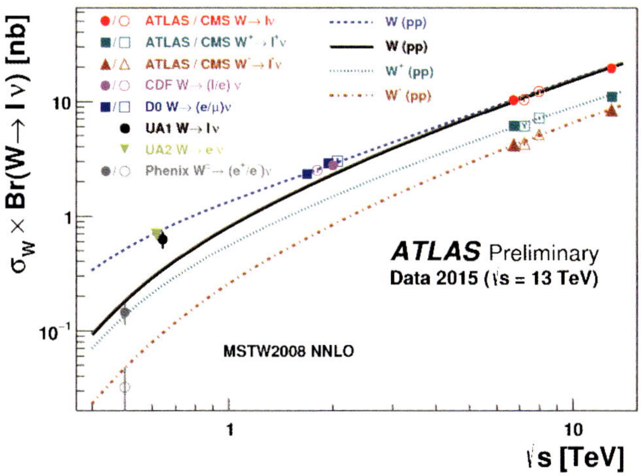

Fig. 8. W production cross section times leptonic branching ratio as a function of the collision energy; also shown are previous measurements at lower energy colliders.

momentum jets, which are the collimated sprays of particles originating from the hadronization of the initially scattered partons (quarks, gluons) in the colliding protons. At work is the strong interaction described by Quantum ChromoDynamics (QCD). Most commonly two jets emerge at opposite azimuthal angles with balanced transverse momenta, from an initially lowest order parton–parton scattering process. However, higher order QCD corrections alter this picture significantly, and detailed measurements of multi-jet configurations are very important to constrain the QCD description of hadronic processes.

The heaviest known particle in the SM is the top quark with a mass of roughly 175 GeV. It decays almost exclusively into a W and a bottom quark. The measurement of top quark pair production typically uses at least one W that decays leptonically (also needed to trigger the events), and therefore the final states require one or two leptons (electrons or muons), E_T^{miss}, and jets. Some of these jets arising from the b-quarks can be tagged by the displaced secondary vertices due to the finite life times of b-hadrons. Clear top pair signals are observed in ATLAS and CMS, both in the single and two-lepton channels, when considering the correct jet topologies. The resulting cross-sections are shown in Fig. 9 also illustrating the expected rapid rise of the cross-section with increasing collision energy, from 2 TeV at the Tevatron to 13 TeV[22] at the LHC. Good agreement with NLO QCD calculations is observed within the present few % measurement errors. Both ATLAS and CMS have also reported measurements of the rare single top production (events with just one top quark) with a rate in good agreement with QCD expectations, as well as detailed studies of top properties such as the measurement of its mass.

A summary of many examples of such ATLAS SM studies is shown in Fig. 10 where measurements of cross-sections for various selected electroweak and QCD

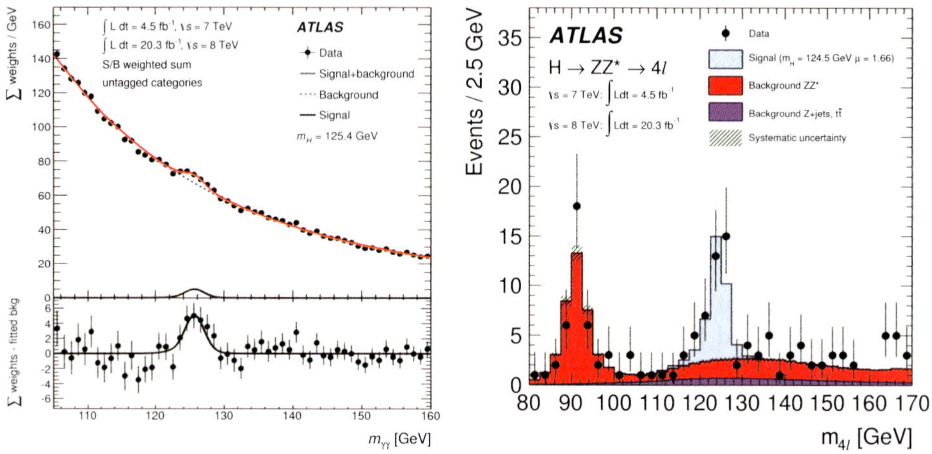

(a) (b)

Fig. 12. (a) Invariant mass distribution of di-photon candidates, with a fit describing a smooth background and a signal component superimposed. (b) Invariant mass distribution of 4 leptons (electrons and muons), background and signal components are indicated.

Table 1. The expected and observed signals in ATLAS and CMS expressed as the corresponding number of standard deviations of the observed excess from the background-only hypothesis, for $m_H = 125\,\text{GeV}$, for various decay modes.

	ATLAS		CMS	
Experiment Decay Mode	Expected (σ)	Observed (σ)	Expected (σ)	Observed (σ)
yy	4.6	5.2	5.3	5.6
ZZ	6.2	8.1	6.3	6.5
WW	5.8	6.1	5.4	4.7
bb	2.6	1.4	2.6	2.0
$\tau\tau$	3.4	4.5	3.9	3.8

W bosons decay leptonically, resulting in a signature with two isolated, oppositely charged, high p_T leptons (electrons or muons) and large E_T^{miss} due to the undetected neutrinos.

4.1.2. *Decays to fermions: The $H \to \tau\tau$ and $H \to bb$ decay modes*

It is important to establish whether the new particle also couples to fermions, and in particular to down-type fermions, since the measurements above mainly constrain the couplings to the up-type top quark (indirectly from the production). Determination of the couplings to down-type fermions requires direct measurement of the Higgs boson decays to bottom quarks and τ leptons. These analyses are complex and require a large amount of data. Evidence of the Higgs boson decaying into two

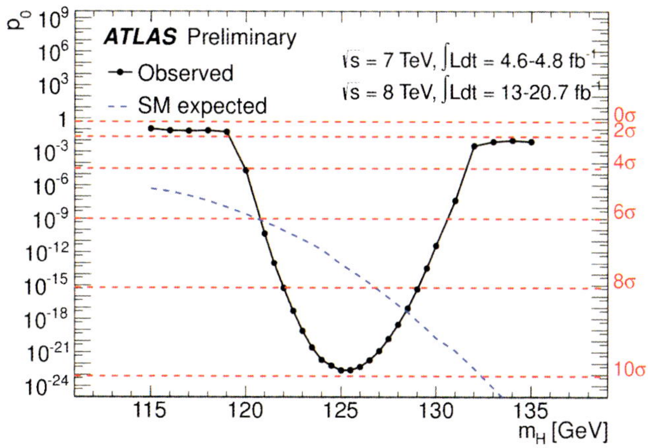

Fig. 13. Combined local probability p0 that a background-only experiment exhibits a fluctuation at least as large as observed in the data, as a function of the Higgs mass.

τ leptons is seen by both experiments. The overall (combining all channels) local probability p0 for a background-only experiment to produce a fluctuation at the level or more signal-like than the data is shown in Fig. 13 as a function of the Higgs mass. ATLAS alone excludes such a fluctuation at the level of 10 standard deviations.

The expected and observed significances for all observed decay modes in both ATLAS and CMS are listed in the Table 1 for the full Run-1 data. It is remarkable, and a success for the overall CERN LHC project, that the two complementary detector designs adopted by ATLAS and CMS have led independently to equivalent results.

4.2. *The combined results from ATLAS and CMS on the Higgs boson properties*

Both the ATLAS and CMS experiments have separately combined their measurements of the mass of the Higgs boson from the two channels that have the best mass resolution, namely $H \rightarrow \gamma\gamma$ and $H \rightarrow ZZ \rightarrow 4l$. The signal in all channels is assumed to be due to a state with a unique mass. The obtained values are from ATLAS m $= 125.36 \pm 0.37 \pm 0.18$ GeV, where the first error is statistical and the second represents the systematic and theoretical uncertainties. The result from CMS is $m = 125.02 \pm 0.27 \pm 0.14$ GeV, in excellent agreement. The two experiments have then combined, in a global analysis, their measurements, and determined the Higgs boson mass to be $m = 125.09 \pm 0.21 \pm 0.11$ GeV, with a remarkable accuracy of 0.2%.[24]

To establish whether or not the newly found state is the Higgs boson of the SM, one needs to precisely measure its other properties and attributes. Several tests of compatibility of the observed excesses with those expected from a SM

Higgs boson have been made. In one comparison, labelled as the signal strength $\mu = \sigma/\sigma_{SM}$, the measured production times decay rate of the signal is compared with the SM expectation, determined for each decay mode individually and for the overall combination of all channels. A signal strength of unity would be indicative of a SM Higgs boson. For all observed decay channels the measurements are compatible with the SM predictions, and the combined result from both experiments is $\mu = 1.09 \pm 0.11$,[25] in good agreement with the expectation from the SM.

Another characteristic fingerprint of the SM Higgs boson is its coupling strength to other elementary particles (i.e. how strongly it interacts with them) in proportion to their masses in a well-defined manner. The reduced coupling modifiers ($\kappa_F.m_F/v$ or $\sqrt{\kappa_V}.m_V/v$) can be expressed in a way such that in a double logarithmic representation there should be a linear proportionality, as shown in Fig. 14. m_F, m_V, v, κ are the masses of the fermion, vector boson, the vacuum expectation value of the Brout–Englert–Higgs field, and the effective coupling modifiers of the Higgs boson. The results show the correct tendency, within the errors, but much more data are needed to test the predicted behaviour more accurately. The accuracy of these types of measurements is one of the strong motivations for the future LHC operation and the intensity upgrade foreseen in the next two decades. Predictions for theories Beyond the SM result in deviations that could be as small as 1% in this representation.

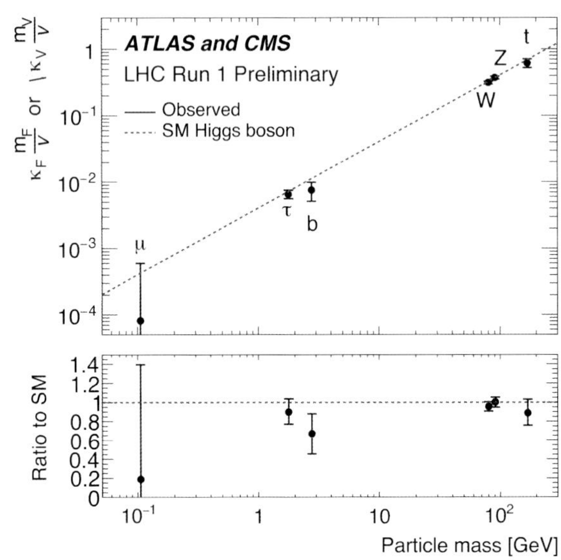

Fig. 14. Couplings of the Higgs boson to other elementary particles, as measured by a combined analyses of the ATLAS and CMS Collaborations with their full data set from LHC Run-1. The line is the prediction from the SM, the lower panel shows the ratio of the measurements to the SM prediction.

Another key to the identity of the new boson is its quantum numbers amongst which is the spin-parity (J^P). The angular distributions of the decay particles can be used to test various spin hypotheses. Detailed analyses show that the expected spin-parity $J^P = 0^+$ hypothesis for a SM Higgs boson is strongly favoured by both experiments, with the alternative hypotheses rejected with confidence levels larger than 99.9%.

5. Search for Physics Beyond the Standard Model and Outlook

Besides the quest to elucidate the mechanism of the electroweak symmetry breaking by searching for the Higgs boson, the major excitement for LHC comes from the great potential to explore uncharted territory of physics Beyond the Standard Model (BSM), thanks to its highest collision energy available in a laboratory so far. Since the beginning of the project, the search for Supersymmetry (SUSY) was a strong motivation, and besides the Higgs boson it has been the other main benchmark physics that was guiding the detector designs. However, many other hypothetical new processes can be searched for, and indeed ATLAS and CMS have already reported in many publications a very broad spectrum of searches for BSM signatures (mass peaks for new particles or kinematical distributions with deviations from the expectations of known physics processes). No such new effect has yet been found, and all of these searches result in highly-improved, stringent exclusion limits, often well beyond the one TeV scale already. Only a few examples are mentioned below.

The most popular searches concern SUSY, which predicts additional fundamental particles. The search for SUSY is motivated in part by the prospect that the lightest stable neutral SUSY particle (LSP) could be an excellent candidate for explaining the Dark Matter (DM) in the Universe. The mysterious existence of DM was postulated by Knut Lundmark, Fritz Zwicky, and rather convincingly evidenced by Vera Rubin, all astronomers, in the 1930s and 1970s respectively.

The SUSY searches at LHC are very complex as they must be sensitive to many (model-dependent) decay chains, implying a large variety of possible final state topologies. A common feature for most of them is the existence of significant missing transverse energy E_T^{miss} due to the escaping LSPs (an experimental signature similar to that of the neutrinos in the W decays mentioned before). Furthermore, the SUSY signatures often include high transverse momentum jets, some tagged as b-jets for third-generation squarks as particularly motivated by naturalness arguments, and leptons. The expected topologies depend not only on the model parameters, but also on the mass relations between various squarks and gluinos (the SUSY partners of the SM quarks and the gluons). A summary of 95% CL mass exclusion regions from many SUSY searches is shown in Fig. 15, with similar results available from CMS as well.

Many other searches aimed at exploring BSM physics have been conducted, at this stage all without finding anywhere an excess of observed event rates over the

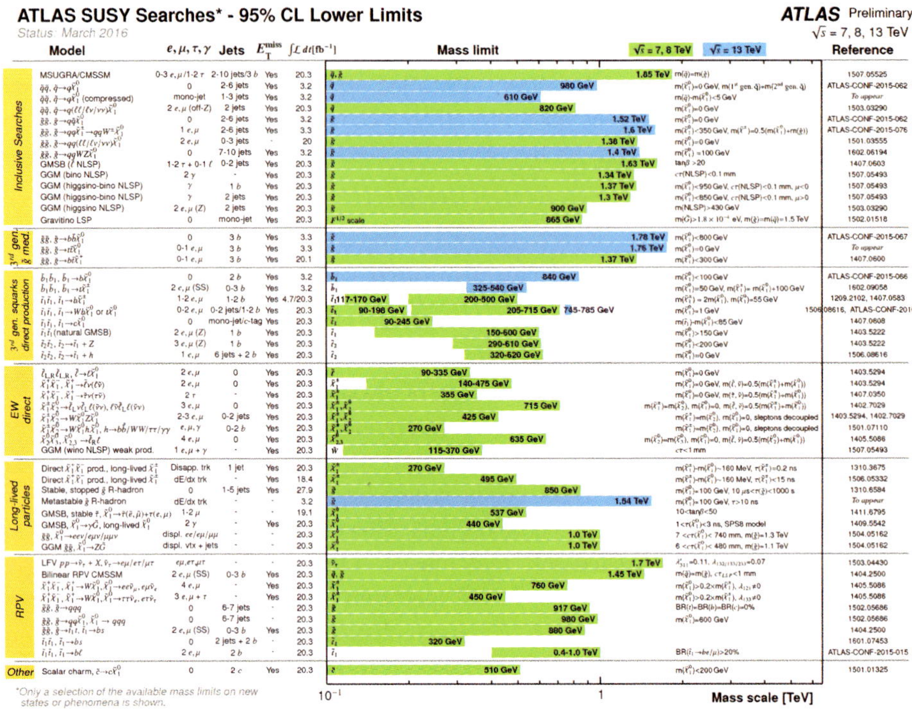

Fig. 15. A summary of 95% CL mass exclusion limits from many SUSY searches.

expected backgrounds from the known SM processes. However, much more stringent limits and constraints could be established than what were available up to now. A collection of them is shown in Fig. 16.

Of very topical interest is the recently reported small excess of events in the search for high-mass di-photon states with the 2015 LHC data at 13 TeV, where ATLAS[26] and CMS[27] see accumulation of events around a mass of 750 GeV, albeit with local significances p0 well below 4 standard deviations, and global significances at the level of 2 standard deviations only. The ATLAS result is shown in Fig. 17. Only more data will clarify the status of this search.

6. Outlook

After the first long shut-down during 2013–2014 the LHC started operation again in 2015 with Run-2 at the collision energy of 13 TeV, and will eventually reach its full design energy of 14 TeV in the coming years. Together with the increase in energy there will be also an increase in the luminosity, bringing the LHC to its full design performance (14 TeV, $2 \times 10^{34}\,\mathrm{cm}^{-2}\mathrm{s}^{-1}$). The projected integrated luminosity by 2022 is about $300\,\mathrm{fb}^{-1}$. The increased energy means larger cross-sections, particularly striking for heavy objects. It is therefore with great expectations that the

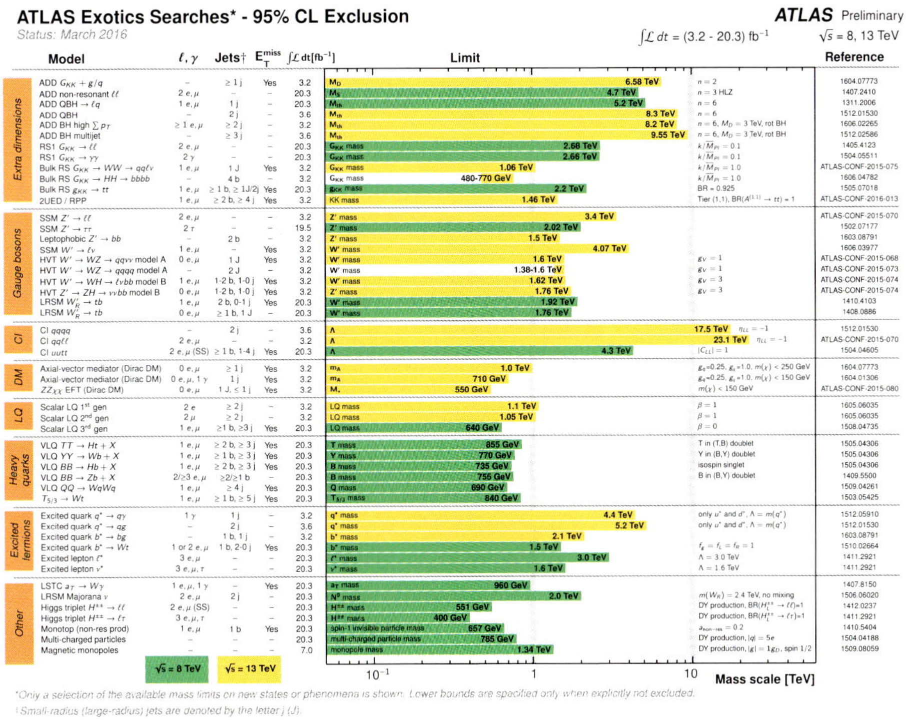

Fig. 16. A summary of 95% CL exclusion limits from many searches (other than SUSY, see Fig. 15) for physics Beyond the Standard Model.

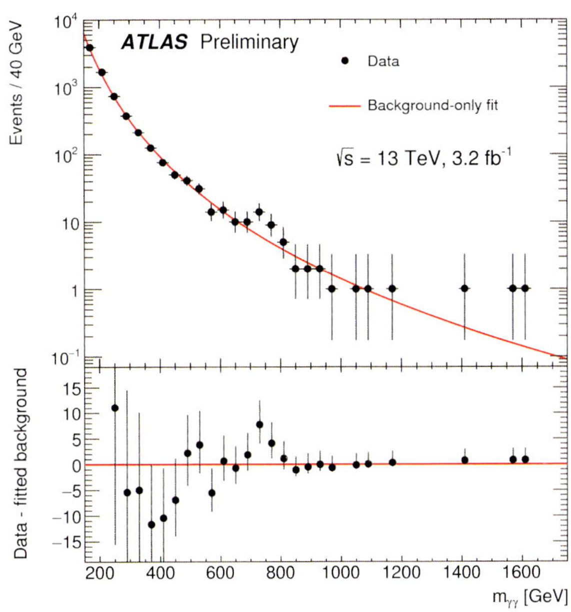

Fig. 17. Di-photon invariant mass distribution from the preliminary data accumulated at 13 TeV in 2015 during the first year of LHC Run-2.

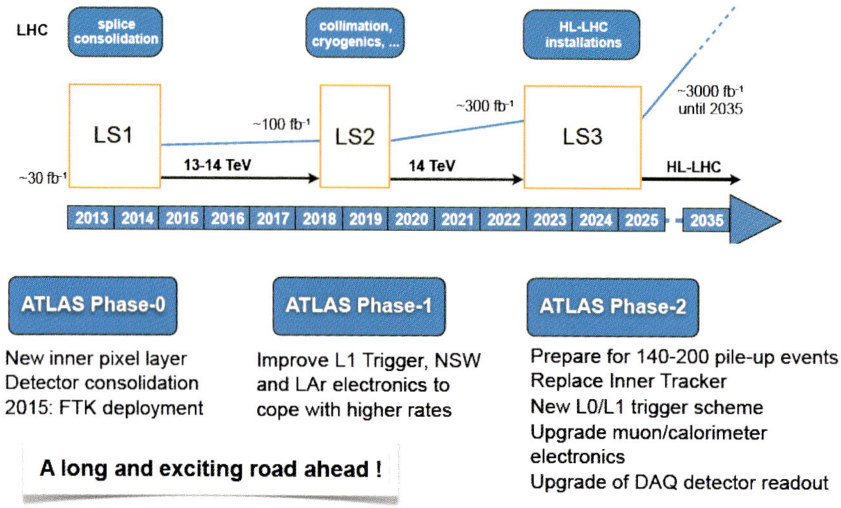

Fig. 18. Sketch of the medium- and long-term planning for ATLAS and the LHC.

experiments are looking forward to collect data in these Run-2 and Run-3 periods, covering the initial LHC project planning.

The roadmap of physics at the LHC beyond its initial design phase has significantly changed with the discovery of the Higgs boson. Not only will there be the unchallenged window for directly observable hypothetical heavy mass particles, messengers of new physics beyond the Standard Model, but also a clear task to investigate in greatest details the properties of the new boson. Needless to say, this basic scenario could well be strongly enriched further if the forthcoming data of the current decade would reveal any new BSM physics, which would then be of course exploited best with the highest available integrated luminosity.

These prospects have strongly motivated to launch a very mayor high luminosity upgrade project planning, both for the experiments and the LHC machine, called the HL-LHC, with the goal to integrate a tenfold luminosity (3000 fb^{-1}) by the mid-2030s, as indicated in Fig. 18.

Acknowledgments

It was a great honor for me to present the ATLAS road map from its conception to the joint Higgs boson discovery with the sister-experiment CMS at this Memorial Meeting for Nobel Laureate Professor Abdus Salam's 90th birthday at the Nanyang Technological University Singapore. Professor Salam has been a giant for the core physics of LHC, the Standard Model, and a shining example of promoting to share physics with all over the world, a spirit well present in the large collaborations. My deep thanks to the chairpersons of the meeting, Professors Kok Khoo Phua, Lars Brink and Michael Duff for having invited me to this memorable conference.

References

1. Proceedings of the ECFA-CERN Workshop, *Large Hadron Collider in the LEP Tunnel*, Lausanne March 1984, CERN 84-10, 1984.
2. Proceedings of the Workshop, *On Physics at Future Accelerators*, La Thuile January 1987, CERN 87-07, 1987.
3. Proceedings of the ECFA-CERN Workshop, *Large Hadron Collider Workshop*, Aachen October 1990, CERN 90-10, 1990.
4. Proceedings of the General Meeting on LHC Physics and Detectors, *Towards the LHC Experimental Programme*, Evians-les-Bains March 1992, ECFA-CERN, 1992.
5. M. Della Negra, P. Jenni and T. S. Virdee, Journey in the Search for the Higgs Boson: The ATLAS and CMS Experiments at the Large Hadron Collider, *Science* **338**(6114), 1560–1568 (2012).
6. ATLAS Collaboration, The ATLAS Experiment at the LHC, *JINST* **3**, S08003 (2008).
7. P. Grannis and P. Jenni, The Evolution of Hadron-Collider Experiments, *Phys. Today* **66**(6), 38 (2013).
8. L. Evans (ed.), *The Large Hadron Collider, a Marvel of Technology*, EPFL Press, 2009.
9. WLCG comprehensive information: http://wlcg.web.cern.ch/
10. S. L. Glashow, *Nucl. Phys.* **22**, 579 (1961).
11. S. Weinberg, *Phys. Rev. Lett.* **19**, 1264 (1967).
12. A. Salam, Proceedings of the eighth Nobel symposium, ed. N. Svartholm, p. 367. Almqvist & Wiskell, 1968.
13. F. Englert and R. Brout, *Phys. Rev. Lett.* **13**, 321 (1964).
14. P. W. Higgs, *Phys. Lett.* **12**, 132 (1964).
15. P. W. Higgs, Phys. *Rev. Lett.* **13**, 508 (1964).
16. G. S. Guralnik, C. R. Hagen and T. W. B. Kibble, *Phys. Rev. Lett.* **13**, 585 (1964).
17. P. W. Higgs, *Phys. Rev.* **145**, 1156 (1966).
18. T. W. B. Kibble, *Phys. Rev.* **155**, 1554 (1967).
19. ATLAS publications: https://twiki.cern.ch/twiki/bin/view/AtlasPublic
20. ATLAS Collaboration, *Measurement of the Inelastic Proton-Proton Cross Section at 13 TeV with the ATLAS Detector at LHC*, ATLAS-CONF-2015-03 (2015).
21. ATLAS Collaboration, *Measurement of W and Z Boson Production Cross Sections in pp Collisions at 13 TeV with the ATLAS Detector*, arXiv:1603.09222[hep-ex], (2016), submitted to Phys. Lett. B.
22. Measurements of the tt production cross sections at 13 TeV pp collisions reported by the ATLAS and CMS Collaborations, ATLAS-CONF-2015-033 and CMS PAS TOP-15-003 (2015).
23. ATLAS Collaboration, Observation of a new particle in the search for the Standard Model Higgs boson with the ATLAS detector at the LHC, *Phys. Lett. B* **716**, 1 (2012); CMS Collaboration, Observation of a new boson at a mass of 125 GeV with the CMS experiment at the LHC, *Phys. Lett. B* **716**, 30 (2012).
24. ATLAS and CMS Collaborations, Combined Measurement of the Higgs Boson Mass in pp Collisions at 7 and 8 TeV with the ATLAS and CMS Experiments, *Phys. Rev. Lett.* **114**, 191803 (2015).
25. ATLAS and CMS Collaborations, *Measurements of the Higgs Boson Production and Decay Rates and Constraints on its Couplings from a Combined ATLAS and CMS analysis of the LHC pp Collision Data at 7 and 8 TeV*, ATLAS-CONF-2015-044 and CMS-PAS-HIG-15-002 (2015).

Before going there, we continue to look at the quantum version of the Maxwell theory, QED:

$$S = \frac{1}{\hbar}\int [F_{\mu\nu}F^{\mu\nu} + \overline{\psi}(i\gamma_\mu \partial^\mu + m + e\gamma_\mu A^\mu)\psi].$$

By using

$$\frac{A_\mu}{\sqrt{\hbar}} \to A_\mu, \ \frac{\psi_\mu}{\sqrt{\hbar}} \to \psi_\mu$$

we get

$$S = \int \left[F_{\mu\nu}F^{\mu\nu} + \overline{\psi}\left(i\gamma_\mu \partial^\mu + \frac{m}{\sqrt{\hbar}} + \frac{e}{\sqrt{\hbar}}\gamma_\mu A^\mu\right)\psi\right].$$

We cannot totally absorb $\hbar$ in QED although it can be absorbed in m and e. The situation is instructive. We do not regard m or e as fundamental in the "final theory." Therefore, they must become dynamical. Then, it seems that we can absorb $\hbar$ into these dynamical variables. This is what is going to happen in 11-dimenstional supergravity.

Let me next discuss the case of gravity

$$R_{\mu\nu} - \frac{1}{2}Rg_{\mu\nu} = 8\pi G T_{\mu\nu}.$$

The question here is whether we can absorb G entirely into some dynamical variables. It can be done if $T_{\mu\nu}$ is expressed in some homogeneous expression of the dynamical variables. Unfortunately, this does not happen in existing theories with interactions among particles or fields. Remarkably, it happens in the case of 11-dimensional supergravity theory even though it contains interactions among fields.

11-dimensional supergravity is defined as the quantum system with

$$S = \exp\left\{\frac{i}{\hbar}\int L\right\}$$

where L is given by[1]

$$\begin{aligned}
L &= L_B + L_F \\
L_B &= -\frac{1}{\kappa^2}eR - \frac{1}{48}G^{M_1N_1}G^{M_2N_2}G^{M_3N_3}G^{M_4N_4}F_{M_1M_2M_3M_4}F_{N_1N_2N_3N_4} \\
&\quad - \frac{\kappa}{3456}\varepsilon^{M_1\cdots M_{11}}F_{M_1M_2M_3M_4}F_{M_5M_6M_7M_8}A_{M_9M_{10}M_{11}}, \\
L_F &= -\frac{1}{2}e\overline{\psi}_{M_1}\Gamma^{M_1M_2M_3}D_{M_2}\psi_{M_3} \\
&\quad - \frac{\kappa}{384}e(\overline{\psi}_{M_1}\Gamma^{M_1M_2M_3M_4M_5M_6}\psi_{M_6} + 12g^{M_2N_1}g^{N_2M_5}\overline{\psi}_{N_1}\Gamma^{M_3M_4}\psi_{N_2}) \\
&\quad \times(F + \hat{F})_{M_2M_3M_4M_5}.
\end{aligned}$$

We can easily prove that $\hbar$ and κ can be absorbed in dynamical variables and will disappear from the definition of the physical system.

The outstanding question is: What is the symmetry implied by this procedure? In the case of light velocity, time and space coordinates were identified and the symmetry was Lorenz group symmetry. What are the dynamical variables identified by absorbing $\hbar$ and κ? And, what is the resulting symmetry?

First, it cannot be the supersymmetry, which is already explicit in the above Lagrangian. Identification of boson and fermion leads to the supersymmetry and for that we need one parameter that has the dimension of length or energy by combining the parameter with $\hbar$ and κ. A popular idea in this case was that this length parameter shows up in the physical process in a "hidden sector"[2] and that a much larger length parameter shows up in our "visible sector." But this idea is now on the verge of extinction because there has been no sign of superparticles in LHC experiments.[3]

Anyway, the symmetry in question seems to unify space time variables with field variables as occurs in matrix models.[4] The problem with matrix models seems to be that only once $N \to \infty$ does the 11-dimensional character and the sought-after symmetry become available.

Quite independently of above considerations, Peter West in his presentation at this symposium and in several papers by him and his collaborators[5] showed that the non-linear realization of the "very extended algebra" of E_8, namely E_{11}, contains all the degrees of freedom of M-theory.

Can E_{11} or the even further extended algebra E_{12} be the symmetry we are talking about? If so, can we work directly on the algebra without invoking the non-linear realization? In this connection I observe that the non-linear realization of an affine Lie algebra (det $A = 0$, where A is the Cartan matrix) leads to a sigma model.[6] Peter West's proof that a non-linear realization of the Lorenzian algebra E_{11} contains all the string spectra is very encouraging to consider directly with the algebra, presumably with the energy momentum tensor as a homogeneous function of the generators of the algebra.[6]

2. Is There Any Indication of the Planck Scale in Nature?

The above argument implies that the constants $\hbar$ and κ are vacuum values of some dynamical variables which we may call moduli, implying a certain geometric significance that should be realized in the dynamical system defined by the Lorenzian Kac–Moody algebra E_{11}. At least in case of affine algebra there is a connection between the space of vacua and that of the moduli space of some geometric object, as presented by M. S. Narasimhan in this symposium and as he also showed in several papers with his collaborators.[7]

In such situation the numerical values of $\hbar$ and κ should be interpreted as the low energy limit of certain moduli and there is no guarantee that they take the same values at high energy. This means that we are not sure whether the length unit called the "Planck scale" actually exists in nature.

Here I would like to examine whether there is any indication of the existence of a "Planck scale."

First, let us check if the early universe indicates such existence. In the standard model of cosmology (Friedman universe) the initial stage of Big Bang is radiation dominated, the theory is completely classical, and there is no way of introducing the Planck scale ($l_{pl} = \sqrt{\hbar G/C^3}$) into the model and the actual time scale of the Big Bang is much greater than the Planck time. In modern cosmology the Big Bang energy was accumulated in the period preceding the Big Bang called the inflation[8] period. Here again the entire theory is classical and the only way to introduce the Planck scale is by defining the vacuum potential (energy density) as

$$V = \frac{MC^2}{l_{pl}^3}.$$

But there is no need to put M around the Planck mass given by

$$m_{pl} = \sqrt{\frac{\hbar C}{G}}.$$

All that is needed for sufficient energy to be provided for Big Bang is

$$\frac{\dot{V}}{V} \ll \Delta t \quad \text{and} \quad \frac{\ddot{V}}{V} \ll (\Delta t)^2,$$

where Δt is the duration of the inflation period. Slow rolling will yield enough energy for the Big Bang.

The way to obtain a Planck scale is with the quantum gravity version of the initial state of the universe, such as the no-boundary theory of S. Hawking. But a consistent quantum gravity theory is given only by string theory, and within the framework of string theory, the model yielded so far is the inflation model based on dynamics in moduli space, and there is no way to introduce a Planck scale there.[9]

I conclude that there is no evidence for the existence of a Planck scale from the analysis of the early universe.

Some time ago there was a report from CERN's LEP experiment[10] that there is an indication of a grand unified scale in the high energy behavior of the electroweak and QCD coupling constants. But the lack of evidence of proton decay so far[11] casts some doubt about its existence.

A much more interesting indication of a short distance parameter comes from the observation and the analysis of the high energy behavior of the quartic Higgs self-coupling.

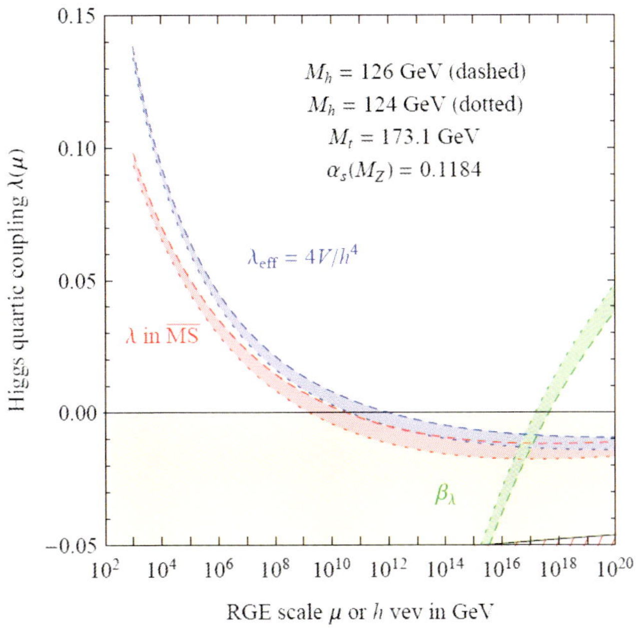

Fig. 1. Behavior of Higgs self-coupling as a function of energy.[12]

The above figure shows the value of the Higgs self-coupling λ implied by the smallness of the Higgs mass obtained from ATLAS and CMS experiments in CERN. It is very sensitive to the mass of the top quark, but interestingly λ seems to be either:

(1) asymptotically free, in which case a mechanism for the hierarchy suggested by W. Bardeen[13] may work; or
(2) negative at some very high energy above which the standard model loses its validity.

The second case gives one some high energy parameter which could be either the Planck scale or something else. Either way of (1) or (2) is extremely interesting and, therefore, the precise measurement of a top quark mass is tantalizing and it should be one of the main goal of the future lepton collider as will be discussed later in this report.

3. Past and Present of Big Accelerators

Here I would like to look back at the past, discuss the present and then in the next section I will offer some advice on the future of big machines to perform high energy experiments.

In the early 1980's, leaders of the high energy physics community formed a committee to discuss future accelerators, ICFA (International Committee for Future Accelerators), in order to promote world-wide collaboration on the construction of future big accelerators. Unfortunately, the US decided to proceed unilaterally in the construction of SSC (Superconducting Super Collider).

I quote from the 10,13 August 1983 ICFA meeting:

DISCUSSION ON THE RECOMMENDATIONS OF THE WOODS HOLE SUB-PANEL TO HEPAP.

Several speakers expressed their unhappiness that the recent US development had taken place without any previous discussion in ICFA or sufficient consultation between the regions. A long discussion then ensued on the possible future role of ICFA in promoting more inter-regional collaboration and consultation.

ICFA was not able to function as it was intended.

This was the time of the Reagan administration, and the US high energy physics community was allegedly promised by the President his support of SSC in exchange for the community's incorporation into his SDI (Strategic Defense Initiative) project. This was still a time when US high energy physics was intimately connected to the US defense policy. Unfortunately, SDI eventually faded away and so did SSC. Of course, the process was not that simple.

Anyway, ICFA maintained its stance of emphasizing the importance of world-wide collaboration. I quote:

In the report by ICFA Chair Telegdi to the 12 October 1984 ICFA meeting:

It had in fact been recognized at the ICFA meeting held at Fermilab in August 1983 that ICFA's original charge, i.e. to promote the construction of a "World Accelerator," the so-called VBA, could not be realistically pursued at present, since some current regional projects (e.g. the SSC in the USA) were already of the size originally envisaged for the VBA, and hence appeared to banish the "real" VBA into a future too distant for constructive discussion at present.

I quote another report of ICFA much later in 1989:

RESOLUTION IN SUPPORT OF INTERNATIONAL PARTICIPATION IN THE SSC (L. Pondrom, paper dated 1 June 1989)

After a frank discussion, during which several members pointed out that it was not ICFA's role to comment on the merits of regional plans for future accelerators, L. Pondrom agreed to withdraw his draft resolution. However, ICFA reiterated its belief that international participation in the construction and exploitation of future accelerators in the different regions should be strongly encouraged.

The above statement is a reflection of the saga of SSC-LHC conflict. ICFA was powerless to resolve the conflict and just kept insisting on the principle of world collaboration. I quote:

In a summary of ICFA activities in March 1991, ICFA Chair Skrinsky said:

ICFA expresses its strong support for the ongoing efforts towards extensive international collaboration for the construction and exploitation of the next generation of high energy hadron accelerators...

In a summary of ICFA Activities in July 1992, ICFA Chair Skrinsky said:

ICFA further recommends that the design and use of future large high-energy physics facilities, including the appropriate R&D, should have international participation from the start to ensure that the full intellectual capabilities of the international community are utilized.

I summarize the reason of the failure of SSC in the following way:

(1) When SDI faded away, so did the influence of the high energy physics community in US defense policy; the SSC was regarded as the symbol of its lost influence.
(2) Failure to resolve the SSC-LHC conflict. Because of its rival LHC, the US could not lower the energy of SSC to reduce the cost.
(3) Failure to effectively persuade participation by other countries because it was started as a US national project.

The world high energy physics community learned a lesson and became united in support of the LHC project at CERN because Europe was still not ready to approve the LHC. With worldwide support in which the ICFA played a key role, the LHC was finally approved. I regard this as the first victory of worldwide collaboration in high energy physics.

The next problem was what to do about a complementary lepton machine to LHC. Deeply influenced by the lessons learned from the demise of SSC, ICFA took the initiative to define ILC and to proceed with the plan.

It was started as a fully international project and still it is. Some of the measures taken by ICFA include:

- ICFA has been taking the initiative in defining ILC;
- Under ICFA we created ILCSC (steering committee);
- We also created FALC (funding agency for LC);
- Under ILCSC we created ITRP (International Technology Recommendation Panel).

For example the ITRP was created by ICFA to decide which technology was to be adopted for ILC, warm or cold. Warm technology (X-band microwave with copper structure) had been developed mainly in KEK and in SLAC. Cold technology had been developed in DESY with strong push by the late B. Wiik. The report was submitted to ICFA and cold technology was selected. SLAC lost momentum but KEK survived because it was engaging in the development of NbTi cavity as an option to X-band.

The members of the ITRP are the following:

Jean-Eudes Augustin (France)
Jonathan Bagger (USA)
Barry Barish (USA) – Chair
Giorgio Bellettini (Italy)
Paul Grannis (USA)
Norbert Holtkamp (USA)
George Kalmus (UK)
Gyung-Su Lee (Korea)
Akira Masaike (Japan)
Katsunobu Oide (Japan)
Volker Soergel (Germany)
Hirotaka Sugawara (Japan)
David Plane – Scientific Secretary

Every member was given a task and I recall that my task was to assess the positron source of each technology, which was vastly different between the two cases.

The final Recommendation reads:

We recommend that the linear collider be based on super-conducting rf technology (from Exec. Summary)

This recommendation is made with the understanding that we are recommending a technology, not a design. We expect the final design to be developed by a team drawn from the combined warm and cold linear collider communities, taking full advantage of the experience and expertise of both (from the Executive Summary).

We submit the Executive Summary today to ILCSC & ICFA

Details of the assessment will be presented in the body of the ITRP report to be published around mid September

The superconducting technology has features that tipped the balance in its favor. They follow in part from the low rf frequency.

<u>This recommendation had a huge effect on the LC activities:</u>

- SLAC stopped all LC activities based on X-band technology;
- DESY continued but switched to FEL;
- KEK changed its policy from warm to cold (painful, but KEK was technically advanced in "cold" also).

Upon this recommendation international effort began and the GDE (Global Design Effort) was formed with B. Barish leading the team. After five years of hard work they have completed their design and made two reports to ICFA:

- The ILC Technical Design Report;
- The ILC Detailed Baseline Design Report.

Currently a new international team lead by Lin Evans is engaging in the following activities, working closely with ICFA and ILCSC:

1. Continue the current R/D effort;
2. Concretely shape the International Linear Collider Laboratory;
3. Determine the ILC site;
4. Finalize the ILC technology;
5. Determine the initial ILC energy;
6. Work hard to finance the ILC project.

As you see the process taken to realize the ILC contrasts sharply to that for SSC or LHC. Everything is done in worldwide collaboration.

Finally I would like to report my correspondence with former CERN Director General R. Aymar on the role CERN will play in realizing the ILC project. I hope the CERN management to continue to follow what the former DG promised to do.

Dear Prof. Sugawara,

Thank you for your cordial letter of May 21st about future accelerators, with which I am largely in agreement.

You are surely correct that the best way to ensure a healthy future for our field is to adopt and maintain a unified approach. In that respect, I am very glad that representatives of funding agencies, including that of Japan, were able to take important steps towards such a unified view at their recent meeting in London.

As you say, an important aspect of this unified view is that the LHC will need to be complemented by a linear electron-positron collider. It was generally recognized at the London meeting that final approval for its construction is likely to be realistic only after the first results from the LHC become available, and that the desirable energy scale may need to be reviewed in light of LHC results. It seems that you agree with me on at least the first part of this point, in view of the American and Japanese situations, even if you would have preferred an earlier go-ahead. In the mean time, thanks in large part to your personal efforts, Japan has an exciting programme of neutrino physics in front of it.

You may be assured that CERN will do its best to help secure approval of a linear collider, first by completing the LHC and extracting key physics information from it, and secondly by striving to convince the world's political authorities of the importance of our common endeavor. I look forward to working together to ensure the continuing vitality of our field.

Yours sincerely,

Robert Aymar

4. Future High Energy Physics Projects

Here I would like to concentrate on the Chinese proposal CEPC (Circular Electron Positron Collider)-SPPC (Super Proton Proton Collider), but most of my advice also applies to the similar CERN proposal, FCC-hh(ee).

First, let us compare SSC and SPPC parameters:

Table 1. Comparison of SSC and SPPC crucial parameters (SPPC parameters taken from Ref. 14).

	SPPC	**SSC**
CM energy	71.2 TeV	40 TeV
Circumference	54.4 km	83 km
Dipole field FCC-hh(ee)	20 T	6.6 T
Luminosity	$1.2 \times 10^{35}\ \mathrm{cm}^2\ \mathrm{sec}^{-1}$	$10^{33}\ \mathrm{cm}^2\ \mathrm{sec}^{-1}$

Interestingly both machines are unilaterally proposed by one country and they were (are) in competition with CERN's project. But, here I am interested in the technological aspect of the project. Remarkable difference between SSC and SPPC is the magnitude of the field strength of the dipole magnets. 20 Tesla of SPPC cannot be achieved by NbTi wires. The only practical material we can currently think of is Nb_3Sn. The technology (U.S. contribution) will be provided for hl-LHC insertion quadrapole magnets and, although what matters in this case is the gradient of the field, 13 to 14 Tesla was already achieved very locally. Achieving 20 Tesla for large dipole magnets is another matter.

Next, we compare CEPC and ILC:

CEPC and ILC

- A 240 GeV circular machine is not far from LEP (207 GeV). Therefore, not many technical issues are involved unless we try to deploy some new feature like a single ring for e^+ and e^-, in which case the issue of the pretzel scheme will be formidable.
- The big difference between ILC and CEPC is the power consumption:
  CEPC (240 GeV) — 500 MW;
  ILC (500 GeV) — 160 MW.
- This makes it almost impossible for a circular machine to go beyond 240 GeV. We have to keep in mind that the radiation loss is E^4.

Whereas, physics required by a lepton collider can be summarized as:

Physics

- Understanding of the Higgs potential is the most important issue.
- For this purpose, we must know the precise top quark mass, requiring at least 350 GeV e^+e^- energy.

- Higgs self-coupling must be studied which also cannot be done by 240 GeV machine.
- Many Higgs parameters can be better studied at higher than 240 GeV.

Based on the above technical and physics considerations, I would like to advise both the Chinese and the CERN project advocates of the following:

My friendly advice

1. ~100 TeV hadron machine is more or less the consensus of the community. Global effort towards the 20 T Nb(3)Sn superconducting magnet must be started as soon as possible to finish the work in 10 years. IHEP (Beijing) and CERN should take the initiative under the supervision of ICFA.
2. US, Japan and other countries must be involved.
3. Test stands may be constructed in IHEP, CERN, Fermilab, KEK, etc. (Going from 13 T (HL-LHC), 16 T (FCC-hh) to 20 T is an increasingly steeper up-hill process.)
4. As for the CEPC (FCC-ee), I advise ICFA to form an international panel (like ITRP) to discuss whether it is a good idea to have a circular lepton machine before going to a ~100 TeV hadron machine.
5. The panel must be composed of theorists, experimentalists and accelerator physicists, as in the case of ITRP.

(What I am afraid of is that 240 GeV e^+e^- circular machine may not produce much physics output, but instead simply delay the construction of ~100 TeV hadron machine.)

Acknowledgments

My first encounter with Professor Salam was through his famous paper on the overlapping divergence. I was a graduate student at Tokyo University and my impression was that he must be a very bright person to tackle such an intricate problem.

I met him personally in ICTP in Trieste in 1968 and he was very kind to me and I was very much impressed by his strong determination to contribute to the science of developing countries. I often contemplate how different my life could have been if I stayed in Trieste rather than coming back to Japan.

I want to thank H. Ooguri for reading my presentation before the symposium, P. West and M. S. Narasimhan for the discussions on their presentations, A. Yamamoto and K. Yokoya for providing materials on SPPC-CEPC, R. Rubinstein for providing some materials from ICFA and L. Pondrom for explaining to me his role in 1989 ICFA meeting at FERMILAB.

Lastly, I would like to thank the organizers of this Symposium for inviting me, especially to Professor Brink and Mr. Hua.

References

1. E. Cremmer, B. Julia and J. Scherk, *Phys. Let.* **76B**, 409 (1978).
2. There are number of papers which suggested this possibility summarized by H. P. Nilles as BBDDEFFIKNPRRSW. See H. P. Nilles, *Phys. Rep.* **110**, 1–162 (1984).
3. For example, gluino search by ATLAS is given in arXiv:1605.09318 at 13 TeV run, May 31, 2016.
4. N. Ishibashi *et al.*, *Nucl. Phys. B* **498** 467 (1997); T. Banks *et al.*, *Phys. Rev. D* **55** 5112 (1997).
5. P. West, Presentation at this symposium, arXiv: 1512.01644v3 [hep-th], 22 Jan 2016 and papers cited in this paper.
6. H. Sugawara and M. Yoshimura, *Phys. Rev.* **173**, 1419 (1968); H. Sugawara, *Phys. Rev.* **170**, 1659 (1968).
7. M. S. Narasimhan, Presentation at this symposium and S. Kumar, M. S. Narasimhan and A. Ramanathan, *Math. Ann.* **300**, 41 (1994).
8. A. H. Guth, *Phys. Rev.* **D23**, 347 (1981); A. D. Linde, *Phys. Let.* **B108**, 389 (1982); A. Albrecht and P. Sternhardt, *Phys. Rev. Let.* **48**, 1220 (1982).
9. For a review see, for example, U. H. Danielsson, *Class. Quant. Grav.* **22**, S1 (2005), arXiv:hep-th/0409274.
10. OPAL Collab., *Phys. Let.* **B253**, 511 (1991).
11. The most recent result on kaon-neutrino mode is reported by KamLAND collaboration in arXiv:1505.03612v3, 24 Sep 2015.
12. G. Degrassi *et al.*, arXiv: 1205.6497v1, 29 May 20 (2012).
13. W. Bardeen, On Naturalness in the Standard Model, FERMILAB-CONF-95-391-T.
14. CEPC-SPPC, *Preliminary Conceptual Design Report* Volume II – Accelerator, The CEPC-SPPC Study Group, March 2015.

Chern–Simons Theories with Fundamental Matter: A Brief Review of Large N Results Including Fermi–Bose Duality and the S-matrix

Spenta R. Wadia
International Centre for Theoretical Sciences (ICTS-TIFR),
Tata Institute of Fundamental Research, Bangalore, 560089, India
spenta.wadia@icts.res.in
www.icts.res.in/people/spenta-wadia

We begin with a few words about Salam's contribution to the growth of String Theory in India. In the technical talk we review results in $SU(N)$ Chern–Simons plus vector matter theories in 2+1 dim in the large N limit. The dressing of charged matter by Chern–Simons gauge fields leads to anyons that interpolate between fermions and bosons and lead to a duality symmetry between fermionic and bosonic theories. The S-matrix (defined in the large N limit) besides exhibiting this duality, also exhibits novel properties due to the presence of anyons. The S-matrix is not analytic, like in Aharonov–Bohm scattering, and satisfies modified crossing symmetry relations.

1. Abdus Salam, ICTP and String Theory in India

Abdus Salam is one of the creators of the Standard Model of elementary particles. In 1979 he shared the Nobel Prize for Physics with Sheldon Glashow and Steven Weinberg. Besides his extensive contributions to physics he created the International Centre for Theoretical Physics (ICTP) in Trieste, Italy in 1964. ICTP is a unique institution with the mission of promoting science in the developing world.

ICTP has contributed to the growth of science in India and in particular to the area of String theory. Salam gave enormous support and encouragement to the fledgling string community in India since we began working in 1984. ICTP became a place to go every summer and meet with peers from all over the world. Here we not only went to listen to lectures by peers from the developed world, but also as lecturers. In the days without the internet these meetings contributed to our being part of the global community of scholars. I vividly remember the excitement at ICTP during the spring and summer schools, and how Salam (till his health permitted) participated in most of them and added life to these meetings by his very generous and sensitive personality and a splendid sense of oriental humour.

Another point worth mentioning is that Salam would support our travel to go and present our work at major conferences elsewhere. This gave visibility to the work we were doing in India all through the 1980s. Without doubt Salam and

ICTP made a solid contribution to the growth of String Theory in India, which has today flowered into one of the strongest communities in this subject in the world.

2. ICTS-TIFR

The International Centre for Theoretical Sciences (ICTS-TIFR) of the Tata Institute of Fundamental Research in Bangalore was established in 2007. It was inspired by theory institutions like ICTP, KITP Santa Barbara and IAS Princeton, but is different and uniquely adapted to the needs of Indian science. It is a new type of institution in India where top quality research across various science disciplines happens side by side with various visitor driven activities like advanced schools, discussion meetings and conferences. ICTS is a hub of scientific activity and a centre for world-class research and education in the Indian subcontinent.

To commemorate the contribution of Abdus Salam to science in India one of the three academic blocks of ICTS is named after him, and every year ICTS hosts a *Abdus Salam Memorial Lecture* that is delivered by a distinguished scientist.

3. Introduction to the Talk on Chern–Simons Theory

This talk is about $U(N)$ Chern–Simons theories at level k coupled with fundamental matter in the limit N, $k \to \infty$ with $\lambda = \frac{N}{k}$ kept fixed. Correlation functions of Wilson loops in pure Chern–Simons gauge theories in 2+1 dims provide a beautiful and powerful mathematical framework to discuss topological properties of knots.[1] The addition of dynamical matter fields changes the scope of the questions that can be pursued. The fact that these theories are soluble in the large N limit enables us to explore and confirm some fascinating dualities in 2+1 dims and also enables us to unravel some novel features of the S-matrix in the presence of the Chern–Simons term.[3]

From the view point of the AdS/CFT correspondence Chern–Simons theories with fundamental (vector) matter turn out to be dual to Vasiliev-type theories of higher spin gauge fields. Besides their intrinsic interest they can serve as a warm up to the much more difficult and important problem when matter is in the adjoint representation of the gauge group, as is the case in the ABJM theory[2] that is dual to type IIA string theory in AdS_4. ABJM theory is a nonperturbative definition of quantum gravity in a 4-dim AdS space–time.

4. Chern–Simons (Fundamental) Matter Theories[4–14]

This report is about the following two $SU(N)$ Chern–Simons theories at level k, coupled with fundamental matter in the limit N, $k \to \infty$ with $\lambda = \frac{N}{k}$ kept fixed.

(1) $SU(N)$ Chern–Simons gauge field coupled to bosons

$$S = \int d^3x\, i\epsilon^{\mu\nu\rho} \frac{k}{4\pi} \mathrm{Tr}\left(A_\mu \partial_\nu A_\rho - \frac{2i}{3} A_\mu A_\nu A_\rho \right)$$

$$+ D_\mu \bar{\phi} D^\mu \phi + b_4 (\bar{\phi}\phi)^2 + m_B^2 (\bar{\phi}\phi). \tag{1}$$

(2) $SU(N)$ Chern–Simons gauge field coupled to fermions

$$S = \int d^3x\, i\epsilon^{\mu\nu\rho} \frac{k}{4\pi} \text{ trace}\left(A_\mu \partial_\nu A_\rho - \frac{2i}{3} A_\mu A_\nu A_\rho \right)$$

$$+ \bar{\psi}\gamma^\mu D_\mu \psi + m_F^{\mathrm{reg}} \bar{\psi}\psi. \tag{2}$$

4.1. *Summary of results on CS + matter theories*

Before we discuss the S-matrix in detail we summarise some previous results in the subject.

4.1.1. *The spectrum*

In the conformal limit of the above two theories (bosons at the Wilson–Fisher point and massless fermions) one can use the conformal group $SO(3,2)$ in $2+1$ dims to show that the spectrum of single trace operators in both theories includes one scalar and one current of spin s for $s = 1, 2, 3, \ldots, \infty$. The dimension of the scalar operator is given, at every value of λ by $\Delta = 2 + o(\frac{1}{N})$, while the dimension of the spin s current $J^{(s)}$ is given by $\Delta = (s+1) + o(\frac{1}{N})$.[4] Since the leading order 'free spectrum' is identical, there is a hint here that the two theories are different avatars of the same underlying theory.

4.1.2. *Partially conserved higher spin currents*

In the conformal limit the higher spin currents $J^{(s)}$ in both theories satisfy identical equations in the large k limit. The operator $J^{(s)}$ comes closest to a conserved current, in that it is the unique current that obeys the exact schematic equation[4]

$$\partial \cdot J^{(s)} \sim \frac{1}{k} JJ + \frac{1}{k^2} JJJ \tag{3}$$

(on the r.h.s. of the equation above, the symbol J refers either to a current or to a descendent of a current; the important point in this equation is that the r.h.s. contains no single trace pieces). Since k is large the higher spin symmetry violation is small. This is why their bulk dual gauge fields in AdS_4 are massless. These equations also indicate that the two theories (including interactions) are different avatars of the same underlying theory.

The presence of currents of higher spin also reflects the fact that the basic gauge invariant degrees of freedom (in a fixed gauge) are bi-local matter fields and their bi-local Fourier conjugates, which at the large N saddle point satisfy the

Schwinger–Dyson equations.[15] See Ref. 6 for an explicit summary of the technique of bi-local fields and their conjugates.

4.1.3. *The 3-point function of currents*

The partial conservation equations for the higher spin currents (appropriately interpreted within correlation functions) constrain multi-point function of the currents. In particular Maldacena and Zhiboedov[7] used these equations to calculate the 3-point function of currents, and exhibited a parametric symmetry, which was interpreted by Aharony, Gur-Ari and Yacoby[8] as a level-rank duality. In summary their result can be stated as a duality of the Fermi–Bose theories

$$\langle J_F^{s_1} J_F^{s_2} J_F^{s_3} \rangle_{U(N_F),k_F} = \langle J_B^{s_1} J_B^{s_2} J_B^{s_3} \rangle_{U(N_B),k_B} \tag{4}$$

provided $N_B = |k|_F - N_F$, $k_B = -k_F$ or equivalently $\lambda_B = \lambda_F - \mathrm{sgn}(\lambda_F)$.

4.1.4. *Thermal partition function*

The thermal partition functions (on $S^2 \times S^1$) in both theories can be exactly calculated in the large N limit with result for all allowed values of λ

$$Z(U(N_F), k_F) = Z(U(N_B), k_B) \tag{5}$$

provided once again $N_B = |k|_F - N_F$, $k_B = -k_F$ or equivalently $\lambda_B = \lambda_F - \mathrm{sgn}(\lambda_F)$.

Here the key insight is to include the holonomy of the gauge field around the thermal circle[9] as relevant degrees of freedom at large N, when the temperature T times the volume of S^2 scales as N.

The partition function is related to the expectation value of a 'potential' function of the S^1 holonomy in pure Chern–Simons theory in several examples the holonomy potential is computed as a function of λ.[10] Level rank duality of pure Chern–Simons theory implies the equality of thermal partition functions of dual pairs of theories as a function of the temperature. The partition function reduces to a matrix integral over holonomies. The summation over flux sectors quantizes the eigenvalues of this matrix in units of $2k$ and the eigenvalue density of the holonomy matrix is bounded from above by $12\pi\lambda$. This introduces a new type of matrix model where the density of eigenvalues is bounded from above. The corresponding matrix integrals generically undergo two phase transitions as a function of temperature. For several Chern–Simons matter theories the relevant matrix models are exactly soluble in the low temperature phase, and the phase transition temperature can be determined as a function of λ. At low temperatures the partition function smoothly matches onto the N and λ independent free energy of a gas of non-renormalized multi-trace operators. An exact solution subject to an upper bound on eigenvalue density can be found for the large N Gross–Witten–Wadia matrix integral.

It is remarkable that the above result is implied by the *level-rank duality of pure Chern–Simons theory* which is simply stated in terms of Wilson loop operators as follows: $\langle W\rangle_{N,k,R} = \langle W\rangle_{|k|-N,-k,\tilde{R}}$, where $\langle W\rangle$ is the expectation value of a Wilson loop operator in pure CS theory. The Young tableaux of $\tilde{R}$ is obtained from that of R by a transposition, i.e. an interchange of rows and columns.

In what follows we discuss the S-matrix in the above bosonic and fermionic theories deformed away from their conformally invariant points by their masses.

5. The S-Matrix[3]

We will now focus on $2 \to 2$ scattering of bosons/fermions in representations R_1 and R_2 of $U(N)$ at large N that is exact for all allowed values of the 't Hooft coupling λ. (The definition of a scattering matrix of coloured particles in a non-Abelian gauge theory is a subtle issue and will be one of our main concerns in the following.)

The tensor product of these two representations decompose as

$$R_1 \times R_2 = \sum_m R_m. \tag{6}$$

It follows from $U(N)$ invariance that the S-matrix for the process takes the schematic form

$$S = \sum_m P_m S_m, \tag{7}$$

where P_m is the projector onto the mth representation, and S_m is the scattering matrix in the 'mth' channel.

Since we study the scattering matrices of the elementary quanta of theories with only fundamental matter, R_1 and R_2, are either both fundamentals, or one fundamental and one anti-fundamental. The scattering of two anti-fundamentals is simply related to the scattering of two fundamentals, and will not be discussed separately. In the case of fundamental–fundamental scattering, R_m is either the 'symmetric' representation with two boxes in the first row (and no boxes in any other row) of the Young Tableaux, or the 'antisymmetric' representation with two boxes in the first column and no boxes in any other column. In the case of fundamental–anti-fundamental scattering, R_m is either the singlet or the adjoint representation.

The non-relativistic limit (i.e. in the limit in which the masses of the scattering particles and the center-of-mass energy are both taken to infinity at fixed momentum transfer) of the scattering matrices of interest have been worked out in the past.[16] We briefly review this.

Gauss's law in Chern–Simons theory ensures that each charged particle carries magnetic flux. The Aharonov–Bohm effect then ensures that the particle R_1 picks up the phase $2\pi\nu_m$ as it circles around the particle R_2, where

$$2\pi\nu_m = \frac{4\pi T_1^a T_2^a}{k} = 2\pi\frac{C_2(R_m) - C_2(R_1) - C_2(R_2)}{k}, \tag{8}$$

$T^a_{1/2}$ are the representation matrices for the group generators in representations R_1 and R_2 and $C_2(A)$ is the quadratic Casimir in representation A. It follows as a consequence[16] that the non-relativistic scattering amplitude in the R_m exchange channel is given by the Aharonov–Bohm scattering amplitude of a $U(1)$ particle of unit charge of a point like magnetic flux of strength $2\pi\nu_m$.[a]

It is easily verified that $\nu_m = \mathcal{O}(\frac{1}{N})$ or smaller in the symmetric, antisymmetric or adjoint channels. In the singlet channel, however, it turns out that to leading order in the large N limit $\nu_m = \frac{N}{k} = \lambda$. It follows that the rotation by π which interchanges the two scattering particles is accompanied by a phase $e^{-i\pi\lambda_B}$ in the bosonic theory and $(-1)e^{-i\pi\lambda_F} = e^{-i\pi(-\mathrm{sgn}(\lambda_F)+\lambda_F)}$ in the fermionic theory.[b] Note that these phases are identical when

$$\lambda_B = \lambda_F - \mathrm{sgn}(\lambda_F). \tag{9}$$

However (9) is precisely the map between λ_B and λ_F induced by the level rank duality transformation described earlier in this report. In the singlet channel, in other words the bosons and conjecturally dual fermions are both effectively anyonic, with the same anyonic phase. This observation provides a partial physical explanation for the duality map (9).

We observe in passing that the anyonic phase $\pi\lambda_B$ is precisely twice the phase of the bulk interaction term in the conjectured Vasiliev duals to these theories.[4] Indeed the first speculation of the bosonization duality for matter Chern–Simons[4] was motivated by argument very similar to that presented in the previous paragraph but in the context of Vasiliev theories (deformations of the bosonic and fermionic theory that lead to the same interaction phase ought to be the same theory). It would certainly be very interesting to find a logical link between the phase of interactions in Vasiliev theory and the anyonic phase of the previous paragraph.

5.1. *Summary of relativistic scattering*

In the following we will sketch the treatment of the relativistic theory. There are two points we will encounter that are different from usual S-matrix theory.

- The forward scattering amplitude has a delta function exactly for the same reason as in the non-relativistic case. This is because a particle with charge carries flux due to Gauss's law: $j_0(x) = \frac{k}{2\pi}F_{12}(x)$, where x is a point in a 2-dim manifold. Hence forward scattering off such a particle has a non-trivial amplitude due to the presence of flux on it, and hence it is required by unitarity. The delta-function in the forward scattering amplitude implies that that S-matrix is not analytic!

---

[a] Readers familiar with the relationship between Chern–Simons theory and WZW theory may recognize this formula in another guise. $\frac{C_2(R)}{k}$ is the holomorphic scaling dimension of a primary operator in the integrable representation R, and $e^{2\pi i\nu_m}$ is the monodromy of the four point function $\langle R_1, R_2, \bar{R}_1, \bar{R}_2\rangle$ in the conformal block corresponding to the OPE $R_1R_2 \to R_m$.

[b] The additional -1 in the fermionic theory comes from Fermi statistics. We have used $-1 = e^{\pm i\pi} = e^{-i\pi\mathrm{sgn}(\lambda_F)}$.

- The standard LSZ formalism of QFT fails because due to Gauss's law the CS gauge field never decouples from the coloured matter! Hence one needs to adopt a manifestly gauge invariant definition of the 4-point function to extract the S-matrix for the scattering of coloured particles. We could make a unique prescription only in the limit of large N (or k). This modification leads to modified crossing relations for the analytic part of the S-matrix. This a new and surprising result perhaps characteristic of the Chern–Simons interaction irrespective of their matter content.

As a consequence of the above points, while a unitary t and u channel S-matrix is calculable at large N, the s-channel S-matrix cannot be obtained from t/u channels by crossing symmetry, because the S-matrix is not analytic. Nevertheless in the following we propose a formula for the s-channel S-matrix that passes many tests, like unitary, duality and a correct non-relativistic limit, but as yet has not been calculated from first principles for technical reasons.

5.1.1. *Proposed s-channel S-matrix*

Introducing the T-matrix by the usual definition: $S = 1 + iT$, we have

$$T_s = \frac{\sin(\pi\lambda)}{\pi\lambda} T_s^{\text{Trial}} - 8\pi i\sqrt{s}(\cos(\pi\lambda) - 1)\delta(\theta), \tag{10}$$

where

$$T_s^{\text{Trial}} = 4i\pi\lambda E(p_1, p_2, p_3)\sqrt{\frac{st}{u}} + 4\pi\lambda\sqrt{s}\frac{1 + e^{-2\lambda\tanh^{-1}\frac{\sqrt{s}}{2|c_B|}}}{1 - e^{-2\lambda\tanh^{-1}\frac{\sqrt{s}}{2|c_B|}}}, \tag{11}$$

where, T_s^{Trial} is obtained using the standard rules of crossing symmetry from the t-channel.

6. Sketch of the Calculation

Consider the scattering of massive coloured particles:

(a) Particles (P) transform in the fundamental representation of $U(N)$,
(b) Anti-particles (A) transform in the anti-fundamental representation of $U(N)$.

Hence there are three distinct scattering processes:

$$\begin{aligned} &P_i + P_j \to P_m + P_n \quad \text{(Sym and Asym channel)},\\ &P_i + A^j \to P_m + A^n \quad \text{(Adj and Singlet(s) channel)},\\ &A^i + A^j \to A^m + A^n \quad \text{(CPT conjugate of case 1)}. \end{aligned} \tag{12}$$

All these S-matrices are contained in appropriate on shell limits of the single correlator

$$\langle \phi_i \phi_m \bar{\phi}^j \bar{\phi}^n \rangle.$$

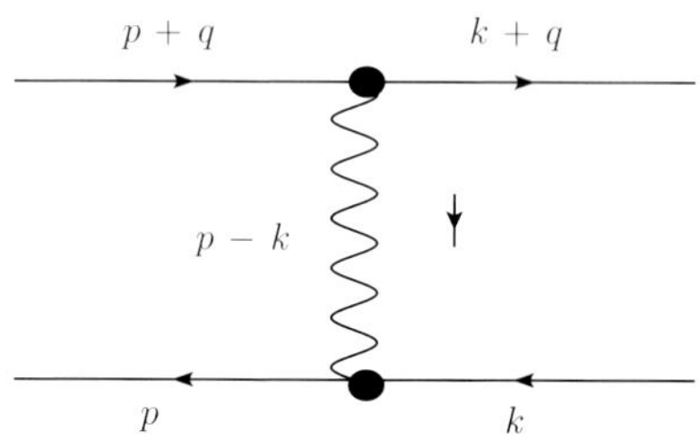

6.1. *Diagrammatic illustration of various channels*

Consider the process

$$P_i(p_1) + A^j(p_2) \to P_m(p_3) + A^n(p_4). \tag{13}$$

- Adjoint channel:

$$p_1 = p + q, \quad p_2 = -(k + q), \quad p_3 = -p, \quad p_4 = k. \tag{14}$$

- Singlet channel:

$$p_1 = p + q, \quad p_2 = -p, \quad p_3 = -(k + q), \quad p_4 = k, \\ \text{C.M. momentum} = q_\mu. \tag{15}$$

6.2. *Schwinger–Dyson equation for 4-point function*

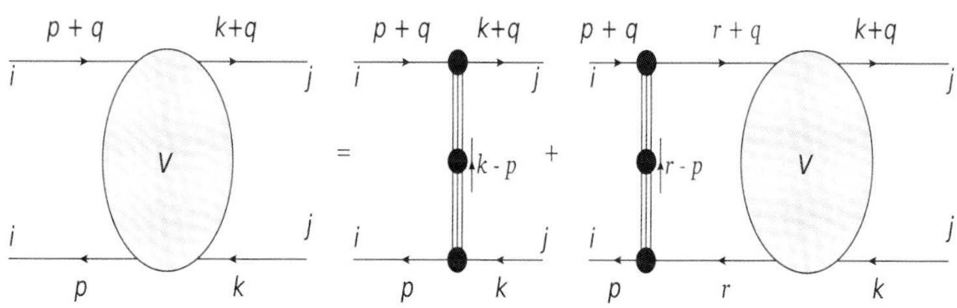

$$V(p,k,q) = V_0(p,k,q) - i \int \frac{d^3 r}{(2\pi)^3} V_0(p,r,q_3) \frac{N V(r,k,q_3)}{(r^2 + c_B^2 - i\epsilon)((r+q)^2 + c_B^2 - i\epsilon)}, \tag{16}$$

where c_B is the exact pole mass of the elementary boson. The choice of gauge is $A_- = 0$ and hence $V(p,k,q)$ depends on this choice.

6.3. *Solving the Schwinger–Dyson equation*

V can be calculated exactly for adjoint (T_{adj}), symmetric (T_{sym}) and antisymmetric T_{asym}, for the bosonic and fermionic theories, using a Lorentz frame with external momenta $q_\pm = 0$ (where $q_\pm = q_0 \pm q_1$). The full answer which is unitary is enabled by Lorentz covariance. It turns out that the fermionic and bosonic S-matrices in these channels are dual to each other under the usual identification of parameters: $N_B = |k|_F - N_F$, $k_B = -k_F$, $\lambda_B = \lambda_F - \text{sgn}\, \lambda_F$ and $m_F = -m_B^{\text{cri}}\lambda$.

6.4. *The singlet channel*

For the singlet channel, since q is the center-of-mass (CM) momentum, setting $q_\pm = 0$ implies the CM momentum is space-like, which is absurd. For this reason we are unable to compute the singlet channel S-matrix. We can attempt to obtain the singlet answer for T_{singlet} from T_{adj} using crossing symmetry but this violates unitarity!

There are two sources of unitarity violation:

(1) There is a delta function for forward scattering and hence the S-matrix is not analytic.
(2) Hence the standard formulas of crossing symmetry cannot be used and one needs a new treatment. It turns out that the S-matrix in say the t- and s-channels are related by a normalization factor $f(\lambda) = \frac{\sin(\pi\lambda)}{\pi\lambda}$ that is dictated by gauge invariance.

7. The Source of the Delta-Function in Aharonov–Bohm Scattering

The Aharonov–Bohm problem involves the scattering of a particle of unit $U(1)$ charge off a point like magnetic vortex. The fact that unitarity requires a delta function in the forward scattering was discovered decades later by S. N. M. Ruijsenaars in 1983![17] As already mentioned the non-Abelian case with a Chern–Simons coupling was studied in Ref. 16. We recall their result again to make this section self-contained. They considered scattering of two particles in representation R_1 and R_2, and the exchange channel is in the representation R. They concluded that the scattering amplitude in this case is the same as the scattering of a particle of unit $U(1)$ charge off a point like magnetic vortex with flux

$$\nu = \frac{C_2(R_1) + C_2(R_2) - C_2(R)}{k}, \tag{17}$$

where $C_2(A)$ is the quadratic Casimir in the representation A.

The scattering amplitude is given by

$$T_{NR} = -8\pi i\sqrt{s}(\cos(\pi\nu) - 1)\delta(\theta) + 4\sqrt{s}|\sin(\pi\nu)|$$
$$+4i\sqrt{s}\sin(\pi\nu)P\nu\left(\cot\left(\frac{\theta}{2}\right)\right), \tag{18}$$

where θ is the scattering angle and $s = 4C_B^2 + \frac{p^2}{4C_B^2}$, p being the incident momentum It is easy to check that for large k and large N

$$\nu_{\text{Adj}} \sim \nu_{\text{sym}} \sim \nu_{\text{as}} \sim \frac{1}{k}, \tag{19}$$

where as

$$\nu_s = \frac{N}{k} = \lambda. \tag{20}$$

Note the $\delta(\theta)$ in the S-matrix which violates analyticity. For T_{Adj}, T_{sym}, T_{as} the delta function coefficient is $\mathcal{O}\left(\frac{1}{k^2}\right)$ and hence vanish for large k. For T_s coefficient of $\delta(\theta)$ is $\mathcal{O}(1)$ and is proportional to $(\cos(\pi\lambda) - 1)\delta(\theta)$ and its coefficient can be interpreted on very general grounds that make no reference to the non-relativistic limit. There are paths in the forward direction which 'see' the magnetic vortex. Hence we expect this term to be present in the relativistic case as well.

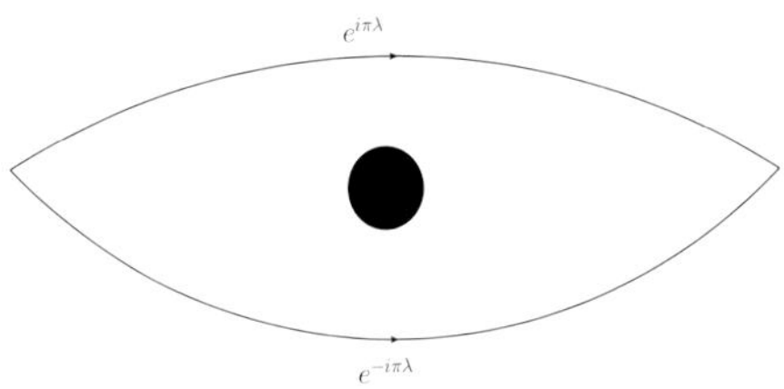

8. Gauge Invariance, Crossing Symmetry and Origin of $f(\lambda)$

We are discussing the S-matrix of coloured particles in a non-Abelian gauge theory. Due to Gauss's law since it is not possible to 'switch off' the gauge field asymptotically we need to extract this S-matrix from a manifestly gauge invariant construction by contracting the on-shell external legs with Wilson lines. However there are many ways in which Wilson lines could be attached to the external legs consistent with gauge invariance. This can lead to inequivalent answers for the S-matrix. This ambiguity in the definition of the S-matrix needs to be addressed.[c] However in the large N limit the skein relations of Jones, which say that the crossing of two Wilson lines costs only a phase, enable a unique gauge invariant construction and hence

[c]We thank Zohar Komargodski for raising this point and discussions.

the S-matrix we can extract is also unique. Let us illustrate the main idea by the following figures:

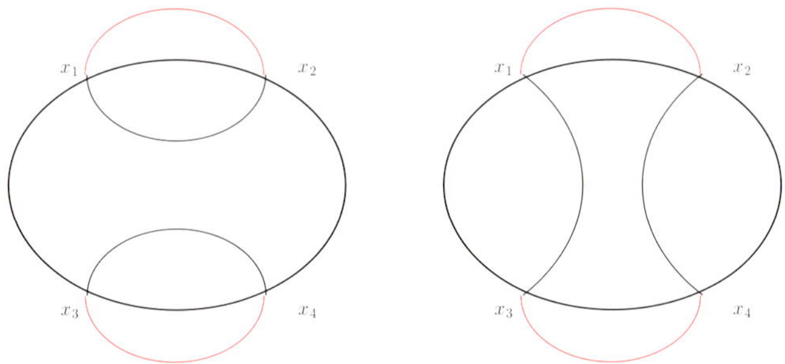

- The left/right figures represent t (adjoint)/s (singlet) channels respectively. The points x_i $i = 1, 2, 3, 4$ are marked on a large 2-sphere, and particles starting from x_i travel over to x_j. The Wilson lines are denoted in red (color online).
- In the t-channel a particle worldline starting at x_1 ends at x_2 and one beginning at x_3 ends at x_4.
- In the s-channel a particle worldline starting at x_1 ends at x_3 and one beginning at x_2 ends at x_4.
- In general the worldlines knot around each other, but at large N the skein relations (Jones) enable them to cross upto a phase: $L_+ = e^{i\phi} L_-$ where $\phi = -2\pi \frac{\lambda}{1+\lambda}$.
- The key point is that the topology of paths in the t/s channels is different and hence the ratio of the S-matrix in the s- and t-channels is $\langle W^2 \rangle / \langle W \rangle$ in pure Chern–Simons theory. This number is precisely $\frac{\sin(\pi\lambda)}{\pi\lambda}$.[1]

9. Bosonic Theory Formulas

Let us now present formulas in the relativistic boson theory described by the action:

$$S = \int d^3x i\epsilon^{\mu\nu\rho} \frac{k}{4\pi} \mathrm{Tr} \left(A_\mu \partial_\nu A_\rho - \frac{2i}{3} A_\mu A_\nu A_\rho \right)$$
$$+ D_\mu \bar{\phi} D^\mu \phi + b_4 (\bar{\phi}\phi)^2 + m_B^2 (\bar{\phi}\phi). \tag{21}$$

The exact propagator is given by

$$\langle \phi_j(p) \bar{\phi}^i(-q) \rangle = \frac{(2\pi)^3 \delta^i_j \delta^3(-p+q)}{p^2 + c_B^2}. \tag{22}$$

The renormalized mass is easily computed to be $c_B^2 = \frac{\lambda_B^2}{4} c_B^2 - \frac{b_4}{4\pi} c_B + m_B^2$, and the critical limit is defined by $b_4 \to \infty$, $m_B \to \infty$ such that $\frac{m_B^2}{b_4}$ kept fixed.

We have the following results for the various channels: T_{adj}, T_{sym}, T_{as} channel answers are consistent with crossing symmetry. We present here only the answer for T_{Adj}:

$$T_{\mathrm{adj}} = -\frac{4i\pi}{k_B}\sqrt{\frac{ut}{s}}\frac{4i\pi}{k_B}\sqrt{-t}$$
$$\times\frac{(\tilde{b}_4 - 4\pi i\lambda_B\sqrt{-t}) + (\tilde{b}_4 + 4\pi i\lambda_B\sqrt{-t})e^{-2i\lambda_B\tan^{-1}\left(\frac{\sqrt{-t}}{2|c_B|}\right)}}{-(\tilde{b}_4 - 4\pi i\lambda_B\sqrt{-t}) + (\tilde{b}_4 + 4\pi i\lambda_B\sqrt{-t})e^{-2i\lambda_B\tan^{-1}\left(\frac{\sqrt{-t}}{2|c_B|}\right)}}, \tag{23}$$

where $\tilde{b}_4 = -b_4 + 2\pi\lambda_B^2|c_B|$. In the $b_4 \to \infty$, we get

$$T_{\mathrm{adj}} = -\frac{4i\pi}{k_B}\sqrt{\frac{ut}{s}} - \frac{4i\pi}{k_B}\sqrt{-t}\frac{1+e^{-2i\lambda_B\tan^{-1}\left(\frac{\sqrt{-t}}{2|c_B|}\right)}}{1-e^{-2i\lambda_B\tan^{-1}\left(\frac{\sqrt{-t}}{2|c_B|}\right)}}. \tag{24}$$

Unitarity for all the above is easily verified.

9.1. *Trial s-channel answer using crossing symmetry*

Using crossing symmetry and the answer in the adjoint channel the s-channel S-matrix would be

$$T_s^{\mathrm{Trial}} = 4i\pi\lambda\sqrt{\frac{st}{u}} + 4\pi\lambda\sqrt{s}\frac{1+e^{-2\lambda\tanh^{-1}\frac{\sqrt{s}}{2|c_B|}}}{1-e^{-2\lambda\tanh^{-1}\frac{\sqrt{s}}{2|c_B|}}}. \tag{25}$$

Note that, this does not contain the δ function piece. As expected it is not unitary. In the beginning of the talk we proposed s-channel answer to be:

$$T_s = \frac{\sin(\pi\lambda)}{\pi\lambda}T_S^{\mathrm{Trial}} - 8\pi i\sqrt{s}(\cos(\pi\lambda)-1)\delta(\theta), \tag{26}$$

where T_s^{Trial} as discussed above. In order to check for unitarity it is convenient to rewriting the above formula as

$$T_s(\sqrt{s},\theta) = H(\sqrt{s})T(\theta) + W_1(\sqrt{s}) - iW_2(\sqrt{s})\delta(\theta), \tag{27}$$

where

$$T(\theta) = i\cot\left(\frac{\theta}{2}\right), \quad H(\sqrt{s}) = 4\sqrt{s}\sin(\pi\lambda),$$

$$W_1(\sqrt{s}) = 4\sqrt{s}\sin(\pi\lambda)\mathrm{sgn}(\lambda)\frac{1+e^{-2\lambda\tanh^{-1}\frac{\sqrt{s}}{2|c_B|}}}{1-e^{-2\lambda\tanh^{-1}\frac{\sqrt{s}}{2|c_B|}}}, \tag{28}$$

$$W_2(\sqrt{s}) = 8\pi\sqrt{s}(\cos(\pi\lambda)-1).$$

The unitarity equations $T - T^\dagger = iTT^\dagger$ now take the form

$$
\begin{aligned}
H - H^* &= \frac{1}{8\pi\sqrt{s}}(W_2 H^* - H W_2^*),\\
W_2 + W_2^* &= -\frac{1}{8\pi\sqrt{s}}(W_2 W_2^* + 4\pi^2 H H^*),\\
W_1 - W_1^* &= \frac{1}{8\pi\sqrt{s}}(W_2 W_1^* - W_2^* W_1) - \frac{i}{4\sqrt{s}}(H H^* - W_1 W_1^*).
\end{aligned}
\tag{29}
$$

These equations are indeed satisfied by $T_s(\sqrt{s}, \theta)$. Since the unitarity equations are non-linear this is a non-trivial check for the proposed answer to satisfy these equations. In the non-relativistic limit the answer reduces to the Aharonov–Bohm–Ruijsenaars result.

10. Fermionic Theory Formulas

Now let us present the S-matrix in the fermionic theory. $SU(N)$ Chern–Simons gauge field coupled to regular fermions is described by

$$
\begin{aligned}
S = \int d^3x\, i\epsilon^{\mu\nu\rho} \frac{k}{4\pi} \operatorname{trace}\left(A_\mu \partial_\nu A_\rho - \frac{2i}{3} A_\mu A_\nu A_\rho \right)\\
+ \bar{\psi}\gamma^\mu D_\mu \psi + m_F^{\text{reg}} \bar{\psi}\psi.
\end{aligned}
\tag{30}
$$

Consider $N \to \infty$ and $k \to \infty$ keeping $\lambda = \frac{N}{k}$ fixed. The renormalized mass $c_f = \frac{m_F^{\text{reg}}}{1-\lambda}$.

10.1. *Schwinger–Dyson equation for fermions*

Fermionic 4-point function satisfies the integral equation:

$$
\begin{aligned}
V_{\alpha\beta\gamma\delta}(p,k,q) = -\frac{1}{2}\gamma^\mu_{\alpha\beta} G_{\nu\mu}(p-k)\gamma^\nu_{\gamma\delta} - \frac{1}{2}\int \frac{d^3q'}{(2\pi^3)}\\
\times [\gamma^\mu G(q'-P)]_{\alpha\sigma} V_{\sigma\beta\gamma\tau}(p_1, p_2, q')[G(q')\gamma^\nu]_{\tau\delta} G_{\nu\mu}(q'-q).
\end{aligned}
\tag{31}
$$

This equation can be solved explicitly for all orders in λ in the frame $q_\pm = 0$.

10.1.1. *Explicit results for t-channel*

In order to evaluate the on-shell S-matrix, we need to contract the above four point function by the Dirac wave functions u and v and put the external momenta on-shell.

In the adjoint channel (t-channel) the result is

$$T_t^F(k_F, \lambda_F, c_F) = \frac{4i\pi}{k_F} E(p_1, p_2, p_3)\sqrt{\frac{ut}{s}} + \frac{4i\pi}{k_F}\sqrt{-t}\frac{1 + e^{-2i(\lambda_F - \mathrm{sgn}(m_F))\tan^{-1}\left(\frac{\sqrt{-t}}{2|c_F|}\right)}}{1 - e^{-2i(\lambda_F - \mathrm{sgn}(m_F))\tan^{-1}\left(\frac{\sqrt{-t}}{2|c_F|}\right)}}. \tag{32}$$

The other channel answers are consistent with crossing symmetry.

11. Fermion–Boson Duality

Given our answers we observe that adjoint channel critical boson S-matrix maps to the fermion S-matrix: $T_T^{B\infty}(-k_F, \lambda_F - \mathrm{sgn}(\lambda_F), c_F) = T_T^F(k_F, \lambda_F, c_F)$. On physical grounds we expect that, $T^B_{\mathrm{sym}} = T^F_{\mathrm{asym}}$, $T^F_{\mathrm{sym}} = T^B_{\mathrm{asym}}$. This is because, the sign picked up by the exchange of two identical fermions is seen on the bosonic side as antisymmetry with respect to the colour indices. The proposed s-channel answer satisfies duality as well as unitarity.

12. $N = 1, 2$ Supersymmetric Theories[18]

We now briefly state results for $\mathcal{N} = 1$ supersymmetric matter. The S-matrix can be considered as a superfield $S(p_1, \theta_1, p_2, \theta_2, p_3, \theta_3, p_4, \theta_4)$ and the expansion in terms of the θ parameters consists of 8 terms. The identity term is the four boson scattering S_B, the four θ term is the fermion scattering S_F. SUSY ward identities determine all other 6 terms in terms of S_B and S_F. Here too the proposed s-channel S-matrix is unitary and consistent with duality. The $\mathcal{N} = 2$ theories have additional constraints which relate S_B and S_F.

13. Conclusions and Future Problems

- The $2 \to 2$ S-matrix of critical bosons and fermions can be calculated exactly and unambiguously in the large N limit in the t and u channels. The answers are unitary and are consistent with fermion–boson duality.
- We proposed a s-channel S-matrix with modified crossing symmetry rules and a delta function for forward scattering that is unitary and consistent with gauge invariance and fermion–boson duality.
- These results, obtained for large N if valid for the case $N = 1$, can have implications for the understanding of the fractional quantum Hall effect at filling fraction $\nu = 1/2$.
- Studying an odd number of Majorana fermions coupled to CS theory would be an interesting problem to explore.
- It would be useful to be able to formulate the gauge theory in terms of variables that make the duality manifest.
- The definition of the S-matrix at finite N and k needs to be explored.

- What are the implications of our results for ABJ theories? In the non-relativistic limit one can clearly see a delta function with fixed coefficients in the forward scattering.[d]

Acknowledgments

I would like to thank the organizers of this conference to celebrate the contribution of Abdus Salam to science and society, and for bringing together a lot of people whose lives were touched by him.

References

1. E. Witten, Quantum field theory and the Jones polynomial, *Commun. Math. Phys.* **121**, 351 (1989).
2. O. Aharony, O. Bergman, D. L. Jafferis and J. Maldacena, $N = 6$ superconformal Chern–Simons-matter theories, M2-branes and their gravity duals, *J. High Energy Phys.* **0810**, 091 (2008), arXiv:0806.1218 [hep-th].
3. S. Jain, M. Mandlik, S. Minwalla, T. Takimi, S. R. Wadia and S. Yokoyama, Unitarity, crossing symmetry and duality of the S-matrix in large N Chern–Simons theories with fundamental matter, *J. High Energy Phys.* **1504**, 129 (2015), doi:10.1007/JHEP04(2015)129, arXiv:1404.6373 [hep-th].
4. S. Giombi, S. Minwalla, S. Prakash, S. P. Trivedi, S. R. Wadia and X. Yin, Chern–Simons theory with vector fermion matter, *Eur. Phys. J. C* **72**, 2112 (2012), doi:10.1140/epjc/s10052-012-2112-0, arXiv:1110.4386 [hep-th].
5. O. Aharony, G. Gur-Ari and R. Yacoby, $d = 3$ bosonic vector models coupled to Chern–Simons gauge theories, *J. High Energy Phys.* **1203**, 037 (2012), doi:10.1007/JHEP03(2012)037, arXiv:1110.4382 [hep-th].
6. S. Jain, S. P. Trivedi, S. R. Wadia and S. Yokoyama, Supersymmetric Chern–Simons theories with vector matter, *J. High Energy Phys.* **1210**, 194 (2012), doi:10.1007/JHEP10(2012)194, arXiv:1207.4750 [hep-th].
7. J. Maldacena and A. Zhiboedov, Constraining conformal field theories with a slightly broken higher spin symmetry, *Class. Quantum Grav.* **30**, 104003 (2013), doi:10.1088/0264-9381/30/10/104003, arXiv:1204.3882 [hep-th].
8. O. Aharony, G. Gur-Ari and R. Yacoby, Correlation functions of large N Chern–Simons-matter theories and bosonization in three dimensions, *J. High Energy Phys.* **1212**, 028 (2012), doi:10.1007/JHEP12(2012)028, arXiv:1207.4593 [hep-th].
9. O. Aharony, S. Giombi, G. Gur-Ari, J. Maldacena and R. Yacoby, *J. High Energy Phys.* **1303**, 121 (2013), doi:10.1007/JHEP03(2013)121, arXiv:1211.4843 [hep-th].
10. S. Jain, S. Minwalla, T. Sharma, T. Takimi, S. R. Wadia and S. Yokoyama, Phases of large N vector Chern–Simons theories on $S^2 \times S^1$, *J. High Energy Phys.* **1309**, 009 (2013), doi:10.1007/JHEP09(2013)009, arXiv:1301.6169 [hep-th].
11. S. Jain, S. Minwalla and S. Yokoyama, Chern–Simons duality with a fundamental boson and fermion, *J. High Energy Phys.* **1311**, 037 (2013), doi:10.1007/JHEP11(2013)037, arXiv:1305.7235 [hep-th].
12. Y. Dandekar, M. Mandlik and S. Minwalla, Poles in the S-matrix of relativistic Chern–Simons matter theories from quantum mechanics, *J. High Energy Phys.* **1504**, 102 (2015), doi:10.1007/JHEP04(2015)102, arXiv:1407.1322 [hep-th].

[d]S. Minwalla and S. R. Wadia, unpublished.

13. O. Aharony, P. Narayan and T. Sharma, On monopole operators in supersymmetric Chern–Simons-matter theories, *J. High Energy Phys.* **1505**, 117 (2015), doi:10.1007/JHEP05(2015)117, arXiv:1502.00945 [hep-th].
14. O. Aharony, Baryons, monopoles and dualities in Chern–Simons-matter theories, *J. High Energy Phys.* **1602**, 093 (2016), doi:10.1007/JHEP02(2016)093, arXiv:1512.00161 [hep-th].
15. S. R. Wadia, On the Dyson–Schwinger equations approach to the large N limit: Model systems and string representation of Yang–Mills theory, *Phys. Rev. D* **24**, 970 (1981), doi:10.1103/PhysRevD.24.970.
16. D. Bak, R. Jackiw and S. Y. Pi, NonAbelian Chern–Simons particles and their quantization, *Phys. Rev. D* **49**, 6778 (1994), doi:10.1103/PhysRevD.49.6778, arXiv:hep-th/9402057.
17. S. N. M. Ruijsenaars, The Aharonov–Bohm effect and scattering theory, *Ann. Phys.* **146**, 1 (1983), doi:10.1016/0003-4916(83)90051-9.
18. K. Inbasekar, S. Jain, S. Mazumdar, S. Minwalla, V. Umesh and S. Yokoyama, Unitarity, crossing symmetry and duality in the scattering of $\mathcal{N}=1$ SUSY matter Chern–Simons theories, *J. High Energy Phys.* **1510**, 176 (2015), doi:10.1007/JHEP10(2015)176, arXiv:1505.06571 [hep-th].

Lorentz Invariant CPT Breaking in the Dirac Equation

Kazuo Fujikawa
RIKEN Nishina Center, Wako 351-0198, Japan

If one modifies the Dirac equation in momentum space to $[\not{p} - m - \Delta m(\theta(p_0) - \theta(-p_0))\theta(p_\mu^2)]\psi(p) = 0$, the symmetry of positive and negative energy eigenvalues is lifted by $m \pm \Delta m$ for a small Δm. The mass degeneracy of the particle and antiparticle is thus removed in a Lorentz invariant manner since the combinations $\theta(\pm p_0)\theta(p_\mu^2)$ with step functions are manifestly Lorentz invariant. We explain an explicit construction of this CPT breaking term in coordinate space, which is Lorentz invariant but nonlocal at the distance scale of the Planck length. The application of this Lorentz invariant CPT breaking mechanism to the possible mass splitting of the neutrino and antineutrino in the Standard Model is briefly discussed.

1. Introduction

It is timely to have a memorial meeting for the 90th birthday of the late Professor Abdus Salam when those who directly worked with him are still active in physics. On my part, I first met Salam at a small gathering in the suburb of Oxford in 1972. This was a nice gathering, and M. Veltman talked on a work with G. 't Hooft and I also talked on the so-called R_ξ gauge written with B. W. Lee and A. I. Sanda. Salam gave a talk on his famous unification paper. At that time, people were talking about the mass value of the W boson at least as heavy as 37.5 GeV; this value, which is not large by the present standard, was regarded to be absurdly large at that time. Note that this was before the discovery of the charmed quark. To defend this large mass needed in his model, Salam was saying, "if you insist on simplicity, Newton's theory of gravity which requires a single scalar potential is much simpler than Einstein's theory which requires 10 metric variables. But we prefer Einstein's theory because of its beautiful symmetry. The decisive factor is not superficial simplicity but rather deep symmetry behind any theory." I was very impressed by his verdict and I still remember it.

The greatest contribution of Salam is probably the proposal of the Standard Model, namely, the Weinberg–Salam–Glashow Model or Weinberg–Salam theory with Higgs mechanism,[1–3] among his numerous contributions to physics. The Standard Model is very successful so far, and people discuss the possible extensions of the model. I would like to discuss a breaking scheme of CPT symmetry in a Lorentz invariant non-local theory which produces a particle–antiparticle mass splitting, and as an application of this CPT breaking scheme, I briefly discuss the possible mass splitting between the neutrino and antineutrino in a minimal extension of the Standard Model.

I start with the **CPT Theorem:** This theorem states that CPT symmetry is valid for any Lorentz invariant local theory defined by a hermitian Lagrangian with normal spin-statistics relation.

The original references on the CPT theorem are:[4,5] W. Pauli, in *Niels Bohr and the Development of Physics*, W. Pauli (ed.) (Pergamon Press, New York, 1955), and G. Lüders, *On the Equivalence of Invariance under Time-Reversal and under Particle-Antiparticle Conjugation for Relativistic Field Theories*, Det. Kong. Danske Videnskabernes Selskab, Mat.-fys. Medd. **28** (5) (1954).

The title of this second reference shows clearly what we mean by the CPT theorem, namely, the equivalence of PT symmetry (*strong reflection*) with C symmetry. More generally, CPT symmetry implies the degeneracy of the masses of the particle and antiparticle. The CPT theorem, which is valid for any sensible local Lorentz invariant theory, is very general. Nevertheless the possible breaking of CPT theorem has been discussed by many people in the past. One may count two representative necessary conditions to break the CPT theorem:

1. Non-local theory,
2. Lorentz symmetry breaking.

The Lorentz symmetry breaking scheme is more common in literature, but we are interested in the possible breaking of CPT in a Lorentz invariant non-local theory. To be specific, we mainly discuss the modified Dirac equation

$$[\not{p} - m - \Delta m(\theta(p_0) - \theta(-p_0))\theta(p_\mu^2)]\psi(p) = 0 \tag{1}$$

with the step function $\theta(x) = 1$ for $x > 0$ and $\theta(x) = 0$ for $x < 0$. The combinations $\theta(\pm p_0))\theta(p_\mu^2)$ are manifestly Lorentz invariant since the signature of the time-component of any time-like vector has a Lorentz invariant meaning. Obviously, this modified Dirac equation produces the mass splitting between the particle and antiparticle, $m \pm \Delta m$ for small Δm, and thus leads to the breaking of CPT symmetry. In passing, we emphasize that the breaking of CPT symmetry is a necessary condition but not sufficient for the particle–antiparticle mass splitting. For example, T breaking with C and P intact leads to CPT breaking but no mass splitting between the particle and antiparticle.

We explain how to reproduce Eq. (1) as a consequence at energy scale $E/M_P \ll 1$ of the Lorentz invariant CPT breaking at the Planck scale M_P.[6] As is clear in the statement of the CPT theorem, we have to sacrifice some basic assumptions in current physics to realize CPT symmetry breaking. We discuss the interesting aspects of the present CPT symmetry breaking scheme as well as the remaining issues to be resolved. In this connection, it may be appropriate to mention that the more common CPT symmetry breaking on the basis of Lorentz symmetry breaking also encounters severe difficulties once one attempts to realize the mass splitting between the particle and antiparticle.[7]

2. Lorentz Invariant CPT Breaking with Particle–Antiparticle Mass Splitting

We have studied the models of Lorentz invariant CPT breaking which give rise to the particle and antiparticle mass splitting in the Dirac equation in a series of papers.[8,96] We would like to discuss the essence of these works.

We start with a free Dirac action

$$S = \int d^4x\bar{\psi}(x)[i\gamma^\mu\partial_\mu - m]\psi(x), \tag{2}$$

and consider to add a small Lorentz invariant non-local term to break CPT symmetry. We examine the hermitian combination with a real μ,[8]

$$\int d^4x d^4y[\theta(x^0-y^0)-\theta(y^0-x^0)]\delta((x-y)^2-l^2)[i\mu\bar{\psi}(x)\psi(y)]. \tag{3}$$

The transformation property of the operator part in this expression is given using *spin-statistics theorem* by

$$\begin{aligned}
&\mathrm{C}:\ i\mu\bar{\psi}(x)\psi(y)\to i\mu\bar{\psi}(y)\psi(x),\\
&\mathrm{P}:\ i\mu\bar{\psi}(x^0,\vec{x})\psi(y^0,\vec{y})\to i\mu\bar{\psi}(x^0,-\vec{x})\psi(y^0,-\vec{y}),\\
&\mathrm{T}:\ i\mu\bar{\psi}(x^0,\vec{x})\psi(y^0,\vec{y})\to -i\mu\bar{\psi}(-x^0,\vec{x})\psi(-y^0,\vec{y}),
\end{aligned} \tag{4}$$

and the overall transformation property of the combination in (3) is confirmed to be

$$C=-1,\ P=1,\ T=1. \tag{5}$$

Namely,

$$C=CP=CPT=-1, \tag{6}$$

and thus all symmetries which may protect the equality of the masses of the particle and antiparticle are broken.

It is thus interesting to examine a Lorentz invariant and hermitian action[8]

$$\begin{aligned}
S = \int d^4x\{&\bar{\psi}(x)i\gamma^\mu\partial_\mu\psi(x) - m\bar{\psi}(x)\psi(x)\\
&-\int d^4y[\theta(x^0-y^0)-\theta(y^0-x^0)]\delta((x-y)^2-l^2)[i\mu\bar{\psi}(x)\psi(y)]\}.
\end{aligned} \tag{7}$$

The Dirac equation is replaced by

$$i\gamma^\mu\partial_\mu\psi(x) = m\psi(x) + i\mu\int d^4y[\theta(x^0-y^0)-\theta(y^0-x^0)]\delta((x-y)^2-l^2)\psi(y). \tag{8}$$

By inserting an ansatz for the possible solution

$$\psi(x) = e^{-ipx}U(p), \tag{9}$$

we have

$$\not{p}U(p) = mU(p) + i\mu\int d^4y[\theta(x^0 - y^0) - \theta(y^0 - x^0)]\delta((x-y)^2 - l^2)e^{-ip(y-x)}U(p) = mU(p) + i\mu[f_+(p) - f_-(p)]U(p), \quad (10)$$

where

$$f_\pm(p) \equiv \int d^4z_1 e^{\pm ipz_1}\theta(z_1^0)\delta((z_1)^2 - l^2), \quad (11)$$

is the Lorentz invariant form factor.

For the *space-like* p, we go to the frame where $p_0 = 0$, and we have $f_+(p) = f_-(p)$ and no mass splitting, and thus no tachyons.

For the *time-like* p, we go to the frame where $\vec{p} = 0$, and the eigenvalue equation

$$p_0 = \gamma_0\{m + i\mu[f_+(p_0) - f_-(p_0)]\}, \quad (12)$$

is written as

$$p_0 = \gamma_0\left[m - 4\pi\mu\int_0^\infty dz\frac{z^2\sin[p_0\sqrt{z^2+l^2}]}{\sqrt{z^2+l^2}}\right], \quad (13)$$

where we used the explicit form in (11).

This eigenvalue equation under $p_0 \to -p_0$ becomes

$$-p_0 = \gamma_0\left[m + 4\pi\mu\int_0^\infty dz\frac{z^2\sin[p_0\sqrt{z^2+l^2}]}{\sqrt{z^2+l^2}}\right]. \quad (14)$$

By sandwiching this equation by γ_5, we have

$$p_0 = \gamma_0\left[m + 4\pi\mu\int_0^\infty dz\frac{z^2\sin[p_0\sqrt{z^2+l^2}]}{\sqrt{z^2+l^2}}\right], \quad (15)$$

which is not identical to the original equation in (13).

In other words, if p_0 is the solution of the original equation, $-p_0$ cannot be the solution of the original equation for $\mu \neq 0$. The last term in the Lagrangian (7) with C=CP=CPT=-1 thus splits the particle and antiparticle masses. As a crude estimate of the mass splitting, one may assume that the term with μ gives a much smaller contribution than m and solve these equations iteratively. If the particle mass is chosen at

$$p_0 \simeq m - 4\pi\mu\int_0^\infty dz\frac{z^2\sin[m\sqrt{z^2+l^2}]}{\sqrt{z^2+l^2}}, \quad (16)$$

then the antiparticle mass is estimated at

$$p_0 \simeq m + 4\pi\mu\int_0^\infty dz\frac{z^2\sin[m\sqrt{z^2+l^2}]}{\sqrt{z^2+l^2}}. \quad (17)$$

We here mention that the factor $f_{\pm}(p)$ in (11) is mathematically related to the two-point Wightman function,

$$\langle 0|\phi(x)\phi(y)|0\rangle = \int d^4p e^{i(x-y)p}\theta(p^0)\delta(p^2 - m^2), \tag{18}$$

if one replaces the coordinate and momentum, $x^\mu \leftrightarrow p^\mu$. This knowledge is useful to analyze our problem. For example, the Wightman function is quadratically divergent at short distances, $(x-y)^\mu \to 0$, which implies that our form factor is quadratically divergent at infrared, $p^\mu \to 0$. To avoid this infrared divergence, we replace the non-local factor in (3) to[6]

$$\delta\left((x-y)^2 - l^2\right) \Rightarrow \Delta_l(x-y) \equiv \delta\left((x-y)^2 - l^2\right) - \delta\left((x-y)^2 - l'^2\right) \tag{19}$$

with $l'^2 \to 0$ in practical applications.

With this replacement, for time-like $p^2 > 0$, one may obtain at the frame $\vec{p} = 0$,

$$p_0 = \gamma^0[m + f(p_0)], \tag{20}$$

with

$$\begin{aligned} f(p_0) &\equiv i[f_+(p_0) - f_-(p_0)] \\ &= -4\mu\pi \int_0^\infty dz \left\{ \frac{z^2 \sin[p_0\sqrt{z^2+l^2}]}{\sqrt{z^2+l^2}} - \frac{z^2 \sin[p_0\sqrt{z^2}]}{\sqrt{z^2}} \right\}. \end{aligned} \tag{21}$$

For space-like $p^2 < 0$, one can confirm that the CPT violating term vanishes, $f(p) = 0$, by choosing $p_\mu = (0, \vec{p})$.

Evaluation of mass splitting:

The CPT breaking factor in (21), which is now written as $f(p)$ for simplicity, is rewritten as[6]

$$\begin{aligned} f(p) &= 4\pi\mu l^2[\theta(p_0) - \theta(-p_0)]\theta(p^2) \\ &\quad \times \left\{ \int_1^\infty du \frac{1}{2u(\sqrt{u^2-1}+u)^2} \sin(|p_0|lu) \right. \\ &\quad \left. - \frac{1}{2}\int_0^1 du \frac{\sin(|p_0|lu)}{u} + \int_0^1 du u \sin(|p_0|lu) + \frac{1}{2}\int_0^\infty du \frac{\sin(u)}{u} \right\}, \end{aligned} \tag{22}$$

namely, the CPT violating term is characterized by the quantity

$$\mu l^2, \tag{23}$$

which has the dimensions of mass. If one chooses the non-locality length l at the Planck length, we have $|p_0|l \ll 1$ for the energy scale at laboratory, and the formula (22) gives a manifestly Lorentz invariant

$$f(p) \simeq \pi^2\mu l^2[\theta(p_0) - \theta(-p_0)]\theta(p^2). \tag{24}$$

Note that no term linear in $|p_0|l$ arises. Thus the particle–antiparticle mass splitting is given by

$$2\Delta m \simeq 2\pi^2 \mu l^2. \tag{25}$$

Our CPT violating term $f(p_0)$ is odd in p_0 and $f(\pm 0) = \pm\Delta m$ but $f(0) = 0$.

The Lorentz invariant non-local factor in (3) after the modification in (19),

$$\left[\theta(x^0 - y^0) - \theta(y^0 - x^0)\right]\left[\delta((x-y)^2 - l^2) - \delta((x-y)^2)\right], \tag{26}$$

cancels out the infinite time-like volume effect and eliminates the quadratic infrared divergence completely. In effect, the non-locality is limited within the fluctuation around the tip of the light-cone characterized by the length scale l, which we choose to be the Planck length.

By setting

$$l = 1/M_P, \quad \mu = M^3, \tag{27}$$

with the Planck mass M_P, the particle–antiparticle mass splitting is given by

$$2\Delta m = 2\pi^2 \mu l^2 = 2\pi^2 M (M/M_P)^2. \tag{28}$$

If one chooses $M \sim 10^9$ GeV, the particle–antiparticle mass splitting becomes of the order of the observed neutrino mass (difference) $\sim$0.1 eV.[10]

3. Quantization

As for the quantization, we adopted the path integral formulation on the basis of Schwinger's action principle in Ref. 8, which formally integrates the equations of motion.[11] This scheme which emphasizes the equations of motion is analogous to Yang–Feldman formalism[12] in the operator formulation. We adopt this path integral formulation throughout this paper, since the conventional canonical quantization is not defined for theory non-local in time.

For the non-locality of the order of the Planck length, our Lorentz invariant non-local CPT breaking factor (24) is written as

$$f(p) = \Delta m[\theta(p_0) - \theta(-p_0)]\theta(p^2). \tag{29}$$

The propagator of the fermion in path integral on the basis of Schwinger's action principle is then given by,[11]

$$\langle T^\star \psi(x)\bar{\psi}(y)\rangle = \int \frac{d^4p}{(2\pi)^4} e^{-ip(x-y)} \frac{i}{\not{p} - m + i\epsilon - \Delta m[\theta(p_0) - \theta(-p_0)]\theta(p^2)}, \tag{30}$$

where $T^\star$ stands for the covariant T-product which avoids the precise coincident time $x^0 = y^0$. This propagator shows that the extra terms are ignored for $p = p_0 \rightarrow \pm\infty$. In fact one can confirm that[6]

$$f(p) = i[f_+(p) - f_-(p)] \rightarrow 0, \tag{31}$$

for $p = p_0 \to \pm\infty$ in Minkowski space without any approximation. The propagator for Minkowski momentum is thus well behaved and the effects of non-locality are mild and limited. If one recalls the vanishing of

$$\lim_{p_0 \to \text{large}} \frac{i}{\not{p} - m + i\epsilon - \Delta m[\theta(p_0) - \theta(-p_0)]\theta(p^2)} = 0 \tag{32}$$

is the criterion of the T-product in (30) in *ordinary theory*, our non-local action almost defines the canonical T-product. In the Bjorken–Johnson–Low prescription, however, one generally requires the stronger condition of vanishing for $|p_0| \to \infty$ in the complex plane of p_0.[11]

If one considers the Euclidean amplitude obtained from the Minkowski amplitude by Wick rotation, our propagator, which contains trigonometric functions, has undesirable behavior after the Wick rotation such as

$$\sin p_0 z \to i \sinh p_4 z, \tag{33}$$

and the exponentially divergent behavior is generally induced in the extra terms. In this sense, the effects of non-locality become non-negligible. However, one might still argue that higher order effects in field theory defined in Minkowski space are in principle analyzed in Minkowski space and, if that is the case, our propagator suggests the ordinary renormalizable behavior.

This issue is left for the future study.

4. Neutrino–Antineutrino Mass Splitting

The neutrino masses are not precisely specified by the original Standard Model and thus may provide a window to "brave New World". It may be allowed to entertain the idea of the possible CPT breaking in the neutrino mass sector and apply our scheme of Lorentz invariant CPT breaking. Since the Standard Model is very successful, we incorporate the conditions:

(a) Lorentz invariance
(b) SU(2)xU(1) gauge invariance
(c) C, CP and CPT breaking
(d) Non-local within a distance scale of the Planck length.

The Standard Model Lagrangian relevant to our discussion of the electron sector is given by

$$\begin{aligned}\mathcal{L} &= i\overline{\psi}_L\gamma^\mu\left(\partial_\mu - igT^aW^a_\mu - i\frac{1}{2}g'Y_LB_\mu\right)\psi_L \\ &\quad + i\overline{e}_R\gamma^\mu(\partial_\mu + ig'B_\mu)e_R + i\overline{\nu}_R\gamma^\mu\partial_\mu\nu_R \\ &\quad + \left[-\frac{\sqrt{2}m_e}{v}\overline{e}_R\phi^\dagger\psi_L - \frac{\sqrt{2}m_D}{v}\overline{\nu}_R\phi_c^\dagger\psi_L - \frac{m_R}{2}\nu_R^TC\nu_R + \text{h.c.}\right],\end{aligned} \tag{34}$$

The test of CPT symmetry in neutrino oscillation has been discussed in some detail in Ref. 13. More concrete analyses of a possible sizable neutrino–antineutrino mass splitting have been performed in connection with the LSND experiment in,[14] but such a large mass splitting of the order of 1eV appears to be disfavored by experiments.[15] An implication of the possible neutrino mass splitting on the baryon asymmetry was also discussed.[6,16] At this moment, it is known that there is a small mass discrepancy between the antineutrino in the reactor experiment and the neutrino from the sun, which is about 2σ discrepancy.[17] This may indicate that the neutrino oscillation is a good testing ground of CPT symmetry.

References

1. S. Weinberg, *Phys. Rev. Lett.* **19**, 1264 (1967).
2. A. Salam, in *Elementary Particle Theory*, ed. N. Svartholm (Almquist and Wiksell, Stockholm, 1968), p. 367.
3. S.L. Glashow, *Nucl. Phys.* **22**, 579 (1961).
4. W. Pauli, *Niels Bohr and the Development of Physics*, W. Pauli (ed.), (Pergamon Press, New York, 1955).
5. G. Lüders, *Mat. Fys. Medd. Dan. Vid. Selsk.* **28**(5), 1 (1954).
6. K. Fujikawa and A. Tureanu, *Phys. Lett. B* **743**, 39 (2015); *Mod. Phys. Lett. A* **30**, 1530016 (2015).
7. O. M. Del Cima, D. H. T. Franco, A. H. Gomes, J. M. Fonseca, O. Piguet, *Phys. Rev. D* **85**, 065023 (2012).
8. M. Chaichian, K. Fujikawa and A. Tureanu, *Phys. Lett. B* **712**, 115 (2012).
9. M. Chaichian, K. Fujikawa and A. Tureanu, *Phys. Lett. B* **718**, 178 (2012); *Phys. Lett. B* **718**, 1500 (2013).
10. K. A. Olive *et al.* (Particle Data Group), *Chin. Phys. C* **38**, 090001 (2014), and 2015 update.
11. K. Fujikawa, *Phys. Rev. D* **70**, 085006 (2004). As for the Bjorken–Johnson–Low method, see also Appendix in K. Fujikawa and P. van Nieuwenhuizen, *Annals Phys.* **308**, 78 (2003).
12. C. N. Yang and D. Feldman, *Phys. Rev.* **79**, 972 (1950).
13. V. D. Barger, S. Pakvasa, T. J. Weiler and K. Whisnant, *Phys. Rev. Lett.* **85**, 5055 (2000).
14. H. Murayama and T. Yanagida, *Phys. Lett. B* **520**, 263 (2001).
    G. Barenboim, L. Borissov and J. Lykken, *Phys. Lett. B* **534**, 106 (2002).
    S. M. Bilenky, M. Freund, M. Lindner, T. Ohlsson and W. Winter, *Phys. Rev. D* **65**, 073024 (2002).
    G. Barenboim and J. D. Lykken, *Phys. Rev. D* **80**, 113008 (2009).
15. P. Adamson *et al.* (MINOS Collaboration), *Phys. Rev. Lett.* **108**, 191801 (2012).
16. G. Barenboim, L. Borissov, J. D. Lykken and A. Y. Smirnov, *JHEP* **0210**, 001 (2002).
17. Private communications from A. Suzuki and A.Y. Smirnov.

Cosmology and Supergravity

S. Ferrara
Theoretical Physics Department, CERN,
CH-1211 Geneva 23, Switzerland
INFN-Laboratori Nazionali di Frascati,
Via Enrico Fermi 40, 00044 Frascati, Italy
Department of Physics and Astronomy, U.C.L.A.,
Los Angeles, CA 90095-1547, USA
sergio.ferrara@cern.ch

A. Kehagias
Physics Division, NTU Athens,
15781 Zografou, Athens, Greece
kehagias@central.ntua.gr

A. Sagnotti
Scuola Normale Superiore and INFN,
Piazza dei Cavalieri 7 56126 Pisa, Italy
sagnotti@sns.it

Abdus Salam was a true master of 20th Century Theoretical Physics. Not only was he a pioneer of the Standard Model (for which he shared the Nobel Prize with S. Glashow and S. Weinberg), but he also (co)authored many other outstanding contributions to the field of Fundamental Interactions and their unification. In particular, he was a major contributor to the development of supersymmetric theories, where he also coined the word "Supersymmetry" (replacing the earlier "Supergauges" drawn from String Theory). He also introduced the basic concept of "Superspace" and the notion of "Goldstone Fermion"(Goldstino). These concepts proved instrumental for the exploration of the ultraviolet properties and for the study of spontaneously broken phases of super Yang–Mills theories and Supergravity. They continue to play a key role in current developments in Early-Universe Cosmology. In this contribution we review models of inflation based on Supergravity with spontaneously broken local supersymmetry, with emphasis on the role of nilpotent superfields to describe a de Sitter phase of our Universe.

Keywords: Supersymmetry; supergravity; cosmology; inflation; string theory.

1. Introduction

Supergravity[1] combines Supersymmetry with General Relativity (GR). This brings about scalar fields, some of which can play a natural role in the Early Universe. Nowadays it is well established that inflationary Cosmology is accurately described via the evolution of a single real scalar field, the *inflaton*, in a Friedmann, Lemaître,

Robertson, Walker (FLRW) geometry.[2] A scalar field associated with the Higgs particle was also recently discovered at LHC,[3] confirming the interpretation of the Standard Model as a spontaneously broken phase (BEH mechanism) of a non-Abelian Yang–Mills theory.[4] There is thus some evidence that Nature is inclined to favor, both in Cosmology and in Particle Physics, theories with scalar degrees of freedom, albeit in diverse ranges of energy scales.

Interestingly, there is also a cosmological model where *inflaton* and *Higgs* fields are identified: this is the Higgs inflation model of Ref. 5, which rests on a non-minimal coupling $h^2 R$ of the Higgs field h to gravity. Another well-known example rests on an $R + R^2$ extension of General Relativity (GR). This is the Starobinsky model of inflation,[6,7] which is also conformally equivalent to GR coupled to a scalar field, the *scalaron*,[8] with the special scalar potential

$$V = V_0 \left(1 \; - \; e^{-\sqrt{\frac{2}{3}}\,\phi}\right)^2 , \quad V_0 \sim 10^{-9} \text{ in Planck units.} \tag{1}$$

These two models (and also a more general class) give identical predictions[9] for the slow-roll parameters ϵ and η, which are determined by the potential according to

$$\epsilon \;=\; \frac{M_P^2}{2}\left(\frac{V'}{V}\right)^2 , \quad \eta \;=\; M_P^2\,\frac{V''}{V} \,. \tag{2}$$

The spectral index of scalar perturbations (scalar tilt) and the tensor-to-scalar ratio turn out to be

$$n_s \;=\; 1 \;-\; 6\,\epsilon \;+\; 2\,\eta \;\simeq\; 1 \;-\; \frac{2}{N} , \quad r \;=\; 16\,\epsilon \;\simeq\; \frac{12}{N^2} , \tag{3}$$

where

$$N \;=\; \frac{1}{M_P^2}\int_{\phi_{\rm end}}^{\phi} \frac{V}{V'}\, d\phi \tag{4}$$

is the total number of e-folds of inflation.

An interesting modification of the Starobinsky potential, suggested by its embedding in $R + R^2$ Supergravity,[10,11] involves a deformation parameter α and reads[11,12]

$$V_\alpha = V_0 \left(1 \; - \; e^{-\sqrt{\frac{2}{3\alpha}}\,\phi}\right)^2 \,. \tag{5}$$

It gives the same result of Eq. (3) for n_s, but the tensor-to-scalar ratio is now

$$r \;\simeq\; \frac{12\,\alpha}{N^2} \,. \tag{6}$$

This family of models provides an interpolation between the Starobinsky model (for $\alpha = 1$) and Linde's chaotic inflation model[13] with a quadratic potential (in the limit $\alpha \to \infty$). The chaotic inflation model leads again to the scalar tilt (3), but now the tensor-to-scalar ratio becomes

$$r \;\simeq\; \frac{8}{N} \,. \tag{7}$$

The recent 2015 data analysis from Planck[14] and BICEP2[15] favors $n_s \approx 0.97$ and $r < 0.1$, and thus the Starobinsky model, which lies well within the allowed parameter space due to the additional $1/N$ suppression factor r present in Eq. (3) as compared to Eq. (7).

The form (5) for V_α can be further generalized, allowing for an arbitrary, monotonically increasing function $f\left(\tanh\frac{\varphi}{\sqrt{6\alpha}}\right)$, such that

$$V_\alpha = V_0\, f\left(\tanh\frac{\varphi}{\sqrt{6\alpha}}\right)^2, \quad \varphi \to \infty: \quad f\left(\tanh\frac{\varphi}{\sqrt{6\alpha}}\right) \to 1 - e^{-\sqrt{\frac{2}{3\alpha}}\,\phi} + \cdots. \tag{8}$$

These modifications led to the concept of α-attractors.[12]

This contribution is organized as follows. In Sec. 2 we describe the single-field inflation in Supergravity, in Sec. 3 we discuss inflation and supersymmetry breaking and in Sec. 4 we present some minimal Supergravity models of inflation. Nilpotent superfields and sgoldstino-less models are reviewed in Sec. 5, in Sec. 6 we discuss higher-curvature Supergravity and its dual standard Supergravity description, in Sec. 7 orthogonal nilpotent superfields are explored and Sec. 8 contains our conclusions and outlooks. Finally in App. A we briefly review constraint superfields which preserve $\mathcal{N} = 1$ supersymmetry.

2. Single-Field Inflation in Supergravity

We can now describe how $\mathcal{N} = 1$ Supergravity can accommodate these "single-field" inflationary models, explaining how to embed the *inflaton* φ in a general Supergravity theory coupled to matter in an FLRW geometry and the role of its superpartners. Under the assumption that no additional Supersymmetry ($\mathcal{N} \geq 2$) is restored in the Early Universe, the most general $\mathcal{N} = 1$ extension of GR is obtained by coupling the graviton multiplet $(2, 3/2)$ to a certain number of chiral multiplets $(1/2, 0, 0)$, whose complex scalar fields are denoted by $z^i, i = 1, \ldots, N_s/2$ and to (gauge) vector multiplets $(1, 1/2)$, whose vector fields are denoted by A^Λ_μ ($\Lambda = 1, \ldots, N_V$). These multiplets can acquire supersymmetric masses, and in this case the massive vector multiplet becomes $(1, 2(1/2), 0)$, eating a chiral multiplet in the supersymmetric version of the BEH mechanism.

For Cosmology, the relevant part of the Lagrangian[16,17] is the sector that couples the scalar fields to the Einstein–Hilbert action, described by

$$\mathcal{L} = -R - \partial_i \partial_{\bar{\jmath}} K\, D_\mu z^i\, D_\nu \bar{z}^{\bar{\jmath}}\, g^{\mu\nu} - V(z, \bar{z}) + \cdots, \tag{9}$$

where K is the Kähler potential of the σ-model scalar geometry and the "dots" hide fermionic terms and gauge interactions. The scalar covariant derivative is

$$D_\mu z^i = \partial_\mu z^i + \delta_\Lambda z^i A^\Lambda_\mu, \tag{10}$$

where the $\delta_\Lambda z^i$ are Killing vectors. This term allows to write massive vector multiplets *à la* Stueckelberg. The scalar potential is

$$V(z^i, \overline{z}^{\overline{i}}) = e^G \left[G_i \, G_{\overline{\jmath}} \, (G^{-1})^{i\overline{\jmath}} - 3 \right] + \frac{1}{2} (Ref_{\Lambda\Sigma})^{-1} D_\Lambda \, D_\Sigma, \tag{11}$$

where, in terms of the superpotential $W(z^i)$,

$$G = K + \log |W|^2, \quad G_{i\overline{j}} = \partial_i \, \partial_{\overline{j}} \, K. \tag{12}$$

The first and third non-negative terms in Eq. (11) are usually referred to as "F" and "D" term contributions: together with the second, negative term, they encode the option of attaining unbroken Supersymmetry in Anti-de Sitter space. Alternatively, the potential can be recast in the more compact form

$$V(z^i, \overline{z}^{\overline{\imath}}) = F_i F^i + D_\Lambda D^\Lambda - 3 \, |W|^2 \, e^K, \tag{13}$$

where

$$F_i = e^{\frac{K}{2}} \left(W K_{,i} + W_{,i} \right), \quad D_\Lambda = G_{,i} \, \delta_\Lambda z^i. \tag{14}$$

The D-term potential can endow a vector multiplet with a supersymmetric mass term, and can also give rise to a de Sitter phase, thanks to its non-negative contribution to the potential. Only F-breaking terms can thus give AdS phases. The (field dependent) matrices $\mathrm{Re}f_{\Lambda\Sigma}, \mathrm{Im}f_{\Lambda\Sigma}$ provide the normalization of the terms quadratic in Yang–Mills curvatures. Their role in Cosmology deserves to be investigated further, since they give direct couplings of the *inflaton* to matter, which are relevant for the epoch of reheating.

3. Inflation and Supersymmetry Breaking

In a given phase, unbroken Supersymmetry requires

$$F_i = D^\Lambda = 0, \tag{15}$$

so that

$$V = -3 \, |W|^2 \, e^K. \tag{16}$$

These are Minkowski or AdS phases depending on whether or not W vanishes. On the other hand, supersymmetry is broken if at least one of the F_i or D^Λ does not vanish. In phases with broken Supersymmetry one can have maximally symmetric AdS, dS or Minkowski vacua, so that one can accommodate both the inflationary phase (dS) and the subsequent Particle Physics (Minkowski) phase. However, it is not trivial to construct corresponding models, since the two scales are very different if Supersymmetry is at least partly related to the Hierarchy problem.

In view of the negative term present in the scalar potential (11), it might seem impossible (or at least not natural) to retrieve a de Sitter phase for large values of a scalar field to be identified with the *inflaton*. The supersymmetric versions of the $R + R^2$ (Starobinsky) model show how this puzzle is resolved: either the theory has (with F-terms) a no-scale structure, which makes the potential positive along the inflationary trajectory,[18] or the potential is a pure D-term and is therefore positive.[19]

These models contain two chiral superfields (T, S),[20,21] as in the old minimal version of $R+R^2$ Supergravity,[18] or one massive vector multiplet,[10,11] as in the new minimal version, and attain unbroken Supersymmetry in a Minkowski vacuum at the end of inflation.

In the framework of nilpotent superfield inflation,[22] some progress was recently made[23,24] on the problem of embedding two different supersymmetry breaking scales in the inflationary potential. The multiplet S, which does not contain the *inflaton* (T multiplet), is replaced by a nilpotent superfield satisfying

$$S^2 = 0. \tag{17}$$

This condition eliminates the *sgoldstino* scalar from the theory, but its F-component still drives inflation, or at least participates in it. This mechanism was first applied to the Starobinsky model, replacing the S field by a Volkov–Akulov nilpotent field[22] and then to general F-term induced inflationary models.[25] Although the examples are so far restricted to the $\mathcal{N} = 1 \to \mathcal{N} = 0$ breaking in four-dimensional supergravity, these types of construction are potentially very instructive for String Theory, where one readily looses control of the vacuum in the presence of broken supersymmetry.[26] Orientifold vacua[27] provide a natural and interesting entry point into this intricate dynamics, via the phenomenon of "brane SUSY breaking".[28] This rests on non-BPS combinations of branes and orientifolds that are individually BPS, and its simplest ten-dimensional setting was related to non-linear supersymmetry in Ref. 29. Recent work, starting from Ref. 25, linked it more clearly to the super-Higgs effect in Supergravity,[16] and also to the KKLT scenario of Ref. 30. Let us conclude this section, however, by recalling that a first attempt to make use of the nilpotent Volkov–Akulov multiplet in Cosmology, identifying the *inflaton* with the *sgoldstino*, was made in Ref. 31.

4. Minimal Models for Inflation and Supergravity

This class includes models where the inflaton is identified with the sgoldstino and only one chiral multiplet T is used. However, the $f(R)$ Supergravity models[32] yield potentials that either have no plateau or, when they do, lead to AdS rather than to dS phases.[34,35] This also reflects a no-go theorem.[33]

A way out of this situation was recently found with "α-scale Supergravity":[36] adding two superpotentials $W_+ + W_-$ which separately give a flat potential along the inflaton (ReT) direction can result in a de Sitter plateau for large ReT. The problem with these models is that the inflaton trajectory is unstable in the ImT direction, but only for small inflaton field: modifications to the superpotential are advocated to generate a satisfactory inflationary potential. For single-field models and related problems, see also Ref. 37. $R+R^2$ Supergravity, D-term inflation,[11,38] α-attractor scenarios,[39] no-scale inflationary models,[20] and α-scale models[36] have a nice $SU(1,1)/U(1)$ hyperbolic geometry for the inflaton superfield, with

$$R_\alpha = -\frac{2}{3\alpha}, \quad n_s \simeq 1 - \frac{2}{N}, \quad r \simeq \frac{12\,\alpha}{N^2}, \tag{18}$$

where R_α is the curvature of the scalar manifold.

4.1. *D-term inflation*

An appealing and economical class of models allows to describe any potential of a single scalar field which is the square of a real function:[11]

$$V(\varphi) = \frac{g^2}{2} P^2(\varphi). \tag{19}$$

These are the D-term models, which describe the self-interactions of a massive vector multiplet whose scalar component is the inflaton. Up to an integration constant (the Fayet–Iliopoulos term), the potential is fixed by the geometry, since the Kähler metric is

$$ds^2 = d\varphi^2 + \Big(P'(\varphi)\Big)^2 da^2. \tag{20}$$

After gauge fixing, the field a is absorbed by the vector, via $da + gA$, giving rise to a mass term $\frac{g^2}{2}\Big(P'(\varphi)\Big)^2 A_\mu^2$ (BEH mechanism). In particular, the Starobinsky model corresponds to

$$P(\varphi) = 1 - e^{-\sqrt{\frac{2}{3}}\,\varphi}, \tag{21}$$

but in all these examples there is no superpotential and only a de Sitter plateau is possible. At the end of inflation $\varphi = 0$, $D = 0$ and Supersymmetry is recovered in Minkowski space, since $V = 0$.

4.2. *$R + R^2$ supergravity*

There are two distinct classes of models, depending on the choice of auxiliary fields: old and new minimal models. The off-shell degrees of freedom contain the $6(= 10 - 4\text{diff})$ degrees of freedom of the graviton $g_{\mu\nu}$ and the $12(= 16 - 4\text{diff})$ degrees of freedom of the gravitino ψ_μ. The $n_B = n_F$ off-shell condition requires six more bosons. There are two choices for the latter, which reflect the two minimal supegravity multiplets of the $\mathcal{N} = 1$ theory:

- old minimal: $A_\mu,\ S,\ P$ (6 DOF's)
- new minimal: $A_\mu, B_{\mu\nu}$ (6 DOF's due to gauge inv. $\delta B_{\mu\nu} = \partial_\mu b_\nu - \partial_\nu b_\mu$) .

These $12_B + 12_F$ degrees of freedom must fill massive multiplets like

$$\text{Weyl}^2 : (2, 2(3/2), 1), \quad R^2_{\text{old}} : 2(1/2, 0, 0), \quad R^2_{\text{new}} : (1, 2(1/2), 0). \tag{22}$$

After superconformal manipulations, these two theories can be turned into standard Supergravity coupled to matter. The new minimal gives D-term inflation as described before, while the old minimal gives F-term inflation with the two chiral superfields T (inflaton multiplet) and S (sgoldstino multiplet). The T submanifold is $SU(1,1)/U(1)$ with scalar curvature $R = -2/3$, and the no-scale structure of the Kähler potential is responsible for the universal expression

$$V = M^2\, M_{\rm Pl}^2 \left(1 - e^{-\sqrt{\frac{2}{3}}\varphi}\right)^2 , \tag{23}$$

along the inflationary trajectory where $F_S \neq 0,\ F_T = 0$, which identifies S with the sgoldstino.

4.3. *Other models*

Several examples exist with two chiral multiplets of the same sort, for which F_S leads to a de Sitter plateau with $F_T = 0$, while at the end of inflation $F_S = F_T = 0$ and Supersymmetry is recovered. A class of models (α attractors) modify the superpotential but not the Kähler geometry of the original $R + R^2$ theory, which now reads:[12]

$$W(S, T) = S\, f(T), \tag{24}$$

with scalar curvature $R_\alpha = -\frac{2}{3\alpha}$. Along the inflationary trajectory the potential is positive since

$$V \sim |f|^2 \geq 0. \tag{25}$$

An alternative class of models with opposite role for Kähler potential and superpotential rest on the choice of Eq. (24), combined however with the trivial Kähler geometry corresponding to

$$K = \frac{1}{2}\left(\Phi + \overline{\Phi}\right)^2 + S\,\overline{S}. \tag{26}$$

The inflaton is now identified with $\varphi = {\rm Im}\Phi$, thus avoiding the dangerous exponential factor e^K in the supersymmetric potential. Along the inflationary trajectory

$$V(\varphi) \sim |f(\varphi)|^2, \tag{27}$$

so that the inflaton potential is fully encoded in the superpotential shape.

5. Nilpotent Superfields and Sgoldstino-less Models

In all the models reviewed so far it is difficult to exit inflation with Supersymmetry broken at a scale much lower than the de Sitter plateau (Hubble scale during inflation). A way to solve this problem is to introduce a nilpotent (Volkov–Akulov) multiplet S satisfying[40–43] the constraint of Eq. (17), so that the goldstino lacks its scalar partner, which is commonly called the sgoldstino. This solves the stabilization problem and gives rise to a de Sitter plateau.

The first cosmological model with a nilpotent sgoldstino multiplet was a generalization of the Volkov–Akulov–Starobinsky supergravity,[22] where

$$W(S,T) = S\, f(T), \quad V = e^{\,K(T)} K^{-1}_{S\overline{S}} |f(T)|^2. \tag{28}$$

Two classes of models which incorporate separate scales of Supersymmetry breaking during and at the exit of inflation were then proposed. They rest on a trivial (flat) Kähler geometry

$$K(\Phi, S) = \frac{1}{2}\left(\Phi + \overline{\Phi}\right)^2 + S\,\overline{S}, \tag{29}$$

but differ in their supersymmetry breaking patterns during and after inflation.

- In the first class of models[23]

$$W(\Phi, S) = M^2\, S\left(1 + g^2(\Phi)\right) + W_0, \tag{30}$$

where $g(\Phi)$ vanishes at $\Phi = 0$ and the inflaton φ is identified with its imaginary part. Along the inflaton trajectory $\mathrm{Re}\Phi = 0$, and the potential reduces to

$$V = M^4 |g(\Phi)|^2 \left(2 + |g(\Phi)|^2\right) + V_0, \quad V_0 = M^4 - 3\, W_0^2. \tag{31}$$

Assuming $V_0 \simeq 0$, one finds

$$m_{3/2} = \frac{H}{\sqrt{3}}, E_{SB} = |F_S|^{\frac{1}{2}} = \sqrt{H M_{\rm Pl}} > H,\ V = F_S F^S - 3W_0^2\,, \tag{32}$$

while $F_\Phi = 0$ during inflation ($\mathrm{Re}\ \Phi = 0$).
- In the second class of models[24] the superpotential is

$$W(\Phi, S) = f(\Phi)\left(1 + \sqrt{3}\, S\right), \tag{33}$$

which combines nilpotency and no-scale structure. Here the function $f(\Phi)$ satisfies the conditions

$$\overline{f}(\Phi) = f(-\overline{\Phi}), \quad f'(0) = 0, \quad f(0) \neq 0. \tag{34}$$

The scalar potential is of no-scale type, and letting $\Phi = (a + i\varphi)/\sqrt{2}$,

$$F_S F^S = 3e^{a^2} |f(\Phi)|^2, V(a,\varphi) = F^\Phi F_\Phi = e^{\,a^2} |f'(\Phi) + a\sqrt{2} f(\Phi)|^2\,. \tag{35}$$

The field a is stabilized at $a = 0$, since f is an even function of a. During inflation a gets a mass $\mathcal{O}(H)$ without mixing with Φ and is rapidly driven to $a = 0$, so that the inflationary potential reduces to

$$V(a = 0, \varphi) = \left| f'\left(\frac{i\varphi}{\sqrt{2}}\right) \right|^2, \quad V(0,0) = 0. \tag{36}$$

These models lack the fine-tuning of the previous class ($V_0 = 0$), and it is interesting to compare the supersymmetry breaking patterns. Here F_S never vanishes, and at the end of inflation

$$F^S\, F_S = 3\, e^{G(0,0)} = 3\, m_{3/2}^2. \tag{37}$$

In particular,

$$\langle F^S \rangle_{\Phi=0} = \sqrt{3}\, \overline{f}(0), \quad m_{3/2} = |f(0)|, \tag{38}$$

and the inflaton potential vanishes at the end of inflation. A choice that reproduces the Starobinsky potential is

$$f(\Phi) = \lambda - i\, \mu_1 \Phi + \mu_2\, e^{i \frac{2}{\sqrt{3}}\, \Phi}. \tag{39}$$

Interestingly, m_a and $m_{3/2}$ depend on the integration constant λ, but V is independent of it, and hence the same is true for m_φ.

6. Higher-Curvature Supergravity and Standard Supergravity Duals

Work in this direction started with the $R + R^2$ Starobinsky model, whose supersymmetric extension was derived in the late 80s[18,19] and was recently revived in view of the new CMB data.[10,11,20,34] Models dual to higher-derivative theories give more restrictions than their bosonic counterparts or standard Supergravity duals. Theories with unconstrained superfields also include the Supergravity embedding of R^2 duals, whose bosonic counterparts describe standard Einstein gravity coupled to a massless scalar field in de Sitter space. These theories were recently resurrected in Refs. 44 and 45. The R^2 higher curvature Supergravity was recently obtained in both the old and new minimal formulations.[46] In the old-minimal formulation, the superspace Lagrangian is

$$\alpha\, \mathcal{R}\overline{\mathcal{R}}\Big|_D - \beta\, \mathcal{R}^2\Big|_F, \tag{40}$$

where

$$\mathcal{R} = \frac{\Sigma(\overline{S}_0)}{S_0}, \quad \overline{\mathcal{D}}_{\dot\alpha} \mathcal{R} = 0 \tag{41}$$

is the scalar curvature multiplet, with Weyl and chiral weights $(w = 1, n = 1)$. The dual standard Supergravity has Kähler potential and superpotential

$$K = -3\, \log(T + \overline{T} - \alpha\, S\, \overline{S}), \quad W = T\, S - \beta\, S^3, \tag{42}$$

and the Kählerian manifold is $SU(2,1)/U(2)$. Note the rigid scale invariance of the action under

$$T \to e^{2\,\lambda}\, T, \quad S \to e^{\lambda}\, S, \quad S_0 \to e^{-\lambda}\, S_0. \tag{43}$$

If $\alpha = 0$ S is not dynamical, and integrating it out gives an $SU(1,1)/U(1)$ σ-model with Kähler potential and superpotential

$$K = -3\, \log\left(T + \overline{T}\right), \quad W = \frac{2\, T^{\frac{3}{2}}}{3\, \sqrt{3\, \beta}}. \tag{44}$$

Higher-curvature supergravities can be classified by the nilpotency properties of the chiral curvature $\mathcal{R}$. Such nilpotency constraints give rise to dual theories with nilpotent chiral superfields.[22] In particular, the constraint

$$\mathcal{R}^2 = 0, \tag{45}$$

in $R + R^2$ generates a dual theory where the inflaton chiral multiplet T (scalaron) is coupled to the Volkov–Akulov multiplet S

$$S^2 = 0, \quad \overline{\mathcal{D}}_{\dot\alpha}\, S = 0. \tag{46}$$

For this theory (the V-A-S Supergravity), the Kähler potential and superpotential are

$$K = -3\,\log\left(T + \overline{T} - S\,\overline{S}\right), \quad W = M\,S\,T + f\,S + W_0, \tag{47}$$

respectively, and due to its no-scale structure the scalar potential is semi-positive definite

$$V = \frac{|M\,T + f|^2}{3\,(T + \overline{T})^2}. \tag{48}$$

In terms of the canonically normalized field

$$T = e^{\sqrt{\frac{2}{3}}\,\phi} + i\,a\,\sqrt{\frac{2}{3}}, \quad (\phi, a) \in \frac{SU(1,1)}{U(1)}, \tag{49}$$

the potential Eq. (48) becomes

$$V = \frac{M^2}{12}\left(1 - \,e^{-\sqrt{\frac{2}{3}}\,\phi}\right)^2 + \frac{M^2}{18}\,e^{-\,2\sqrt{\frac{2}{3}}\,\phi}\,a^2. \tag{50}$$

Here a in the axion, which is much heavier than the inflaton during inflation

$$m_\phi^2 \simeq \frac{M^2}{9} e^{-\,2\,\sqrt{\frac{2}{3}}\,\phi_0} \ll m_a^2 = \frac{M^2}{9}. \tag{51}$$

There are then only two natural supersymmetric models with genuine single-field ϕ inflation. One is the new-minimal $R + R^2$ theory, where the inflaton has a massive vector as bosonic partner, and the V-A-S (sgoldstino-less) Supergravity just described.

Another interesting example is the sgoldstino-less version of the $\mathcal{R}\overline{\mathcal{R}}$ theory described before. This is obtained imposing the same constraint $\mathcal{R}^2 = 0$ as for the V-A-S Supergravity,[47] and is dual to the latter with

$$f = W_0 = 0. \tag{52}$$

The corresponding potential

$$V = M^2\,\frac{|T|^2}{3\,(T+\overline{T})^2} = \frac{M^2}{12} + \frac{M^2}{18}e^{\,-\,2\sqrt{\frac{2}{3}}\,\phi}\,a^2, \tag{53}$$

is positive definite and scale invariant. This model results in a de Sitter vacuum geometry with a positive vacuum energy

$$V(a=0) = \frac{M^2}{12}M_{\rm Pl}^4. \tag{54}$$

On the other hand, the Volkov–Akulov model coupled to Supergravity involves two parameters, and its vacuum energy has an arbitrary sign. The pure V-A theory coupled to Supergravity has indeed a superfield action determined by[22]

$$K = 3\,S\,\overline{S}, \quad W = f\,S + W_0, \quad S^2 = 0. \tag{55}$$

Moreover, the cosmological constant turns out to be

$$\Lambda = \frac{1}{3}|f|^2 - 3\,|W_0|^2. \tag{56}$$

The full-fledged component expression of the model, including all fermionic terms, was recently worked out.[48,49] The higher-curvature supergravity dual[50,51] is the standard (anti-de Sitter) supergravity Lagrangian augmented with the nilpotency constraint

$$\left(\frac{\mathcal{R}}{S_0} - \lambda\right)^2 = 0. \tag{57}$$

This is equivalent to adding to the action the term

$$\sigma\left(\frac{\mathcal{R}}{S_0} - \lambda\right)^2 S_0^3\Big|_F, \tag{58}$$

where σ is a chiral Lagrange multiplier. A superfield Legendre transformation and the superspace identity

$$\left[(\Lambda+\overline{\Lambda})S_0\overline{S}_0\right]_D = \left[\Lambda\mathcal{R}S_0^2\right]_F + \text{h.c.}, \tag{59}$$

which holds up to a total derivative for any chiral superfield Λ, turn indeed the action into the V-A superspace action coupled to standard Supergravity with

$$f = \lambda - 3\,W_0. \tag{60}$$

Hence, supersymmetry is broken whenever

$$3\,W_0 \neq \lambda \neq 0. \tag{61}$$

In the higher-derivative formulation, the goldstino G is encoded in the Rarita-Schwinger field. At the linearized level around flat space

$$G = -\frac{3}{2\lambda}\left(\gamma^{\mu\nu}\,\partial_\mu\,\psi_\nu - \frac{\lambda}{2}\gamma^\mu\,\psi_\mu\right), \quad \delta G = \frac{\lambda}{2}\epsilon. \tag{62}$$

The linearized equation of motion for the gravitino reads

$$\gamma^{\mu\nu\rho}\,\partial_\nu\,\psi_\rho - \frac{\lambda}{6}\gamma^{\mu\nu}\,\psi_\nu - \frac{1}{3}\left(\gamma^{\mu\nu}\,\partial_\nu - \frac{\lambda}{2}\gamma^\mu\right)G = 0, \tag{63}$$

and is gauge invariant under

$$\delta\psi_\mu = \partial_\mu\epsilon + \frac{\lambda}{6}\gamma_\mu\,\epsilon. \tag{64}$$

Both the γ-trace and the divergence of the equation of motion yield

$$\gamma^{\mu\nu}\,\partial_\mu\,\psi_\nu - \gamma^\mu\,\partial_\mu G = 0, \tag{65}$$

so that gauging away the Goldstino G one recovers the standard formulation of a massive gravitino.

Tables 1–3 summarize the various dualities linking higher-curvature supergravities in the old-minimal and new-minimal formulations with standard Supergravity.

7. Orthogonal Nilpotent Superfields

We have seen so far that simple models of inflation, and in particular the supersymmetric version of the Starobinsky model, rest on a pair of chiral multiplet, the sgoldstino multiplet S and the inflaton multiplet T. Sgoldstino-less models are obtained by replacing S by a nilpotent superfield ($S_{NL}^2 = 0$), which is the local version of the V-A multiplet. This setting should correspond to a linear model where the scalar partners of the goldstino are infinitely heavy, so that the sgoldstino becomes a non-dynamical composite field. Following Refs. 43 and 52, other types of constraints can be imposed, which remove other degrees of freedom from the T multiplet. The most interesting of them is the orthogonality constraint[53–55]

$$S_{NL}\,(T_{ONL} - \overline{T}_{ONL}) = 0, \quad (S_{NL}^2 = 0), \tag{66}$$

Table 1. Old-minimal dualities: $-\Phi S_0\overline{S}_0\Big|_D + W S_0^3\Big|_F$, $\Phi = \exp\left(-\frac{K}{3}\right)$.

Higher Curvature Supergravity	Standard Supergravity
$\Phi_H = 1 - h\left(\frac{\mathcal{R}}{S_0}, \frac{\overline{\mathcal{R}}}{\overline{S}_0}\right)$	$\Phi_S = 1 + T + \overline{T} - h(S, \overline{S})$
$W_H = W\left(\frac{\mathcal{R}}{S_0}\right)$	$W_S = TS - W(S)$
$\Phi_H = 1$	$\Phi_S = 1 + T + \overline{T}$
$W_H = W\left(\frac{\mathcal{R}}{S_0}\right)$	$W_S = -SW'(S) + W(S)\Big|_{T=-W'(S)}$
$\Phi_H = -\alpha\frac{\mathcal{R}}{S_0}\frac{\overline{\mathcal{R}}}{\overline{S}_0}$	$\Phi_S = T + \overline{T} - \alpha S\overline{S}$
$W_H = -\beta\frac{\mathcal{R}^3}{S_0^3}$	$W_S = TS - \beta S^3$

Table 2. Nilpotent old-minimal dualities: $-\Phi S_0 \overline{S}_0\Big|_D + W S_0^3\Big|_F$, $\Phi = \exp\left(-\frac{K}{3}\right)$.

Higher Curvature Supergravity	Standard Supergravity
$\Phi_H = 1 - \frac{1}{M^2}\frac{\mathcal{R}}{S_0}\frac{\overline{\mathcal{R}}}{\overline{S}_0}$	$\Phi_S = T + \overline{T} - S\overline{S}$
$W_H = W_0 + \xi\frac{\mathcal{R}}{S_0} + \sigma\frac{\mathcal{R}^2}{S_0^2}$	$W_S = MTS + fS + W_0$
	$(S^2 = 0, \quad f = \xi - \frac{1}{2})$
$\Phi_H = -\frac{1}{M^2}\frac{\mathcal{R}}{S_0}\frac{\overline{\mathcal{R}}}{\overline{S}_0}$	$\Phi_S = T + \overline{T} - S\overline{S}$
$W_H = \sigma\frac{\mathcal{R}^2}{S_0^2}$	$W_S = MTS$
	$(S^2 = 0)$
$\Phi_H = 1$	$\Phi_S = 1 - S\overline{S}$
$W_H = W_0 + \sigma\left(\frac{\mathcal{R}}{S_0} - \lambda\right)^2$	$W_S = fS + W_0$
	$(S^2 = 0, \ f = \lambda - 3W_0)$

Table 3. New-minimal dualities: $\Phi = \exp\left(-\frac{K}{3}\right)$.

Higher Curvature Supergravity	Standard Supergravity
$L\log\left(\frac{L}{S_0\overline{S}_0}\right)\Big|_D$	$\Phi_S = -U\exp U$
$W_\alpha\left(\frac{L}{S_0\overline{S}_0}\right)W^\alpha\left(\frac{L}{S_0\overline{S}_0}\right)\Big|_F$	$W_\alpha(U)W^\alpha(U)$
	$\Phi_S = (T + \overline{T})\exp V$
$W_\alpha\left(\frac{L}{S_0\overline{S}_0}\right)W^\alpha\left(\frac{L}{S_0\overline{S}_0}\right)\Big|_F$	$W_\alpha(V)W^\alpha(V)$

which also implies

$$\left(T_{ONL} - \overline{T}_{ONL}\right)^3 = 0. \tag{67}$$

This constraint removes the inflatino (spin-1/2 partner of the inflaton), as well as the sinflaton (spin-0 partner of the inflaton), so that this description should correspond to a regime where the inflatino and the sinflaton are both infinitely heavy.

The new aspect of these "non-chiral orthogonality constraints" is that the T-auxiliary field F_T becomes nilpotent, and therefore fails to contribute to the scalar potential, which takes the form

$$V(\varphi = \text{Re}\, T) = f^2(\varphi) - 3\, g^2(\varphi), \tag{68}$$

for a quadratic Kähler potential and a superpotential of the form

$$W(S_{NL}, T_{ONL}) = S_{NL}\, f(T_{ONL}) + g(T_{ONL}). \tag{69}$$

Table 4. Orthogonality constraints with S_{NL} ($S_{NL}^2 = 0$).

$S_{NL}(T_{ONL} - \overline{T}_{ONL}) = 0$ (implies $(T_{ONL} - \overline{T}_{ONL})^3 = 0$)	sgoldstino-less, inflatino-less, sinflaton-less
$S_{NL}\overline{T}'_{ONL}$ = chiral (implies $S_{NL}\overline{\mathcal{D}}_{\dot{\alpha}}T'_{ONL} = 0$)	sgoldstino-less, inflatino-less
$S_{NL}T''_{ONL} = 0$ (implies $(T''_{ONL})^3 = 0$)	sgoldstino-less, scalar-less
$S_{NL}W_\alpha(V_{ONL} = 0$	sgoldstino-less, gaugino-less

The potential V in Eq. (68) may or may not reproduce the inflaton trajectory for models with a "linear T multiplet". This setting presents an advantage with respect to the linear T model, because it eliminates the sinflaton, thus bypassing the problems related to its stabilization. It also avoids goldstino-inflatino mixing, which makes matter creation in the Early Universe very complicated. In the unitary gauge, the inflatino field simply vanishes, since it is proportional to the goldstino.[53,56]

In Table 4 we collect the various orthogonality constraints. The supergravity model for a matter multiplet T corresponding to the constraint $ST = 0$ was derived in Ref. 56. This model has been recently shown[57] to describe the effective dynamics of a fermion, other than the $\mathcal{N} = 1$ goldstino, which lives on a $\overline{D}3$-brane world volume.

8. Conclusions and Outlook

The orthogonality constraints in Eq. (66) and the resulting scalar potential in Eq. (68) allow the construction of MSIM (minimal supersymmetric inflationary models), which accommodate, with appropriate fine tuning, dark energy (cosmological constant Λ), the supersymmetry breaking scale $m_{3/2}$, and the inflationary Hubble scale H.[54] A simplified class of models is obtained with (in $M_{\rm Pl}$ units)

$$g(\varphi) = g_0 = m_{3/2}, \quad f(\varphi) = H\, f_I(\varphi) + f_0, \tag{70}$$

where φ is the appropriate canonically normalized scalar field, whenever the Kähler potential is not quadratic but has the more general form as in Refs. 53 and 54. Here, $f_I(\varphi)$ is a function with the property $f_I(\varphi) \to 1$, (for φ large) while at the extremum of the potential $\varphi = 0$, $f_I(\varphi) = 0$. Hence, the scalar potential satisfies $V(\varphi = 0) = f_0^2 - 3m_{3/2}^2 = \Lambda$, while for large φ, $V(\varphi) \to H^2$ (as $\varphi \to \infty$), for values of the parameters such that $\Lambda \approx 10^{-120}$, $m_{3/2} \approx 10^{-16}$ and $H = 10^{-5}$, where we took the SUSY breaking scale at the end of inflation (approximate Minkowski spacetime) to be at the TeV scale as a minimal value, which is inspired by the current LHC results.

Finally, we would like to note that microscopic models which may yield in suitable limits the non-linear realisations considered so far have been proposed in Refs. 58–60 and matter couplings to the inflation sector, with and without non-linear superfields, were considered in Refs. 23, 24, 50, 61 and 62.

This review reflects the lecture presented by SF at the 2016 Memorial Meeting for Prof. Salam, and overlaps in part with Ref. 63.

Acknowledgments

We are grateful to I. Antoniadis, G. Dall'Agata, E. Dudas, F. Farakos, P. Fré, R. Kallosh, A. Linde, M. Porrati, A. Riotto, A. Sorin, J. Thaler, A. Van Proeyen, T. Wrase and F. Zwirner for useful discussions and collaboration on related projects. SF is supported in part by INFN-CSN4-GSS, while AS is supported in part by Scuola Normale Superiore and by INFN-CSN4-STEFI.

Appendix A. Constrained Superfields and $\mathcal{N}=1$ Supersymmetry

Non-linear constraints involving a pair of chiral superfields (X, W_α), $(X, U_{\dot{a}})$ can have solutions that differ sharply from the V-A case. Here $W_\alpha, U_{\dot{a}}$ are the chiral superfields

$$W_\alpha = \frac{1}{4}\bar{D}^2 D_\alpha V, \quad \text{gauge field-strength multiplet,} \tag{A.1}$$

$$U_{\dot{a}} = \bar{D}_{\dot{a}} L, \quad L \text{ linear (or tensor) multiplet, } D^2 L = \bar{D}^2 L = 0. \tag{A.2}$$

The (chiral) constraints in question are

$$1)\ X^2 = 0, \quad X W_\alpha = 0, \tag{A.3}$$

$$2)\ X^2 = 0, \quad X U_{\dot{a}} = 0, \quad (\text{or } XL = \text{chiral}). \tag{A.4}$$

$\mathcal{N}=1$ supersymmetry is broken solving $X^2=0$, with the V-A solution

$$X = X_{NL} = S = \left(\frac{G^2}{2F}, G_\alpha, F\right), \tag{A.5}$$

and then solving the second portions of Eqs. (A.3), (A.4),

$$X_{NL} W_\alpha = 0, \quad X_{NL} U_{\dot{a}} = 0. \tag{A.6}$$

Using the fact that $W_\alpha, U_{\dot{a}}$ have components

$$W_\alpha = \left(\lambda_\alpha,\ L_\alpha^\beta = \delta_\alpha^\beta D - \frac{i}{2}(\sigma^\mu \bar{\sigma}^\nu)_\alpha^\beta F_{\mu\nu},\ \partial_{\alpha\dot{a}} \bar{\lambda}^{\dot{a}}\right), \tag{A.7}$$

$$U_{\dot{a}} = \left(\bar{\chi}_{\dot{a}},\ \Lambda_{\dot{a}\beta} = \sigma^\mu_{\dot{a}\beta}(H_\mu + i\partial_\mu \phi),\ \partial_{\alpha\dot{a}} \bar{\chi}^a\right), \tag{A.8}$$

where

$$F_{\mu\nu} = \partial_\mu A_\nu - \partial_\nu A_\mu, \quad H_\mu = \frac{1}{3!}\epsilon_{\mu\nu\rho\sigma}\,\partial^\nu\, b^{\rho\sigma}, \tag{A.9}$$

the constraints leave free the bosonic fields but express the *gaugino* (and *tensorino*) in terms of the V-A G goldstino[43] according to

$$\lambda_\alpha = i\, L_{\alpha\beta}\frac{G^\beta}{\sqrt{2}F} + \mathcal{O}(G^2), \tag{A.10}$$

$$\bar{\chi}_{\dot{a}} = -\, i\, \Lambda_{\dot{\alpha}\beta}\frac{G^\beta}{\sqrt{2}F} + \mathcal{O}(G^2). \tag{A.11}$$

The full solution can be obtained from the last component of the constraints by iteration. The constraint on the linear multiplet was considered in Ref. 60, and has the effect of leaving $(\phi,\ b_{\mu\nu})$ in the spectrum. However, there is another solution to the constraints, where instead the chiral multiplet X is not the V-A multiplet but the constraints in Eqs. (A.3), (A.4) can be used to express X in terms of W_α (or $U_{\dot\alpha}$). This is the case of the supersymmetric Born–Infeld and the non-linear tensor multiplet constraints of Bagger and Galperin[63–66] where

$$1)\ X = \frac{W^\alpha W_\alpha}{m - \bar{D}^2\bar{X}}, \tag{A.12}$$

$$2)\ X = \frac{U^{\dot\alpha} U_{\dot\alpha}}{m - \bar{D}^2\bar{X}}. \tag{A.13}$$

The resulting Lagrangians, which are simply the F-components of X, describe a non-linear theory with $\mathcal{N} = 2$ spontaneously broken to $\mathcal{N} = 1$.

References

1. D. Z. Freedman, P. van Nieuwenhuizen and S. Ferrara, "Progress Toward a Theory of Supergravity," *Phys. Rev. D* **13**, 3214 (1976); S. Deser and B. Zumino, "Consistent Supergravity," *Phys. Lett. B* **62**, 335 (1976). For a review see: D. Z. Freedman and A. Van Proeyen, "Supergravity" (Cambridge University Press, 2012).
2. For recent reviews see: V. Mukhanov, "Physical Foundations of Cosmology," (Cambridge University Press, 2005); S. Weinberg, "Cosmology," (Oxford University Press, 2008).
3. G. Aad *et al.* [ATLAS Collaboration], "Observation of a new particle in the search for the Standard Model Higgs boson with the ATLAS detector at the LHC," *Phys. Lett. B* **716**, 1 (2012) [arXiv:1207.7214 [hep-ex]]; S. Chatrchyan *et al.* [CMS Collaboration], "Observation of a new boson at a mass of 125 GeV with the CMS experiment at the LHC," *Phys. Lett. B* **716**, 30 (2012) [arXiv:1207.7235 [hep-ex]].
4. F. Englert and R. Brout, "Broken Symmetry and the Mass of Gauge Vector Mesons," *Phys. Rev. Lett.* **13**, 321 (1964); P. W. Higgs, "Spontaneous Symmetry Breakdown without Massless Bosons," *Phys. Rev.* **145**, 1156 (1966); G. 't Hooft and M. J. G. Veltman, "Regularization and Renormalization of Gauge Fields," *Nucl. Phys. B* **44**, 189 (1972).
5. F. L. Bezrukov and M. Shaposhnikov, "The Standard Model Higgs boson as the inflaton," *Phys. Lett. B* **659**, 703 (2008) [arXiv:0710.3755 [hep-th]].
6. A. A. Starobinsky, "A New Type of Isotropic Cosmological Models Without Singularity," *Phys. Lett. B* **91**, 99 (1980).
7. V. F. Mukhanov and G. V. Chibisov, "Quantum Fluctuation and Nonsingular Universe. (In Russian)," *J. Exp. Theor. Phys. Lett.* **33**, 532 (1981) [**33**, 549 (1981)].
8. B. Whitt, "Fourth Order Gravity as General Relativity Plus Matter," *Phys. Lett. B* **145**, 176 (1984).
9. A. Kehagias, A. M. Dizgah and A. Riotto, "Remarks on the Starobinsky model of inflation and its descendants," *Phys. Rev. D* **89** no. 4, 043527 (2014) [arXiv:1312.1155 [hep-th]].
10. F. Farakos, A. Kehagias and A. Riotto, "On the Starobinsky Model of Inflation from Supergravity," *Nucl. Phys. B* **876**, 187 (2013) [arXiv:1307.1137 [hep-th]].
11. S. Ferrara *et al.*, "Minimal Supergravity Models of Inflation," *Phys. Rev. D* **88** no. 8, 085038 (2013) [arXiv:1307.7696 [hep-th]]; "Higher Order Corrections in Minimal

Supergravity Models of Inflation," *J. Cosmol. Astropart. Phys.* **1311**, 046 (2013) [arXiv:1309.1085 [hep-th]].
12. R. Kallosh, A. Linde and D. Roest, "Universal Attractor for Inflation at Strong Coupling," *Phys. Rev. Lett.* **112** no. 1, 011303 (2014) [arXiv:1310.3950 [hep-th]], "Superconformal Inflationary α-Attractors," *J. High Energy Phys.* **1311**, 198 (2013) [arXiv:1311.0472 [hep-th]].
13. A. D. Linde, "Chaotic Inflation," *Phys. Lett. B* **129**, 177 (1983).
14. P. A. R. Ade *et al.* [Planck Collaboration], "Planck 2015 results. XX. Constraints on inflation," arXiv:1502.02114 [astro-ph.CO].
15. P. A. R. Ade *et al.* [BICEP2 and Planck Collaborations], "Joint Analysis of BICEP2/*KeckArray* and *Planck* Data," *Phys. Rev. Lett.* **114**, 101301 (2015) [arXiv:1502.00612 [astro-ph.CO]].
16. E. Cremmer *et al.*, "Yang–Mills Theories with Local Supersymmetry: Lagrangian, Transformation Laws and SuperHiggs Effect," *Nucl. Phys. B* **212**, 413 (1983).
17. J. Bagger and E. Witten, "The Gauge Invariant Supersymmetric Non-linear Sigma Model," *Phys. Lett. B* **118**, 103 (1982).
18. S. Cecotti, "Higher Derivative Supergravity Is Equivalent To Standard Supergravity Coupled To Matter. 1.," *Phys. Lett. B* **190**, 86 (1987);
19. S. Cecotti *et al.*, "New Minimal Higher Derivative Supergravity Coupled To Matter," *Nucl. Phys. B* **306**, 160 (1988).
20. J. Ellis, D. V. Nanopoulos and K. A. Olive, "No-Scale Supergravity Realization of the Starobinsky Model of Inflation," *Phys. Rev. Lett.* **111**, 111301 (2013) [Erratum: *ibid* **111** no. 12, 129902 (2013)] [arXiv:1305.1247 [hep-th]].
21. R. Kallosh and A. Linde, "Superconformal generalizations of the Starobinsky model," *J. Cosmol. Astropart. Phys.* **1306**, 028 (2013) [arXiv:1306.3214 [hep-th]].
22. I. Antoniadis *et al.* "The Volkov–Akulov–Starobinsky supergravity," *Phys. Lett. B* **733**, 32 (2014) [arXiv:1403.3269 [hep-th]].
23. R. Kallosh and A. Linde, "Inflation and Uplifting with Nilpotent Superfields," *J. Cosmol. Astropart. Phys.* **1501**, 025 (2015) [arXiv:1408.5950 [hep-th]].
24. G. Dall'Agata and F. Zwirner, "On sgoldstino-less supergravity models of inflation," *J. High Energy Phys.* **1412**, 172 (2014) [arXiv:1411.2605 [hep-th]].
25. S. Ferrara, R. Kallosh and A. Linde, "Cosmology with Nilpotent Superfields," *J. High Energy Phys.* **1410**, 143 (2014) [arXiv:1408.4096 [hep-th]];
26. W. Fischler and L. Susskind, "Dilaton Tadpoles, String Condensates And Scale Invariance," *Phys. Lett. B* **171**, 383 (1986), "Dilaton Tadpoles, String Condensates And Scale Invariance. 2," *Phys. Lett. B* **173**, 262 (1986); E. Dudas *et al.*, "On tadpoles and vacuum redefinitions in string theory," *Nucl. Phys. B* **708**, 3 (2005) [arXiv:hep-th/0410101]; N. Kitazawa, "Tadpole Resummations in String Theory," *Phys. Lett. B* **660**, 415 (2008) [arXiv:0801.1702 [hep-th]]; R. Pius, A. Rudra and A. Sen, "String Perturbation Theory Around Dynamically Shifted Vacuum," *J. High Energy Phys.* **1410**, 70 (2014) [arXiv:1404.6254 [hep-th]].
27. A. Sagnotti, in Cargese '87, "Non-Perturbative Quantum Field Theory", eds. G. Mack et al (Pergamon Press, 1988), p. 521, "Open Strings And Their Symmetry Groups," arXiv:hep-th/0208020; G. Pradisi and A. Sagnotti, "Open String Orbifolds," *Phys. Lett. B* **216**, 59 (1989); P. Horava, "Strings On World Sheet Orbifolds," *Nucl. Phys. B* **327**, 461 (1989), "Background Duality Of Open String Models," *Phys. Lett. B* **231**, 251 (1989); M. Bianchi and A. Sagnotti, "On The Systematics Of Open String Theories," *Phys. Lett. B* **247**, 517 (1990), "Twist Symmetry And Open String Wilson Lines," *Nucl. Phys. B* **361**, 519 (1991); M. Bianchi, G. Pradisi and A. Sagnotti, "Toroidal compactification and symmetry breaking in open string theories," *Nucl. Phys. B* **376**, 365

(1992); A. Sagnotti, "A Note on the Green–Schwarz mechanism in open string theories," *Phys. Lett. B* **294**, 196 (1992) [arXiv:hep-th/9210127]. For reviews see: E. Dudas, "Theory and phenomenology of type I strings and M-theory," *Class. Quant. Grav.* **17**, (2000) R41 [arXiv:hep-ph/0006190]; C. Angelantonj and A. Sagnotti, "Open strings," *Phys. Rept.* **371**, 1 (2002) [Erratum: *ibid.* **376**, 339 (2003)] [arXiv:hep-th/0204089];
28. S. Sugimoto, "Anomaly cancellations in type I D9-D9-bar system and the USp(32) string theory," *Prog. Theor. Phys.* **102**, 685 (1999) [arXiv:hep-th/9905159]; I. Antoniadis, E. Dudas and A. Sagnotti, "Brane supersymmetry breaking," *Phys. Lett. B* **464**, 38 (1999) [arXiv:hep-th/9908023]; C. Angelantonj, "Comments on open-string orbifolds with a non-vanishing B(ab)," *Nucl. Phys. B* **566**, 126 (2000) [arXiv:hep-th/9908064]; G. Aldazabal and A. M. Uranga, "Tachyon-free non-supersymmetric type IIB orientifolds via brane-antibrane systems," *J. High Energy Phys.* **9910**, 024 (1999) [arXiv:hep-th/9908072]; C. Angelantonj *et al.*, "Type I vacua with brane supersymmetry breaking," *Nucl. Phys. B* **572**, 36 (2000) [arXiv:hep-th/9911081].
29. E. Dudas and J. Mourad, "Consistent gravitino couplings in nonsupersymmetric strings," *Phys. Lett. B* **514**, 173 (2001), doi:10.1016/S0370-2693(01)00777-8 [hep-th/0012071].
30. S. Kachru *et al.*, "De Sitter vacua in string theory," *Phys. Rev. D* **68**, 046005 (2003), doi:10.1103/PhysRevD.68.046005 [hep-th/0301240].
31. L. Alvarez-Gaume, C. Gomez and R. Jimenez, *Phys. Lett. B* **690**, 68 (2010) [arXiv:1001.0010 [hep-th]]; "Phenomenology of the minimal inflation scenario: inflationary trajectories and particle production," *J. Cosmol. Astropart. Phys.* **1203**, 017 (2012) [arXiv:1110.3984 [astro-ph.CO]].
32. S. Ketov, "F(R) supergravity," *AIP Conf. Proc.* **1241**, 613 (2010) [arXiv:0910.1165 [hep-th]], "Chaotic inflation in F(R) supergravity," *Phys. Lett. B* **692**, 272 (2010) [arXiv:1005.3630 [hep-th]], "On the supersymmetrization of inflation in f(R) gravity," *PTEP* **2013**, 123B04 (2013) [arXiv:1309.0293 [hep-th]]; S. V. Ketov and S. Tsujikawa, "Consistency of inflation and preheating in F(R) supergravity," *Phys. Rev. D* **86**, 023529 (2012) [arXiv:1205.2918 [hep-th]]. S. V. Ketov, *On the supersymmetrization of f(R) gravity,* arXiv:1309.0293 [hep-th].
33. J. Ellis, D. V. Nanopoulos and K. A. Olive, "Starobinsky-like Inflationary Models as Avatars of No-Scale Supergravity," *J. Cosmol. Astropart. Phys.* **1310**, 009 (2013) [arXiv:1307.3537 [hep-th]].
34. S. Ferrara, R. Kallosh and A. Van Proeyen, "On the Supersymmetric Completion of $R + R^2$ Gravity and Cosmology," *J. High Energy Phys.* **1311**, 134 (2013) [arXiv:1309.4052 [hep-th]].
35. S. Ferrara, A. Kehagias and M. Porrati, "Vacuum structure in a chiral $R+R^n$ modification of pure supergravity," *Phys. Lett. B* **727**, 314 (2013) [arXiv:1310.0399 [hep-th]].
36. D. Roest and M. Scalisi, "Cosmological attractors from -scale supergravity," *Phys. Rev. D* **92**, 043525 (2015) [arXiv:1503.07909 [hep-th]].
37. S. V. Ketov and T. Terada, "Generic Scalar Potentials for Inflation in Supergravity with a Single Chiral Superfield," *J. High Energy Phys.* **1412**, 062 (2014) [arXiv:1408.6524 [hep-th]].
38. S. Ferrara, P. Fre and A. S. Sorin, "On the Gauged Khler Isometry in Minimal Supergravity Models of Inflation," *Fortsch. Phys.* **62**, 277 (2014) [arXiv:1401.1201 [hep-th]]; "On the Topology of the Inflaton Field in Minimal Supergravity Models," *J. High Energy Phys.* **1404**, 095 (2014) [arXiv:1311.5059 [hep-th]].
39. J. J. M. Carrasco *et al.*, "Hyperbolic geometry of cosmological attractors," *Phys. Rev. D* **92** no. 4, 041301 (2015) [arXiv:1504.05557 [hep-th]].
40. M. Rocek, "Linearizing the Volkov–Akulov Model," *Phys. Rev. Lett.* **41**, 451 (1978).

41. U. Lindstrom and M. Rocek, "Constrained Local Superfields," *Phys. Rev. D* **19**, 2300 (1979).
42. R. Casalbuoni *et al.*, "Non-linear Realization of Supersymmetry Algebra From Supersymmetric Constraint," *Phys. Lett. B* **220**, 569 (1989).
43. Z. Komargodski and N. Seiberg, "From Linear SUSY to Constrained Superfields," *J. High Energy Phys.* **0909**, 066 (2009) [arXiv:0907.2441 [hep-th]].
44. C. Kounnas, D. Lüst and N. Toumbas, "R^2 inflation from scale invariant supergravity and anomaly free superstrings with fluxes," *Fortsch. Phys.* **63**, 12 (2015) [arXiv:1409.7076 [hep-th]].
45. L. Alvarez-Gaume *et al.*, "Aspects of Quadratic Gravity," *Fortsch. Phys.* **64** no. 2–3, 176 (2016) [arXiv:1505.07657 [hep-th]].
46. S. Ferrara, A. Kehagias and M. Porrati, "$\mathcal{R}^2$ Supergravity," *J. High Energy Phys.* **1508**, 001 (2015) [arXiv:1506.01566 [hep-th]].
47. S. Ferrara, M. Porrati and A. Sagnotti, "Scale invariant Volkov–Akulov supergravity," *Phys. Lett. B* **749**, 589 (2015) [arXiv:1508.02939 [hep-th]].
48. E. A. Bergshoeff *et al.*, "Pure de Sitter Supergravity," *Phys. Rev. D* **92** no. 8, 085040 (2015) [Erratum: *ibid* **93** no. 6, 069901 (2016)] [arXiv:1507.08264 [hep-th]]; "Construction of the de Sitter supergravity," arXiv:1602.01678 [hep-th].
49. F. Hasegawa and Y. Yamada, "de Sitter vacuum from R^2 supergravity," *Phys. Rev. D* **92** no. 10, 105027 (2015) [arXiv:1509.04987 [hep-th]].
50. E. Dudas *et al.*, "Properties of Nilpotent Supergravity," *J. High Energy Phys.* **1509**, 217 (2015) [arXiv:1507.07842 [hep-th]].
51. I. Antoniadis and C. Markou, "The coupling of Non-linear Supersymmetry to Supergravity," *Eur. Phys. J. C* **75** no. 12, 582 (2015) [arXiv:1508.06767 [hep-th]].
52. A. Brignole, F. Feruglio and F. Zwirner, "On the effective interactions of a light gravitino with matter fermions," *J. High Energy Phys.* **9711**, 001 (1997) [hep-th/9709111], "Four-fermion interactions and sgoldstino masses in models with a superlight gravitino," *Phys. Lett. B* **438**, 89 (1998) [hep-ph/9805282].
53. S. Ferrara, R. Kallosh and J. Thaler, "Cosmology with orthogonal nilpotent superfields," *Phys. Rev. D* **93** no. 4, 043516 (2016) [arXiv:1512.00545 [hep-th]].
54. J. J. M. Carrasco, R. Kallosh and A. Linde, "Minimal supergravity inflation," *Phys. Rev. D* **93** no. 6, 061301 (2016) [arXiv:1512.00546 [hep-th]].
55. G. Dall'Agata and F. Farakos, "Constrained superfields in Supergravity," *J. High Energy Phys.* **1602**, 101 (2016) [arXiv:1512.02158 [hep-th]].
56. G. Dall'Agata, S. Ferrara and F. Zwirner, "Minimal scalar-less matter-coupled supergravity," *Phys. Lett. B* **752**, 263 (2016) [arXiv:1509.06345 [hep-th]].
57. B. Vercnocke and T. Wrase, "Constrained superfields from an anti-D3-brane in KKLT," arXiv:1605.03961 [hep-th].
58. R. Kallosh *et al.*, "Orthogonal Nilpotent Superfields from Linear Models," arXiv:1603.02661 [hep-th].
59. S. Ferrara *et al.*, "Linear Versus Non-linear Supersymmetry, in General," *J. High Energy Phys.* **1604**, 065 (2016) [arXiv:1603.02653 [hep-th]].
60. G. Dall'Agata, E. Dudas and F. Farakos, "On the origin of constrained superfields," arXiv:1603.03416 [hep-th].
61. S. Ferrara and M. Porrati, "Minimal $R+R^2$ Supergravity Models of Inflation Coupled to Matter," *Phys. Lett. B* **737**, 135 (2014) [arXiv:1407.6164 [hep-th]].
62. R. Kallosh, A. Linde and T. Wrase, "Coupling the Inflationary Sector to Matter," *J. High Energy Phys.* **1604**, 027 (2016) [arXiv:1602.07818 [hep-th]].
63. S. Ferrara and A. Sagnotti, "Some Pathways in non-Linear Supersymmetry: Special Geometry Born–Infeld's, Cosmology and dualities," [arXiv:1506.05730 [hep-th]],

"Supersymmetry and Inflation," PoS PLANCK **2015**, 113 (2015) [arXiv:1509.01500 [hep-th]], "Higher Curvature Supergravity and Cosmology," *Fortsch. Phys.* **64**, 371 (2016).
64. S. Ferrara, A. Sagnotti and A. Yeranyan, "Doubly Self-Dual Actions in Various Dimensions," *J. High Energy Phys.* **1505**, 051 (2015) [arXiv:1503.04731 [hep-th]].
65. J. Bagger and A. Galperin, "A New Goldstone multiplet for partially broken supersymmetry," *Phys. Rev. D* **55**, 1091 (1997) [hep-th/9608177].
66. J. Bagger and A. Galperin, "The Tensor Goldstone multiplet for partially broken supersymmetry," *Phys. Lett. B* **412**, 296 (1997) [hep-th/9707061].

A New Look at Newton–Cartan Gravity

E. A. Bergshoeff
Van Swinderen Institute, University of Groningen, Nijenborgh 4, 9722 WB Groningen, The Netherlands
e.a.bergshoeff@rug.nl
www.ericbergshoeff.nl

J. Rosseel
Albert Einstein Center for Fundamental Physics, University of Bern, Sidlerstrasse 5, 3012 Bern, Switzerland
rosseelj@gmail.com

We give a short overview of Newton–Cartan geometry and gravity including its matter couplings. We also present results on a new non-relativistic gravity model in three spacetime dimensions, called extended Bargmann gravity, and show that this model has matter couplings that differ from those of Newton–Cartan gravity.

Keywords: Newton–Cartan gravity, non-relativistic field theories.

1. Introduction

It is well known that in Newtonian gravity free-falling frames are connected by the Galilean symmetries which consist of (constant) time translations, spatial translations, spatial rotations and Galilean boosts. These Galilean boosts rotate space into time but not the other way around since Newtonian time is absolute. In such free-falling frames one does not experience any gravitational force. Such a force is only felt in frames that are not free-falling. For instance, in a (non-rotating) earth-based frame, that is in constant acceleration with respect to a free-falling frame, one experiences a gravitational force that is described by the Newton potential satisfying a Poisson equation. Of course, in a different frame than an earth-based frame, one experiences a different gravitational force that, in principle, can be calculated by relating that frame to a given earth-based or free-falling frame. However, a truly frame-independent formulation of Newtonian gravity was never given by Newton and his followers. The reason for this is that in order to give such a frame-independent formulation one needs a piece of mathematics that was not yet developed around that time. It was only in the middle of the 19th century that Riemann developed the required tool that is now called Riemannian geometry.

When Einstein invented his General Relativity theory in 1915 he achieved two things. First of all, he gave a description of gravity that is consistent with the theory of Special Relativity that he had developed 10 years earlier, by relating inertial frames to each other via the Poincaré transformations and by making use of the geometry of spacetime to give a proper description of gravity. In this way he built in a delay effect that avoided the instantaneous action of Newton's gravitational force. But most importantly, and this took Einstein many years of hard work to achieve, he presented his equations in a frame-independent way. For this, he needed the Riemannian geometry mentioned above and communicated to him by his friend Albert Grossmann. The Poincaré symmetries differ from the Galilean symmetries only as far as the boosts are concerned. Unlike the Galilean boosts the Lorentzian boosts rotate space into time *and* time into space: the concept of time is relative in Einstein's theory. Furthermore, to obtain a frame-independent formulation Einstein introduced a symmetric tensor field to describe the gravitational force. This field replaces the Newton potential and describes geometrical distances in the Riemannian spacetime manifold.

It was only 8 years later that Élie Cartan did for Newtonian gravity what Einstein had achieved for relativistic gravity. The formulation of Newtonian gravity in an arbitrary frame goes under the name of Newton–Cartan (NC) gravity. This NC gravity theory contains more fields than just the Newton potential. Locally, the formulation given by Newton, with a Newton potential in an earth-based frame, can easily be obtained from the general formulation by an appropriate gauge-fixing of the gravitational fields such that one is left with the Newton potential as the only non-zero field. The geometry Cartan was using is called NC geometry. This NC geometry differs from the Riemannian geometry used by Einstein in the sense that it has a *degenerate* metric and a *unique* foliation with an absolute time direction.

Given NC geometry and gravity the question arises: why should we study non-relativistic gravity? There are two main reasons why NC gravity has received a new appreciation in recent years. First of all, it arises in the context of the holographic principle which states that all the information about a gravitational theory in a given volume can be encoded by a different non-gravitational quantum field theory that lives on the surface surrounding this volume. This holographic principle has found a precise mathematical framework in string theory where it goes under the name of the AdS/CFT correspondence. This is a special situation where the gravity theory lives in a maximally symmetric spacetime with a negative cosmological constant, namely an Anti-de Sitter (AdS) spacetime, and where the field theory is a special so-called conformal field theory (CFT). In recent years much research has been done on non-AdS and non-relativistic holography to understand the validity and the basic principles underlying the holographic principle. One of the simplest deviations of AdS, breaking the relativistic isometries, is a Lifshitz spacetime which has less symmetries than AdS. Correspondingly, it has been found that at the field theory side the relativistic scale invariance of the CFT is broken to a non-relativistic scale invariance corresponding to a field theory that couples to an extension of NC geometry with so-called 'twistless torsion'.[1] Independently, NC geometry has recently found applications in the condensed matter physics community. Here one works with an Effective Field Theory (EFT) coupled to NC geometry to describe non-perturbative features of models such as the fractional quantum Hall effect,[2] chiral superfluids and simple fluids. The coupling to NC gravity means that one uses an arbitrary frame formulation in which general features are visible. One can compare this with the Coriolis force that is not visible in a non-rotating earth-based frame but can only be observed in a more general (rotating) frame.

Having the above motivation in mind we will first show in Sec. 2 how NC gravity can be obtained via a kind of gauging procedure from the centrally extended Galilei algebra which is called the Bargmann algebra. In Sec. 3, we will discuss some recent results on matter couplings. Next, in Sec. 4 we discuss NC gravity with torsion while in Sec. 5 we present a new model of gravity in three spacetime dimensions, which we call Extended Bargmann Gravity (EBG). In Sec. 6 we show that this EBG model has matter couplings that differ from the usual matter couplings that occur in NC gravity. Finally, in Sec. 7 we discuss future directions.

2. Newton–Cartan from Gauging Bargmann

Let us first remind ourselves how to obtain Einstein gravity via a kind of gauging procedure from the Poincaré algebra. In General Relativity all free-falling frames are connected by the following Poincaré symmetries:

- space–time translations: $\delta x^\mu = \xi^\mu$,
- Lorentz transformations: $\delta x^\mu = \lambda^\mu{}_\nu\, x^\nu$.

In arbitrary frames the gravitational force is described by the metric field. Instead of a metric, it is convenient to use an equivalent Vierbein formulation, with Vierbein field $E_\mu{}^A$ ($\mu = 0, 1, 2, 3$; $A = 0, 1, 2, 3$) since these Vierbeine are naturally related to the gauge fields of Poincaré translations.

In the non-relativistic case all free-falling frames are connected by the Galilean symmetries:

- time translations: $\delta t = \xi^0$,
- space translations: $\delta x^i = \xi^i$, $\qquad i = 1, 2, 3,$
- spatial rotations: $\delta x^i = \lambda^i{}_j\, x^j$,
- Galilean boosts: $\delta x^i = \lambda^i t$.

They are identical to the Poincaré symmetries except for the Galilean boosts which differ from the Lorentzian boosts as we discussed in the Introduction.

It is important to distinguish Newtonian gravity from NC gravity. Newtonian gravity is valid in frames of constant acceleration with respect to free-falling frames and is described by a single Newton potential $\Phi(x)$. On the other hand, NC gravity is valid in arbitrary frames and needs more fields to describe the gravitational force. To be precise, the required fields are a so-called temporal Vierbein $\tau_\mu(x)$ and a spatial Vierbein $e_\mu{}^a(x)$. Since these two fields together form a 4×4 matrix $\{\tau_\mu, e_\mu{}^a\}$

one would think that these fields suffice. Surprisingly, one needs one more field to describe NC gravity, namely a vector field $m_\mu(x)$.

One way to understand why this extra field is needed is to compare a freely moving relativistic particle with its non-relativistic counterpart. On the one hand a relativistic particle is described by the action

$$S_{\text{relativistic}} = -m \int d\tau \sqrt{-\eta_{\mu\nu}\dot{x}^\mu \dot{x}^\nu} \quad \mu = 0,1,2,3, \tag{1}$$

where $x^\mu(\tau)$ are the embedding coordinates. Clearly, the Lagrangian corresponding to this action is invariant under the Poincaré symmetries. On the other hand, a non-relativistic particle is described by the action

$$S_{\text{non-relativistic}} = \frac{m}{2}\int \frac{\dot{x}^i \dot{x}^j \delta_{ij}}{\dot{t}} d\tau \quad i = 1,2,3. \tag{2}$$

In this case the Lagrangian is <u>not</u> invariant under Galilean boosts. Instead, the Lagrangian transforms with a total derivative as follows:

$$\delta L_{\text{ non-relativistic}} = \frac{d}{d\tau}(m x^i \lambda^j \delta_{ij}). \tag{3}$$

Although the action is invariant, the non-invariance of the Lagrangian leads to modified Noether charges which induce a central extension of the underlying Galilei algebra. One thus ends up with the Bargmann algebra where the gauge field of the extra central charge transformation is the vector field $m_\mu(x)$.

Before gauging the Bargmann algebra it is of interest to compare gaugings and Inönü–Wigner contractions of algebras and taking the non-relativistic limit of gravity. We have indicated the relations between these different manipulations below. We see that, in order to obtain the Bargmann algebra from a contraction of the Poincaré algebra we need first to extend the Poincaré algebra with an additional U(1) generator, in order to account for the central charge generator which is present in the Bargmann algebra on top of the usual Galilei generators. This suggests that the non-relativistic limit of General Relativity can only be taken in the presence of an additional vector field that corresponds to the extra U(1) generator. This non-relativistic limit should mimic the Inönü–Wigner contraction of the algebra.

Poincaré $\otimes$ U(1)	$\overset{\text{‘gauging’}}{\Longrightarrow}$	General Relativity $\otimes$ U(1)
contraction $\Downarrow$		$\Downarrow$ non-relativistic limit
Bargmann	$\overset{\text{‘gauging’}}{\Longrightarrow}$	Newton–Cartan gravity

Fig. 1. This figure indicates the different relations between gaugings, contractions and non-relativistic limits.

We now return to the gauging of the Bargman algebra[3] which is based on a similar gauging procedure developed in the supergravity community many years ago.[4] Our starting point is the set of commutation relations defining the Bargmann algebra

$$[J_{ab}, P_c] = -2\delta_{c[a}P_{b]}, \quad [J_{ab}, G_c] = -2\delta_{c[a}G_{b]},$$
$$[G_a, H] = -P_a, \quad [G_a, P_b] = -\delta_{ab}Z, \quad a = 1, 2, \ldots, d, \tag{4}$$

where $\{H, P_a, J_{ab}, G_a, Z\}$ are the generators of time translations, space translations, spatial rotations, Galilean boosts and central charge transformations, respectively. In this gauging procedure we associate to every generator/symmetry a gauge field, gauge parameters that are arbitrary functions of spacetime and covariant curvatures as indicated in the table below.

From Table 1 we see that besides a timelike Vierbein τ_μ and a spatial Vierbein[a] $e_\mu{}^a$ there are two independent spin-connection fields $\{\omega_\mu{}^{ab}, \omega_\mu{}^a\}$ of spatial rotations and Galilean boosts, respectively, and a gauge field m_μ for the central charge transformations. Following General Relativity, in order to make the spin-connection fields dependent we need to impose constraints on the curvatures. Unlike General Relativity, the curvature $\mathcal{R}_{\mu\nu}(H)$ of time translations cannot play any role here since that curvature does not contain any of the two spin-connections fields. At this point the curvature $\mathcal{R}_{\mu\nu}(Z)$ of the central charge transformations comes to help since that curvature does contain the spin-connection field of the Galilean boosts. Independent of this we do set the curvature of time translations to zero since this defines the foliation of spacetime. We thus arrive at the following set of curvature constraints:

$$\mathcal{R}_{\mu\nu}{}^a(P) = 0, \qquad \mathcal{R}_{\mu\nu}(Z) = 0: \qquad \text{solve for spin-connection fields} \tag{5}$$

$$\mathcal{R}_{\mu\nu}(H) = \partial_{[\mu}\tau_{\nu]} = 0 \;\rightarrow\; \tau_\mu = \partial_\mu\,\tau: \;\text{absolute time ('zero torsion')} \tag{6}$$

Table 1. This table indicates for every symmetry the corresponding generators, gauge fields, local gauge parameters and covariant curvatures.

Symmetry	Generators	Gauge field	Parameters	Curvatures
Time translations	H	τ_μ	$\zeta(x^\nu)$	$\mathcal{R}_{\mu\nu}(H)$
Space translations	P^a	$e_\mu{}^a$	$\zeta^a(x^\nu)$	$\mathcal{R}_{\mu\nu}{}^a(P)$
Galilean boosts	G^a	$\omega_\mu{}^a$	$\lambda^a(x^\nu)$	$\mathcal{R}_{\mu\nu}{}^a(G)$
Spatial rotations	J^{ab}	$\omega_\mu{}^{ab}$	$\lambda^{ab}(x^\nu)$	$\mathcal{R}_{\mu\nu}{}^{ab}(J)$
Central charge transf.	Z	m_μ	$\sigma(x^\nu)$	$\mathcal{R}_{\mu\nu}(Z)$

[a]Even though these Vielbeine are not invertible, one can define projective inverses τ^μ, $e^\mu{}_a$ via the relations $\tau^\mu\tau_\mu = 1$, $\tau^\mu e_\mu{}^a = 0$, $\tau_\mu e^\mu{}_a = 0$, $e^\mu{}_a e_\mu{}^b = \delta_a^b$, $e^\mu{}_a e_\nu{}^a = \delta_\nu^\mu - \tau^\mu\tau_\nu$.

Here $\{P_A, M_{AB}\}$ are the generators of spacetime translations and Lorentz generators, respectively. Next, we decompose $A = (0, a)$ and relate the Poincaré $\otimes$ U(1) generators $\{P_0, P_a, M_{a0}, M_{ab}, \mathcal{Z}\}$ to the Bargmann generators $\{H, P_a, G_a, J_{ab}, Z\}$ as follows:

$$P_0 = \frac{1}{2\omega} H + \omega Z, \quad \mathcal{Z} = \frac{1}{2\omega} H - \omega Z, \quad A = (0, a), \tag{16}$$

$$P_a = P_a, \quad M_{ab} = J_{ab}, \quad M_{a0} = \omega G_a, \tag{17}$$

where we have introduced a contraction parameter ω. In a second step, taking the limit $\omega \to \infty$, we obtain the Bargmann algebra including the following commutator containing the central charge generator Z:

$$\left[P_a, G_b\right] = \delta_{ab} Z. \tag{18}$$

Inspired by the above Inönü–Wigner contraction we now define the non-relativistic limit of General Relativity as follows. We first introduce, on top of the Vierbein field, a vector field M_μ with $\partial_{[\mu} M_{\nu]} = 0$. Next, we relate the relativistic gauge fields $\{E_\mu{}^A, M_\mu\}$ to the non-relativistic gauge fields $\{\tau_\mu, e_\mu{}^a, m_\mu\}$ as follows:

$$E_\mu{}^0 = \omega \tau_\mu + \frac{1}{2\omega} m_\mu, \ M_\mu = \omega \tau_\mu - \frac{1}{2\omega} m_\mu, \ E_\mu{}^a = e_\mu{}^a. \tag{19}$$

This implies for the inverse Vierbein fields the following relation:

$$E^\mu{}_a = e^\mu{}_a - \frac{1}{2\omega^2} \tau^\mu e^\rho{}_a m_\rho + \mathcal{O}\left(\omega^{-4}\right) \text{ and similar for } E^\mu{}_0. \tag{20}$$

The definitions of the non-relativistic inverse fields $\{\tau^\mu, e^\mu{}_a\}$ we have used here can be found in Ref. 3.

In a second step we now take the limit $\omega \to \infty$. In this way we obtain the correct non-relativistic transformation rules (9) and the equations of motion (13). Note that the standard textbooks on General Relativity usually go straight from General Relativity to Newtonian gravity skipping the general frame formulation of NC gravity.

As an example we consider a complex scalar of mass M with Lagrangian given by

$$E^{-1} \mathcal{L}_{\rm rel} = -\frac{1}{2} g^{\mu\nu} D_\mu \Phi^* D_\nu \Phi - \frac{M^2}{2} \Phi^* \Phi \quad \text{with} \tag{21}$$

$$D_\mu \Phi = \partial_\mu \Phi - {\rm i}\, M\, M_\mu\, \Phi, \quad \delta\Phi = {\rm i}\, M\, \Lambda\, \Phi. \tag{22}$$

In a free-falling frame this Lagrangian reduces to the standard Klein–Gordon Lagrangian. Note that M_μ is not an electromagnetic gauge field. The mass M is not equal to the electric charge q. Instead the gauge field M_μ couples to the current expressing the conservation of # particles–# anti-particles.

We now take the non-relativistic limit of General Relativity as defined above together with $M = \omega m, \Phi \to \sqrt{\frac{\omega}{m}}\phi$. This leads us to the following Schrödinger Lagrangian coupled to NC gravity:

$$e^{-1}\mathcal{L}_{\text{Schrödinger}} = \left[\frac{\mathrm{i}}{2}\left(\Phi^*\tilde{D}_0\Phi - \Phi\tilde{D}_0\Phi^*\right) - \frac{1}{2m}\left|\tilde{D}_a\Phi\right|^2\right] \quad \text{with} \tag{23}$$

$$\tilde{D}_\mu\Phi = \partial_\mu\Phi + \mathrm{i}\,m\,m_\mu\,\Phi, \quad \delta\Phi = \xi^\mu\partial_\mu\Phi - \mathrm{i}\,m\,\sigma\,\Phi. \tag{24}$$

In a free-falling frame this is the standard Schrödinger Lagrangian. Note that the non-relativistic gauge field m_μ couples to the current that expresses the conservation of # particles only. Intuitively, the extra vector gauge field takes care of the infinities that occur if you switch between a Lagrangian with two time derivatives and a Lagrangian with one time derivative.

4. Newton–Cartan Gravity with Torsion

When studying non-relativistic holography one of the simplest deviations from AdS spacetime to consider is a Lifshitz spacetime which has non-relativistic Lifshitz isometries.[8] The Lifshitz spacetime metric is given by

$$ds^2 = -\frac{dt^2}{r^{2z}} + \frac{1}{r^2}\left(dr^2 + d\vec{x}^2\right).$$

Here $z \neq 1$ is the dynamical exponent that characterizes the anisotropic scaling

$$t \to \lambda^z t \quad \text{and} \quad \vec{x} \to \lambda\vec{x}.$$

The special thing about this Lifshitz spacetime is that when one approaches the boundary ($r \to \infty$) the relativistic lightcone flattens out and becomes, assuming that $z > 1$, a Galilean lightcone.[1] This is the reason that one finds at the boundary a CFT coupled to a NC background geometry. In fact, one even finds an extended NC geometry with a so-called twistless torsion.[1]

To clarify this extended NC geometry it is convenient to first adapt the so-called conformal method[9] to the non-relativistic case. The relativistic conformal group is the Poincaré group extended with dilatations D and special conformal transformations K_μ:

$$\text{Conformal} = \text{Poincaré} + D\ (\text{dilatations}) + K_\mu\ (\text{special conf. transf.}).$$

The (kinematics of the) corresponding conformal gravity theory can be obtained by gauging the conformal algebra in the same way that the kinematics of General Relativity can be obtained by gauging the Poincaré algebra. The gauge fields b_μ of dilatations and $f_\mu{}^a$ of special conformal transformations are special in the sense that b_μ transforms as a shift under the special conformal transformations with parameter $\Lambda_K^a(x)$ and $f_\mu{}^a$ is a dependent gauge-field that can be expressed in terms of the independent gauge fields $e_\mu{}^a, b_\mu$ and the dependent spin-connection field $\omega_\mu{}^{ab}(e,b)$:

$$\delta b_\mu = \Lambda_K^a(x) e_\mu{}^a, \quad f_\mu{}^a = f_\mu{}^a(e,\omega,b). \tag{25}$$

It is well-known that in the relativistic case there is a relationship between a Poincaré invariant and the CFT of a real scalar. This relationship goes both ways which can be best explained at the hand of the specific example of the Einstein–Hilbert action. Starting from the D-dimensional Poincaré invariant

$$e^{-1}\mathcal{L} = \frac{1}{\kappa^2} R \tag{26}$$

one can 'zip it' to a CFT of a real scalar in two steps. In a first step one writes the Poincaré Vielbein $(e_\mu{}^{\mathrm{A}})^{\mathrm{P}}$ as the product of a compensating scalar φ and a conformal Vielbein $(e_\mu{}^{\mathrm{A}})^{\mathrm{C}}$ as follows:

$$(e_\mu{}^{\mathrm{A}})^{\mathrm{P}} = \kappa^{\frac{2}{D-2}} \varphi \, (e_\mu{}^{\mathrm{A}})^{\mathrm{C}}. \tag{27}$$

The dilatations of the compensating scalar φ cancel the dilatations of the conformal Vielbein such that the left-hand side of the above equation is invariant under dilatations. Next, in a second step one fixes the general coordinate transformations and local dilatations to rigid conformal transformations by imposing the following gauge-fixing condition:

$$(e_\mu{}^{\mathrm{A}})^{\mathrm{C}} = \delta_\mu{}^{\mathrm{A}}. \tag{28}$$

Upon making the redefinition (assuming $D > 2$)

$$\varphi = \phi^{\frac{2}{D-2}} \tag{29}$$

one finally obtains the following CFT for the real scalar ϕ:

$$\text{CFT}: \;\; \mathcal{L} = 4\,\frac{D-1}{D-2}\,\phi\Box\phi. \tag{30}$$

The above procedure also works the other way around. Starting from the CFT (30) one can 'unzip' it by two steps. First we couple the CFT to conformal gravity by replacing the derivatives by conformal-covariant derivatives:

$$e^{-1}\mathcal{L} = 4\,\frac{D-1}{D-2}\,\phi\Box^{\mathrm{C}}\phi. \tag{31}$$

In a second step one fixes the local dilatations by imposing the gauge-fixing condition

$$\phi = \frac{1}{\kappa}, \tag{32}$$

where κ is the gravitational coupling constant. Substituting this gauge condition back into the Lagrangian (31) one re-obtains the Poincaré invariant (26).

We wish to extend the conformal method to the non-relativistic case. However, we will not use the non-relativistic contraction of the conformal algebra which is the Galilean Conformal Algebra (GCA). The reason for this is that the GCA does not allow for a central extension that plays such an important role in the case of the Bargmann algebra. Instead, we will use the so-called Schrödinger algebra which is somewhat smaller than the GCA but does allow a central extension. It is an extension of the Bargmann algebra that is characterized by a dynamical exponent z. For instance, we have the following commutators

$$[H, D] = zH, \quad [P_a, D] = P_a, \tag{33}$$

where D is the generator of dilatations, H of time translations and P_a of space translations. For $z = 2$ the Schrödinger algebra is the Bargmann algebra extended with dilatations D and a single special conformal transformation K:

$$z = 2 \text{ Schrödinger} = \text{Bargmann} + D \text{ (dilatations)} + K \text{ (special conf.)}.$$

For $z = 1$ we obtain the relativistic conformal algebra and for $z \neq 2$ there are no special conformal transformations.

The role of conformal gravity in the relativistic case is taken over by the so-called Schrödinger gravity theory which can be obtained by gauging the $z = 2$ Schrödinger algebra. We find that the independent gauge fields $\{\tau_\mu, e_\mu{}^a, m_\mu\}$ transform as follows:

$$\begin{aligned}
\delta\tau_\mu &= 2\Lambda_D\tau_\mu, \\
\delta e_\mu{}^a &= \Lambda^a{}_b\, e_\mu{}^b + \Lambda^a\tau_\mu + \Lambda_D e_\mu{}^a, \\
\delta m_\mu &= \partial_\mu\sigma + \Lambda_a\, e_\mu{}^a.
\end{aligned}$$

The time projection $\tau^\mu b_\mu$ of b_μ transforms under K as a a shift while the spatial projection $b_a \equiv e_a{}^\mu b_\mu$ is dependent:

$$b_a\,(e, \tau) = e_a{}^\mu \tau^\nu \partial_{[\mu}\tau_{\nu]}. \tag{34}$$

This expression is the solution of the following twistless torsionless or modified foliation condition

$$\partial_{[\mu}\tau_{\nu]} - 2b_{[\mu}\tau_{\nu]} = 0. \tag{35}$$

We are now able to apply the Schrödinger method, i.e. the non-relativistic analog of the conformal method. The simplest case to consider is to 'unzip' the Schrödinger action for a complex scalar Ψ in d spatial dimensions with dilatation weight $w = -d/2$ and central charge weight M, i.e.

$$\delta\Psi = \left(-\frac{d}{2}\lambda_D + iM\sigma\right)\Psi. \tag{36}$$

The Schrödinger Field Theory (SFT) for such a complex scalar is given by the Schrödinger action

$$\text{SFT}: \quad S_{\text{Schrödinger}} = \int dt\, d^dx\, \Psi^\star \left(i\partial_0 - \frac{1}{2M}\partial_a\partial_a\right)\Psi. \tag{37}$$

Unzipping the Schrödinger action leads to a Galilean invariant that has inconsistent equations of motion by itself. This situation can be compared with the cosmological constant in the relativistic case. As a separate Poincaré invariant the cosmological constant has inconsistent equations of motion but it can perfectly be added to the Einstein–Hilbert action. The same applies to the Galilean invariant that corresponds to the Schrödinger action. This invariant, whose explicit expression we refrain from giving here, can be added to a higher-derivative action of the Hořava–Lifshitz type.[1]

Instead of considering these higher-derivative invariants, we will now show that the Schrödinger method can also be applied at the level of the equations of motion. To be specific, we will apply it to the NC equations of motion that cannot be integrated to an action and we will show that the Schrödinger procedure naturally gives rise to the equations of motion of NC gravity with twistless torsion. It is convenient to first consider the case of zero torsion, i.e. $b_a = 0$. The foliation constraint and the equations of motion for this Galilean-invariant case are given by (the index G denotes that we are dealing with the Galilean and not the Schrödinger case)

$$\text{foliation constraint}: \quad \partial_\mu(\tau_\nu)^{\mathrm{G}} - \partial_\nu(\tau_\mu)^{\mathrm{G}} = 0, \tag{38}$$

$$\text{E.O.M.}: \quad (\tau^\mu)^{\mathrm{G}}(e^\nu{}_a)^{\mathrm{G}}\mathcal{R}_{\mu\nu}{}^a(G) = 0, \tag{39}$$

$$(e^\nu{}_a)^{\mathrm{G}}\mathcal{R}_{\mu\nu}{}^{ab}(J) = 0. \tag{40}$$

Applying the Schrödinger method discussed above we now zip these equations to the following SFT of a complex scalar $\Psi = \varphi e^{i\chi}$:

$$\text{SFT}: \quad \partial_0\partial_0\varphi = 0 \quad \text{and} \quad \partial_a\varphi = 0 \quad \text{with} \quad w = 1. \tag{41}$$

Note that the SFT is defined in terms of equations of motion only and that the scalar χ is absent implying that we do not break the central charge transformations. We observe that the first equation is not invariant under the Schrödinger transformations by itself but that it transforms into the constraint $\partial_a\varphi = 0$. This constraint is the zipped version of the foliation constraint (38).

We now consider the case with twistless torsion, i.e. $b_a \neq 0$ and start from the SFT side. It is clear that we cannot use the SFT given in (41) since the foliation constraint has been changed and therefore the constraint $\partial_a\varphi = 0$ is no longer valid anymore. In fact, the new foliation constraint, given in Eq. (35), is invariant under dilatations and therefore there is no zipped version of it, i.e. there is no constraint. This means that the equation $\partial_0\partial_0\phi = 0$ is not invariant under the Schrödinger transformations. To compensate for this lack of invariance the second scalar χ now comes to our help. One can show that the following modified equation, with extra terms containing the scalar χ defines a modified SFT$'$:

$$\text{SFT}': \quad \partial_0\partial_0\varphi - \frac{2}{M}(\partial_0\partial_a\varphi)\partial_a\chi + \frac{1}{M^2}(\partial_a\partial_b\varphi)\partial_a\chi\partial_b\chi = 0. \tag{42}$$

The advantage of the Schrödinger approach is that it is much easier to find the answer in the zipped SFT version than in the unzipped Galilean invariant version. Nevertheless, the answer in terms of the Galilean invariant equations of motion can be reconstructed by unzipping the SFT′ given above. Following the rules (coupling to Schrödinger gravity and gauge-fixing) one finds the following answer:[10]

$$\boxed{-\,\triangle\Phi + \hat{\tau}^{\mu}\partial_{\mu}K + K^{ab}K_{ab} - 8\,\Phi\, b\cdot b - 2\,\Phi\,\mathcal{D}\cdot b - 6\, b^{a}\mathcal{D}_{a}\Phi = 0} \tag{43}$$

plus

$$\boxed{e^{\nu}{}_{a}R_{\mu\nu}{}^{ab}(J) = 0.} \tag{44}$$

Truncating to the case of zero torsion one re-obtains the equations of motion of NC gravity (38). This finishes our discussion of NC gravity with twistless torsion.

5. Extended Bargmann Gravity

It is well-known that taking the limit of matter-coupled General Relativity in three spacetime dimensions is special.[11] Consider, for instance, the action of General Relativity plus a complex scalar in D dimensions:

$$S = \int d^{D}x\sqrt{-g}\left\{\frac{1}{\kappa^{2}}R - \frac{1}{2}\,g^{\mu\nu}\,D_{\mu}\Phi^{*}D_{\nu}\Phi - \frac{M^{2}}{2}\,\Phi^{*}\Phi\right\}. \tag{45}$$

The equations of motion of the metric tensor corresponding to this action are given by

$$R_{\mu\nu} = \kappa^{2}\left(T_{\mu\nu} - \frac{1}{D-2}g_{\mu\nu}T\right). \tag{46}$$

Taking the non-relativistic limit in the equations of motion as before and, furthermore, rescaling the gravitational coupling constant κ with

$$\kappa^{2} \to \kappa^{2}/\omega^{4} \tag{47}$$

yields the following non-relativistic equations of motion

$$R_{0a}{}^{a}(G) = \frac{D-3}{D-2}\kappa^{2}\rho \qquad \text{with} \qquad \rho = m\Phi^{*}\Phi, \tag{48}$$

$$R_{\mu b}{}^{b}{}_{a}(J) = 0. \tag{49}$$

For $D = 4$ the first equation, after gauge-fixing to a frame with constant acceleration, leads to the expected Newton's law

$$\triangle\Phi = 4\pi G_{N}\,\rho \tag{50}$$

but for $D = 3$ the source term vanishes.

Surprisingly, it turns out that there is a different way of taking the non-relativistic limit that only works in $D = 3$ dimensions. The possibility of this limit is due to the fact that for $D = 3$ the Bargmann algebra allows for a second central extension:[12]

$$\text{Galilei} \quad \overset{\text{`Mass'}}{\Longrightarrow} \quad \text{Bargmann} \quad \overset{\text{`Spin'}}{\Longrightarrow} \quad \text{Extended Bargmann.}$$

For a further discussion of this second central charge, see Refs. 13 and 15. We will call the Bargmann Algebra with this second central charge the Extended Bargmann Algebra. Applying a gauging procedure on this Extended Bargmann algebra will lead to the so-called Extended Bargmann Gravity (EBG) model. We will first construct this EBG model and after that show how it can be obtained by defining a special limit of 3D General Relativity, augmented with *two* Abelian gauge fields.

The spacetime symmetry algebra of EBG consists of the generators of time translations H, spatial translations P_a (with $a = 1, 2$), spatial rotations J, Galilean boosts G_a, a central charge M, corresponding to particle mass as well as a second central charge S.[12–15] The generators H, P_a, J, G_a and M form the Bargmann algebra and the inclusion of S leads to the Extended Bargmann Algebra whose non-zero commutation relations are given by

$$\begin{aligned} [H, G_a] &= -\epsilon_{ab} P_b, \qquad [J, G_a] = -\epsilon_{ab} G_b, \\ [J, P_a] &= -\epsilon_{ab} P_b, \quad [G_a, G_b] = \epsilon_{ab} S, \\ [G_a, P_b] &= \epsilon_{ab} M. \end{aligned} \tag{51}$$

Unlike the Bargmann algebra, the extended Bargmann algebra can be equipped with a non-degenerate, invariant bilinear form or ‘trace’, given by[16]

$$\langle G_a, P_b \rangle = \delta_{ab}, \quad \langle H, S \rangle = \langle M, J \rangle = -1. \tag{52}$$

The action of EBG is given by the Chern–Simons action for the gauge algebra (51)

$$S = \frac{k}{4\pi} \int \langle A \wedge dA + \frac{2}{3} A \wedge A \wedge A \rangle, \tag{53}$$

where k is the Chern–Simons coupling constant and the gauge field $A = A_\mu dx^\mu$ is given by

$$A_\mu = \tau_\mu \, H + e_\mu{}^a \, P_a + \omega_\mu \, J + \omega_\mu{}^a \, G_a + m_\mu \, M + s_\mu \, S. \tag{54}$$

Explicitly, one finds the following action for EBG,[16,c]

$$S = \frac{k}{4\pi} \int d^3x \left(\epsilon^{\mu\nu\rho} e_\mu{}^a R_{\nu\rho}(G_a) - \epsilon^{\mu\nu\rho} m_\mu R_{\nu\rho}(J) - \epsilon^{\mu\nu\rho} \tau_\mu R_{\nu\rho}(S) \right), \tag{55}$$

---

[c] We thank Diego Hofman for sharing with us the observation that a particular truncation of this action is related to the lower-spin gravity action introduced in Ref. 17.

where here and in the following we have used the curvatures

$$
\begin{aligned}
R_{\mu\nu}(H) &= 2\partial_{[\mu}\tau_{\nu]},\\
R_{\mu\nu}(P^a) &= 2\partial_{[\mu}e_{\nu]}{}^a + 2\epsilon^{ab}\omega_{[\mu}e_{\nu]b} - 2\epsilon^{ab}\omega_{[\mu b}\tau_{\nu]},\\
R_{\mu\nu}(J) &= 2\partial_{[\mu}\omega_{\nu]},\\
R_{\mu\nu}(G^a) &= 2\partial_{[\mu}\omega_{\nu]}{}^a + 2\epsilon^{ab}\omega_{[\mu}\omega_{\nu]b},\\
R_{\mu\nu}(M) &= 2\partial_{[\mu}m_{\nu]} + 2\epsilon^{ab}\omega_{[\mu a}e_{\nu]b},\\
R_{\mu\nu}(S) &= 2\partial_{[\mu}s_{\nu]} + \epsilon^{ab}\omega_{[\mu a}\omega_{\nu]b}.
\end{aligned} \tag{56}
$$

These curvatures are covariant with respect to the local H, P_a, J, G_a, M and S transformations of τ_μ, $e_\mu{}^a$, m_μ, ω_μ, $\omega_\mu{}^a$ and s_μ, that are found from the gauge algebra (51) following the usual rules of gauge theory. Note that the fields τ_μ, $e_\mu{}^a$, ω_μ, $\omega_\mu{}^a$ and m_μ also appear in the formulation of NC gravity obtained by gauging the Bargmann algebra. As in that case, τ_μ and $e_\mu{}^a$ can be interpreted as Vielbeine for two degenerate timelike and spatial metrics, respectively. The field s_μ is not present in NC gravity and is specific to EBG. We note that the equations of motion for s_μ, ω_μ and $\omega_\mu{}^a$ lead to the curvature constraints

$$
R_{\mu\nu}(H) = 0, \quad R_{\mu\nu}(P^a) = 0, \quad R_{\mu\nu}(Z) = 0, \tag{57}
$$

that are usually imposed by hand in NC gravity. As in NC gravity, these equations imply that EBG is defined on non-relativistic space–times with torsionless NC geometry. The first equation implies that the space–time can be foliated in an absolute time direction, while the last two equations can be used to express ω_μ and $\omega_\mu{}^a$ in terms of τ_μ, $e_\mu{}^a$ and m_μ. Following NC gravity, ω_μ and $\omega_\mu{}^a$ can be seen as appropriate non-relativistic versions of the relativistic spin connection.

Having constructed the EBG model, we will now show how the EBG action (55) can be obtained as the non-relativistic limit of a suitable extension of the three-dimensional Einstein–Hilbert action. In order to show this, we extend the procedure that allows one to obtain the equations of motion of NC gravity from Einstein's equations, that was discussed in the previous section. As a starting point, we take the following Einstein–Hilbert action for the relativistic Vielbein $E_\mu{}^A$ and spin connection $\Omega_\mu{}^{AB}$, written as a Chern–Simons action, plus a Chern–Simons action for two Abelian gauge fields $Z_{1\mu}$ and $Z_{2\mu}$:

$$
S = \frac{k\omega}{4\pi}\int d^3x \left(\epsilon^{\mu\nu\rho}\, E_\mu{}^A\, R_{\nu\rho}(J_A) + 2\,\epsilon^{\mu\nu\rho}\, Z_{1\mu}\,\partial_\nu Z_{2\rho}\right), \tag{58}
$$

where the Riemann tensor $R_{\mu\nu}(J^A)$ reads

$$
R_{\mu\nu}(J^A) = 2\partial_{[\mu}\Omega_{\nu]}{}^A - \epsilon^{ABC}\Omega_{[\mu B}\Omega_{\nu]C}. \tag{59}
$$

Extending the particle-limit procedure of,[6] mimicking the Inönü–Wigner contraction of the underlying Poincaré $\otimes$ $\mathrm{U}(1)^2$ gauge algebra, we express the relativistic gauge fields $E_\mu{}^A$, $\Omega_\mu{}^A$, $Z_{1\mu}$, $Z_{2\mu}$ in terms of the non-relativistic fields

$\tau_\mu, e_\mu{}^a, \omega_\mu{}^a, \omega_\mu, m_\mu, s_\mu$ as follows:

$$E_\mu{}^0 = \omega\,\tau_\mu + \frac{1}{2\omega}\,m_\mu, \qquad Z_{1\mu} = \omega\,\tau_\mu - \frac{1}{2\omega}\,m_\mu,$$
$$\Omega_\mu{}^0 = \omega_\mu + \frac{1}{2\omega^2}\,s_\mu, \qquad Z_{2\mu} = \omega_\mu - \frac{1}{2\omega^2}\,s_\mu,$$
$$E_\mu{}^a = e_\mu{}^a, \qquad \Omega_\mu{}^a = \frac{1}{\omega}\,\omega_\mu{}^a. \tag{60}$$

Using these expressions in the action (58) and taking the limit $\omega \to \infty$[d] it is straightforward to show that the EBG action (55) is obtained.

In the next section we will show that EBG and NC gravity are different theories by comparing their coupling to matter.

6. Matter coupled Extended Bargman Gravity

Introducing matter couplings in EBG can be done by adding one of the matter Lagrangians on arbitrary torsionless NC backgrounds constructed in Ref. 5:

$$S = \frac{k}{4\pi}\int d^3x\,\Big(\epsilon^{\mu\nu\rho}e_\mu{}^a R_{\nu\rho}(G_a) - \epsilon^{\mu\nu\rho}\tau_\mu R_{\nu\rho}(S) - \epsilon^{\mu\nu\rho}m_\mu R_{\nu\rho}(J)\Big) + \int d^3x\, e\,\mathcal{L}_{\rm m}. \tag{61}$$

Here $e = \det(\tau_\mu, e_\mu{}^a)$ denotes the volume element. Since any matter couplings to the s_μ gauge field change the foliation constraint $R_{\mu\nu}(H) = 0$, we do not consider such couplings so that we can stay within the framework of torsionless NC geometry.

Defining the energy current t^μ, the momentum current $t^\mu{}_a$ and the particle number current j^μ by

$$t^\mu = \frac{1}{e}\frac{\delta}{\delta\tau_\mu}\,(e\mathcal{L}_{\rm m}), \quad t^\mu{}_a = \frac{1}{e}\frac{\delta}{\delta e_\mu{}^a}\,(e\mathcal{L}_{\rm m}),$$
$$j^\mu = \frac{1}{e}\frac{\delta}{\delta m_\mu}\,(e\mathcal{L}_{\rm m}), \tag{62}$$

the equations of motion stemming from the action (61) take the form

$$e^{-1}\epsilon^{\mu\nu\rho}R_{\nu\rho}(S) = \frac{4\pi}{k}t^\mu, \quad e^{-1}\epsilon^{\mu\nu\rho}R_{\nu\rho}(J) = \frac{4\pi}{k}j^\mu,$$
$$e^{-1}\epsilon^{\mu\nu\rho}R_{\nu\rho}(G_a) = -\frac{4\pi}{k}t^\mu{}_a. \tag{63}$$

Since the curvatures in these equations obey Bianchi identities, the currents obey various identities for consistency. We distinguish between Bianchi identities 'of the

[d]This limit is well-defined. The term $-2\,\epsilon^{\mu\nu\rho}E_\mu{}^0\,\partial_\nu\Omega_\rho{}^0$ leads to a potentially diverging $-2\,\omega^2\,\epsilon^{\mu\nu\rho}\tau_\mu\,\partial_\nu\omega_\rho$ term, but this term gets cancelled by a contribution coming from the term $2\,\epsilon^{\mu\nu\rho}\,Z_{1\mu}\,\partial_\nu Z_{2\rho}$ that we added to the Einstein–Hilbert action.

first kind' and 'of the second kind'. The identities of the first kind follow from the fact that the equations

$$R_{\mu\nu}(P^a) = 0, \quad R_{\mu\nu}(Z) = 0 \tag{64}$$

are identically satisfied, once one views the spin connections ω_μ and $\omega_\mu{}^a$ as dependent on τ_μ, $e_\mu{}^a$ and m_μ. Substituting equations (64) into the Bianchi identities $D_{[\mu}R_{\nu\rho]}(P^a) = 0$ and $D_{[\mu}R_{\nu\rho]}(Z) = 0$ leads to the following Bianchi identities of the first kind:

$$R_{[\mu\nu}(J)\, e_{\rho]}{}^a = R_{[\mu\nu}(G^a)\, \tau_{\rho]}, \quad \epsilon^{ab} R_{[\mu\nu}(G_a)\, e_{\rho]b} = 0. \tag{65}$$

The remaining Bianchi identities, called of the second kind, are not algebraic in the curvatures and are given by

$$D_{[\mu}R_{\nu\rho]}(G^a) = 0, \quad \partial_{[\mu}R_{\nu\rho]}(J) = 0, \quad D_{[\mu}R_{\nu\rho]}(S) = 0. \tag{66}$$

Combining the equations of motion (63) with the Bianchi identities of the first kind (65) leads to the following algebraic consistency conditions

$$e_\mu{}^a j^\mu = -\tau_\mu t^\mu{}_a, \quad e_\mu{}^{[a} t^{|\mu|b]} = 0. \tag{67}$$

The Bianchi identities of the second kind on the other hand lead to the following current conservation conditions:

$$D_\mu t^\mu = 0, \quad D_\mu t^\mu{}_a = 0, \quad D_\mu j^\mu = 0. \tag{68}$$

The EBG equations of motion (63) are strikingly different from the NC gravity ones. Using the following identity between the Riemann tensor and the curvatures $R(J)$ and $R(G)$

$$R^\sigma{}_{\rho\mu\nu} = \epsilon^{ab} R_{\mu\nu}(J) e^\sigma{}_a e_{\rho b} - \epsilon^{ab} R_{\mu\nu}(G_b) e^\sigma{}_a \tau_\rho \tag{69}$$

it follows from the equations of motion (63) that the purely time-like component of the Ricci tensor $R_{\mu\nu} = R^\rho{}_{\mu\rho\nu}$ is given by $\tau^\mu\tau^\nu R_{\mu\nu} \propto e_\mu{}^a\, t^\mu{}_a$. This is unlike NC gravity, where one rather has $\tau^\mu\tau^\nu R_{\mu\nu} \propto j^0$.[e] Furthermore, in NC gravity only this purely time-like component of the Ricci tensor is non-zero. This is in contrast with EBG where matter sources all components of the Riemann tensor. As a result, three-dimensional EBG admits backgrounds with non-trivial curvature whenever matter is present.

[e]It is instructive to consider this difference in the example of a massive complex Schrödinger field Φ, whose action can be found in Ref. 5. In that case one finds for EBG that $\tau^\mu\tau^\nu R_{\mu\nu} = -\frac{2\pi\mathrm{i}}{k}\left(\Phi^* D_0\Phi - \Phi D_0\Phi^*\right)$ while the case of NC gravity leads to $\tau^\mu\tau^\nu R_{\mu\nu} \propto |\Phi|^2$.

Neutrino Oscillations and Neutrino Masses

Harald Fritzsch
Department für Physik, Universität München,
Theresienstraße 37, München 80333, Germany
fritzsch@mpp.mpg.de

In 1914 James Chadwick discovered that energy and momentum were not conserved in the beta decay of atomic nuclei. For the next 16 years this phenomenon was not understood. In 1930 Wolfgang Pauli suggested in a letter to the participants of a conference in Tuebingen, that in the beta decays not only an electron was emitted, but also a neutral particle, which could not be observed. The energy and momentum of this particle would be the observed missing energy and momentum. Enrico Fermi proposed a name for this hypothetical particle: neutrino.

Pauli assumed, that neutrinos could never be observed directly. For this reason he did not publish his idea. But in 1946 Bruno Pontecorvo suggested that the anti-neutrinos, emitted by a reactor, could be observed. An anti-neutrino interacts with a proton. A neutron and a positron are produced. The positron annihilates with an electron and two photons are produced, which can be observed.

Following the suggestion of Pontecorvo, the neutrino was discovered in 1956 by Clyde Cowan and Frederick Reines. They investigated the neutrino emission of the big reactor near the Savannah River in South Carolina, USA.

Charged pions decay mainly into muons and neutrinos, but it was unknown, whether these neutrinos are the same as the neutrinos which are produced by the beta decay. In 1962 Leon Lederman, Jack Steinberger and Melvin Schwartz investigated a pion beam, produced by the Alternating Gradient Synchroton at the Brookhaven National Laboratory. The muons, produced in the decay of the pions, were stopped by a big steel wall. Only the neutrinos could travel through this wall. They produced muons, not electrons, in a detector behind the steel wall. Thus the neutrinos, produced in the decay of pions, are not identical to the ones, produced in a beta decay. There must be two neutrinos: electron neutrino and muon neutrino.

In 1975 the new tau lepton was discovered at the Stanford Linear Accelerator Center. This heavy lepton (mass about 1777 MeV) is also associated with a new neutrino. Thus there are three different neutrinos: the electron neutrino, the muon neutrino and the tau neutrino.

Pauli assumed, that the mass of the neutrino should be rather small, probably less than the mass of the electron. Today it is known, that the mass of the electron neutrino is less than 0.32 eV. Until 1998 it was assumed, that the neutrinos do not have any mass, as the photon. After the discovery of parity violation Abdus Salam suggested in 1957, that the neutrinos are always left-handed and can be described as massless Weyl fermions.

Pontecorvo assumed, that the neutrinos have a small mass and that they are superpositions of two mass eigenstates. In this case an electron neutrino, emitted from a nucleus, can turn into a muon neutrino after traveling a certain distance. Afterwards it would again become an electron neutrino, etc. Thus neutrinos oscillate. In 1976 Peter Minkowski and I wrote a paper in *Physics Letters*, where the details of such oscillations were described. In 1998 the neutrino oscillations were discovered in Japan.

In the nuclear fusion on the sun many electron neutrinos are produced. In 1963 John Bahcall calculated the flux of the solar neutrinos. He concluded that this flux could be measured by experiments. Raymond Davis and Bahcall started such an experiment, which investigates the solar neutrinos. The experiment was placed in the Homestake Gold Mine in Lead, South Dakota. It took data from 1970 until 1994. One observed only about 1/3 of the flux calculated by Bahcall. Thus there were problems with the solar neutrinos. Today we know, that these problems were due to neutrino oscillations.

In the Japanese Alps, near the small village "Kamioka", a big underground detector was built in 1982 in an old mine. This detector "Kamiokande" was installed in order to find the hypothetical decay of a proton. But thus far no proton decay has been observed, but the detector can also be used to study neutrinos, in particular the atmospheric neutrinos, produced by the decay of pions in the upper atmosphere.

In 1996 a new detector "Superkamiokande" started to investigate these neutrinos. This detector consists of a water tank, containing 50,000 liters of purified water, surrounded by about 11,000 photo multipliers. Two years later, one discovered that the atmospheric neutrinos oscillate mainly with tau neutrinos. The flux of the neutrinos, coming from the atmosphere above Kamioka, was as high as expected, but the flux of the neutrinos, coming from the other side of the earth, was only about 50% of the expected rate. The experiment was carried out by Yoji Totsuka and his group. Totsuka died in 2008. In 2015 his successor Takaaki Kajita received the Nobel prize.

In Canada a neutrino detector was built near Sudbury (Ontario) — the Sudbury Neutrino Observatory (SNO). With this detector one could observe solar neutrinos. An electron neutrino hits a deuteron, which splits into two protons and an electron — this process can be observed. Furthermore it was possible to observe the neutral current interaction of the neutrinos. If a neutrino collides with a deuteron, it splits into a proton and a neutron. This reaction can also be observed.

The neutral current interaction is not affected by oscillations, since all neutrinos have the same neutral current interaction. However oscillations are relevant for the charged current interaction. An electron neutrino, which becomes a muon neutrino, will not produce an electron after colliding with a nucleus. By comparing the interaction rates for the neutral and for the charged current interactions one has observed the oscillations of the solar neutrinos. The director of SNO, Arthur McDonald received the Nobel prize in 2015 together with Kajita.

Before I describe the mixing of the neutrinos, I shall consider the flavor mixing of the quarks. The CKM-matrix describes the weak transitions between the quarks (u, c, t) and the quarks (d, s, b):

$$U = \begin{pmatrix} V_{ud} & V_{us} & V_{ub} \\ V_{cd} & V_{cs} & V_{cb} \\ V_{td} & V_{ts} & V_{tb} \end{pmatrix}. \tag{1}$$

I prefer a parametrization, which Z. Xing and I introduced in 1997, given by the angles θ_u, θ_d and θ:

$$U = \begin{pmatrix} c_u & s_u & 0 \\ -s_u & c_u & 0 \\ 0 & 0 & 1 \end{pmatrix} \times \begin{pmatrix} e^{-i\delta} & 0 & 0 \\ 0 & c & s \\ 0 & -s & c \end{pmatrix} \times \begin{pmatrix} c_d & -s_d & 0 \\ s_d & c_d & 0 \\ 0 & 0 & 1 \end{pmatrix}. \tag{2}$$

Here we used: $c_{u,d} \sim \cos\theta_{u,d}$, $s_{u,d} \sim \sin\theta_{u,d}$, $c \sim \cos\theta$, $s \sim \sin\theta$. The angle θ_u describes the mixing between the u-quark and the c-quark, the angle θ_d the mixing between the d-quark and the s-quark and the angle θ the mixing among the heavy quarks $(t, c/b, s)$. The three angles have been determined by the experiments. Here are the average values:

$$\theta_u \simeq 5.4^\circ, \quad \theta_d \simeq 11.7^\circ, \quad \theta \simeq 2.4^\circ.$$

If the quark masses vanish, the flavor mixing angles would also be zero. Thus the mixing angles must be functions of the quark masses. If the masses change, the mixing angles will also change. For example the Cabibbo angle $\theta_C \simeq 13^\circ$ could be given by the ratio of the quark masses:

$$\tan\theta_C \simeq \sqrt{m_d/m_s}. \tag{3}$$

This relation works very well:

$$\tan\theta_C \simeq 0.23 \simeq \sqrt{m_d/m_s}. \tag{4}$$

Such a relation can be derived, if the quark mass matrices have "texture zeros", as shown by Steven Weinberg and me in 1978. Texture zeros are naturally obtained in the grand unified theory, based on the gauge group SO(10).

Let me discuss a simple example, using only four quarks: u, d, c, s. Their mass matrices have a zero in the $(1, 1)$-position:

$$M = \begin{pmatrix} 0 & A \\ A^* & B \end{pmatrix}. \tag{5}$$

These mass matrices can be diagonalized by a rotation. The rotation angles are:

$$\theta_u \simeq \sqrt{m_u/m_c} \simeq 0.09, \quad \theta_d \simeq \sqrt{m_d/m_s} \simeq 0.23. \tag{6}$$

The Cabibbo angle is given by the difference:

$$\theta_C \simeq \left| \sqrt{m_d/m_s} - e^{i\phi} \sqrt{m_u/m_c} \right|. \tag{7}$$

In the complex plane this relation describes a triangle. The phase parameter is unknown, however it must be close to 90°, since the Cabibbo angle is given by the ratio m_d/m_s:

$$\theta_C \simeq \sqrt{m_d/m_s}. \tag{8}$$

Thus the triangle is close to a rectangular triangle.

For six quarks the "texture zero" mass matrices for the quarks of charge $(2/3)$ and of charge $(-1/3)$ are:

$$M = \begin{pmatrix} 0 & A & 0 \\ A^* & B & C \\ 0 & C^* & D \end{pmatrix}. \tag{9}$$

We can calculate the angles θ_u and θ_d as functions of the mass eigenvalues:

$$\theta_u \simeq \sqrt{m_u/m_c}, \quad \theta_d \simeq \sqrt{m_d/m_s}. \tag{10}$$

Using the observed mass values for the quarks, we find:

$$\theta_u \simeq (5.0 \pm 0.7)^\circ, \quad \theta_d \simeq (13.0 \pm 0.4)^\circ. \tag{11}$$

The experimental values agree very well with the theoretical results:

$$\theta_u \simeq (5.4 \pm 1.1)^\circ, \quad \theta_d \simeq (11.7 \pm 1.1)^\circ. \tag{12}$$

Now we consider the flavor mixing of the leptons. The neutrinos, emitted in weak decays, are mixtures of different mass eigenstates. This leads to neutrino oscillations — at least two neutrinos must have finite masses. Neutrino oscillations give only information about the mass differences of the neutrinos. The masses of the neutrinos are still not known.

Since there are three neutrinos, the mixing of the neutrinos is described by a unitary 3×3-matrix, which is similar to the CKM-matrix for the quarks:

$$U = \begin{pmatrix} V_{1e} & V_{2e} & V_{3e} \\ V_{1\mu} & V_{2\mu} & V_{3\mu} \\ V_{1\tau} & V_{2\tau} & V_{3\tau} \end{pmatrix}. \tag{13}$$

The matrix elements describe the interaction of the charged leptons with the three neutrinos. These are superpositions of three mass eigenstates. The mixing matrix can be parametrized in terms of three angles and three phases. I use a parametrization, introduced in 2009 by Z. Xing and me:

$$U = \begin{pmatrix} c_l & s_l & 0 \\ -s_l & c_l & 0 \\ 0 & 0 & 1 \end{pmatrix} \times \begin{pmatrix} e^{-i\varphi} & 0 & 0 \\ 0 & c & s \\ 0 & -s & c \end{pmatrix} \times \begin{pmatrix} c_\nu & -s_\nu & 0 \\ s_\nu & c_\nu & 0 \\ 0 & 0 & 1 \end{pmatrix} P_\nu. \tag{14}$$

Here we have used: $c_{l,\nu} \sim \cos\theta_{l,\nu}$, $s_{l,\nu} \sim \sin\theta_{l,\nu}$, $c \sim \cos\theta$ and $s \sim \sin\theta$. The angle θ_ν is the solar angle θ_{sun}, the angle θ is the atmospheric angle θ_{at}, and the angle θ_l is the "reactor angle". The phase matrix $P_\nu = \text{Diag}\{e^{i\rho}, e^{i\sigma}, 1\}$ is relevant only if the neutrino masses are Majorana masses which will be discussed later.

The experiments about the neutrino oscillations give the following information about the three mixing angles:

$$\theta_{\text{sun}} = \theta_\nu \simeq 34^\circ, \quad \theta_{\text{at}} = \theta \simeq 38^\circ, \quad \theta_l \simeq 13^\circ.$$

The neutrino oscillations also determine the mass differences between the neutrinos. The difference of the first and the second neutrino mass is about 0.009 eV, the difference of the second and the third neutrino mass is about 0.05 eV.

We assume that the mass matrices of the leptons have "texture zeros":

$$M = \begin{pmatrix} 0 & A & 0 \\ A^* & B & C \\ 0 & C^* & D \end{pmatrix}. \tag{15}$$

In this case we can calculate two leptonic mixing angles as functions of mass ratios:

$$\tan\theta_l \simeq \sqrt{m_e/m_\mu}, \quad \tan\theta_\nu \simeq \sqrt{m_1/m_2}. \tag{16}$$

The solar mixing angle has been measured. Thus we obtain for the neutrino mass ratio:

$$\sqrt{m_1/m_2} \simeq 0.66. \tag{17}$$

This relation and the experimental results for the mass differences of the neutrinos, measured by the neutrino oscillations, allow us to determine the three neutrino masses:

$$\begin{aligned} m_1 &\simeq 0.004 \text{ eV}, \\ m_2 &\simeq 0.010 \text{ eV}, \\ m_3 &\simeq 0.050 \text{ eV}. \end{aligned} \tag{18}$$

These neutrino masses are very small, much smaller than the masses of the charged leptons. The ratio of the mass of the tau neutrino and of the mass of the tauon is only about 2.7×10^{-11}.

Dirac introduced the four-component fermions in 1928, which describe the electron and its antiparticle, the positron. One year later Hermann Weyl discussed two-component fermions without any mass, the "Weyl fermions". In 1937 Ettore Majorana discovered that for neutral fermions he could introduce a mass for a Weyl fermion. Such a particle would be a fermion without the corresponding antiparticle. Such a neutral particle is called a "Majorana particle".

Since the neutrino masses are tiny, one can speculate, whether the neutrinos are normal Dirac particles or Majorana particles. If they are Majorana particles, the smallness of the neutrino masses can be understood by the "seesaw"-mechanism. The mass matrix of the neutrinos is a matrix with one "texture zero" in the $(1,1)$-position. The two off-diagonal terms are given by the Dirac mass term D. A large Majorana mass term appears in the $(2,2)$-position:

$$M_\nu = \begin{pmatrix} 0 & D \\ D & M \end{pmatrix}. \tag{19}$$

After diagonalization one obtains a large Majorana mass M and a small neutrino mass:

$$m_\nu \simeq D^2/M. \tag{20}$$

One expects that the Dirac term D is similar to the corresponding charged lepton mass. For example, let us consider the tau lepton and its neutrino. If D is given by the tau lepton mass, we obtain for the heavy Majorana mass M:

$$M \simeq 6.3 \times 10^{10} \text{ GeV}. \tag{21}$$

The only way to test the nature of the neutrino masses is to study the neutrinoless double beta decay, which violates lepton number conservation. Two neutrons inside an atomic nucleus decay by emitting two electrons and two neutrinos. The two Majorana neutrinos annihilate — only two electrons are emitted. The annihilation rate is a function of the Majorana mass.

In various experiments one has searched for the neutrinoless double beta decay, in particular for the decay of xenon into barium. Thus far the decay has not been observed.

If neutrinos mix, all three neutrino masses will contribute to the decay rate. Their contributions are given by the masses of the neutrinos, by the mixing angles and by the CP-violation phases. Using the neutrino masses and the mixing angles, one finds for the effective neutrino mass, relevant for the neutrinoless double beta decay:

$$\tilde{m} \simeq 0.005 \text{ eV}. \tag{22}$$

The present limit for this effective mass is provided by the Kamland experiment:

$$\tilde{m} < 0.12 \text{ eV}. \tag{23}$$

This value is 24 times larger than the expected value — thus it will be very difficult to find the neutrinoless double beta decay.

The texture zero idea provides a coherent framework to understand the flavor mixing of quarks and leptons. The mixing angles for the quarks are given by quark mass ratios. For the leptons the solar mixing angle is determined by the ratio of the masses of two neutrinos. Thus the absolute masses of the neutrinos can be calculated.

In the near future it should be possible to observe CP-violation for neutrinos, using reactor neutrinos. Furthermore the mixing angles and the mass differences will be determined with higher precision.

The flavor mixing of the quarks is interesting and thus far not understood in detail. The same is true for the physics of neutrinos. Why are the mixing angles of the leptons much larger than the mixing angles for the quarks? Why are the masses of the neutrinos much smaller than the masses of the associated charged leptons? Neutrino physics will remain an interesting field.

Majorana Fermions in Condensed-Matter Physics

A. J. Leggett
*Department of Physics,
University of Illinois at Urbana-Champaign,
Urbana, IL 61801, USA
aleggett@illinois.edu*

It is an honor and a pleasure to have been invited to give a talk in this conference celebrating the memory of the late Professor Abdus Salam. To my regret, I did not know Professor Salam personally, but I am very aware of his work and of his impact on my area of specialization, condensed matter physics, both intellectually through his ideas on spontaneously broken symmetry and more practically through his foundation of the ICTP. Since I assume that most of this audience are not specialized in condensed-matter physics, I thought I would talk about one topic which to some extent bridges this field and the particle-physics interests of Salam, namely Majorana fermions (M.F.s). However, as we shall see, the parallels which are often drawn in the current literature may be a bit too simplistic. I will devote most of this talk to a stripped-down exposition of the current orthodoxy concerning M.F.s. in condensed-matter physics and their possible applications to topological quantum computing (TQC), and then at the end briefly indicate why I believe this orthodoxy may be seriously misleading.

Keywords: Majorana fermions; anyons; topological quantum computing; $p + ip$ state.

As most of you know, the concept of a Majorana fermion (hereafter M.F.) was originally introduced[1] in a particle-physics context, by Ettore Majorana in 1937; it is a fermionic quasiparticle which is its own antiparticle. More formally, the field operator $\widehat{\psi}\,(\mathbf{r})$ which creates a M.F. has the property

$$\widehat{\psi}\,(\mathbf{r}) = \widehat{\psi}^{\dagger}\,(\mathbf{r}), \tag{1}$$

from which follows (with appropriate normalization conventions) the anticommutation relation (ACR)

$$\{\widehat{\psi}\,(\mathbf{r}), \widehat{\psi}\,(\mathbf{r}')\} = \delta(\mathbf{r} - \mathbf{r}'). \tag{2}$$

(instead of the standard fermionic ACR which would have zero on the right-hand side).

How can such a particle arise in condensed matter physics? The standard example is the case of a so-called "topological superconductor" (TS), as described by the standard "mean-field" theory which is a generalization of that developed for a simple Bardeen–Cooper–Schrieffer (BCS) superconductor; for the latter, see e.g.,

Chapter 5 of Ref. 2. Consider a general nonrelativistic Hamiltonian of the form $\hat{H}_0 + \hat{V}$, where

$$\hat{H}_0 \equiv \frac{\hbar^2}{2m} \int \left\{ \sum_\alpha \nabla \hat{\psi}^\dagger_\alpha(\mathbf{r}) \cdot \nabla \hat{\psi}_\alpha(\mathbf{r}) + \sum_{\alpha\beta} U_{\alpha\beta}(\mathbf{r}) \hat{\psi}^\dagger_\alpha(\mathbf{r}) \hat{\psi}_\beta(\mathbf{r}) \right\} d\mathbf{r}. \quad \text{(3a)}$$

$$\hat{V} \equiv \frac{1}{2} \sum_{\alpha\beta\gamma\delta} \iint d\mathbf{r}\, d\mathbf{r}' V_{\alpha\beta\gamma\delta}(\mathbf{r}, \mathbf{r}') \hat{\psi}^\dagger_\alpha(\mathbf{r}) \hat{\psi}^\dagger_\beta(\mathbf{r}') \hat{\psi}_\gamma(\mathbf{r}') \hat{\psi}_\varepsilon(\mathbf{r}). \quad \text{(3b)}$$

Here the suffix $\alpha = \pm 1$ is the spin index, and the quantity $\hat{\psi}_\alpha(\mathbf{r})$ and its Hermitian conjugate $\hat{\psi}^\dagger_\alpha(\mathbf{r})$ are field operators satisfying the standard fermionic ACRs

$$\{\hat{\psi}_\gamma(\mathbf{r}), \hat{\psi}^\dagger_\beta(\mathbf{r}')\} = \delta_{\alpha\beta}\delta(\mathbf{r} - \mathbf{r}'), \quad \{\hat{\psi}_\alpha(\mathbf{r}), \hat{\psi}_\beta(\mathbf{r}')\} = 0. \quad (4)$$

For the moment, we place no particular constraints on the form of the single-particle potential $U_{\alpha\beta}(\mathbf{r})$ or that of the two-particle interaction $V_{\alpha\beta\gamma\delta}(\mathbf{r}, \mathbf{r}')$. Note that the Hamiltonian (3) is automatically invariant under the global $U(1)$ transformation

$$\hat{\psi}_\alpha(\mathbf{r}) \to e^{i\phi} \hat{\psi}_\alpha(\mathbf{r}), \quad \hat{\psi}^\dagger_\alpha(\mathbf{r}) \to e^{-\phi} \hat{\psi}^\dagger_\alpha(\mathbf{r}), \quad (5)$$

where φ is a real position-independent constant. (We could of course introduce a vector potential $\mathbf{A}(\mathbf{r})$ and more general gauge transformations, thereby making explicit contact with Salam's work, but for simplicity of presentation I do not do so here).

The essence of the mean-field (mf) theory of superconductivity[2] consists in introducing the notion of spontaneous breaking of the $U(1)$ symmetry (hereafter abbreviated SBU(1)S) or equivalently the idea that we are free to describe the system by quantum-mechanical states which are not eigenstates of the total particle number operator

$$\hat{N} \equiv \sum_\alpha \int d\mathbf{r}\, \hat{\psi}^\dagger_\alpha(\mathbf{r}) \hat{\psi}_\alpha(\mathbf{r}). \quad (6)$$

Indeed, in mf theory, the ground state of a superconducting system containing an **even** number of fermions is (explicitly or implicitly) taken in the form

$$\Psi_{(\text{even})} = \sum_{N=\text{even}} C_N \Psi_N, \quad (7)$$

where the quantities C_N are complex coefficients. As a result of the ansatz (7), the "anomalous averages", that is, expectation values of the form

$$F_{\alpha\beta}(\mathbf{r}, \mathbf{r}') \equiv \langle \hat{\psi}_\alpha(\mathbf{r}) \hat{\psi}_\beta(\mathbf{r}') \rangle \quad (8)$$

can legitimately be taken to be nonzero in the ground state (and more generally in the superconducting phase). The quantity $F_{\alpha\beta}(\mathbf{r}, \mathbf{r}')$ may be intuitively regarded (apart from its normalization) as the wavefunction of the Cooper pairs which form in the superconductor (on this see e.g., Ref. 3, Chapter 5).

The mf method then proceeds by decoupling the expression for the potential energy operator $\hat{V}$, Eq. (3b), by a "generalized Hartree–Fock" approximation, in which not only operators of the form $\hat{\psi}_\alpha^\dagger(\mathbf{r})\,\hat{\psi}_\beta(\mathbf{r}')$ but also those of the form $\hat{\psi}_\alpha(\mathbf{r})\,\hat{\psi}_\beta(\mathbf{r}')$ are (partially) replaced by their averages. The resulting form of $\hat{V}$ contains not only the standard Hartree and Fock terms, which for simplicity of presentation I will omit, but also the "pairing" ("mf") term

$$\hat{V}_{\text{mf}} = \sum_{\alpha\beta} \iint d\mathbf{r}\, d\mathbf{r}' \{\Delta_{\alpha\beta}(\mathbf{r},\mathbf{r}')\, \hat{\psi}_\alpha^\dagger(\mathbf{r})\, \hat{\psi}_\beta^\dagger(\mathbf{r}') + \text{h.c.}\}, \tag{9}$$

where the $\hat{\psi}_{\alpha,\beta}^\dagger$ are still operators, but the quantity

$$\Delta_{\alpha,\beta}(\mathbf{r},\mathbf{r}') \equiv \sum_{\alpha\beta\gamma\delta} v_{\alpha\beta\gamma\delta}(\mathbf{r},\mathbf{r}')\langle \hat{\psi}_\gamma(\mathbf{r}')\, \hat{\psi}_\delta(\mathbf{r})\rangle \tag{10}$$

is a c-number; Δ is often regarded as a sort of effective mf. While the general expression (9) looks dauntingly complicated, some of you may be familiar with the simpler form it takes in the case of a contact interaction $V_\alpha\delta(\mathbf{r})$ and s-wave pairing:

$$\hat{V}_{\text{mf}} = \int d\mathbf{r}\{\Delta(\mathbf{r})\, \hat{\psi}_\uparrow^\dagger\, V(\mathbf{r})\, \hat{\psi}_\downarrow(\mathbf{r}) + \text{h.c.}\}, \tag{11a}$$

$$\Delta(\mathbf{r}) \equiv V_0\langle \hat{\psi}_\downarrow(\mathbf{r})\, \hat{\psi}_\uparrow(\mathbf{r})\rangle. \tag{11b}$$

Before we proceed further, it is necessary to state one caveat: were we to minimize the sum of $\hat{H}_0$ and $\hat{V}_{\text{mf}}$ as it stands, then since there is now no explicit restriction on the total particle number N the expectation value of the latter would likely turn out to be zero or at most very small (certainly this will be so if the potential is too weak to form a two-particle bound state in free space). Thus, we need to add to the single-particle energy $\hat{H}_0$ a term $-\mu\hat{N}$. With this correction, and continuing to omit the Hartree and Fock terms (which do not affect subsequent arguments qualitatively) the mf [Bogoliubov–de Gennes (BdG)] Hamiltonian is schematically of the form

$$\hat{H}_{\text{mf}} = \sum_{\alpha\beta} \Big\{ \int d\mathbf{r} K_{\alpha\beta}(\mathbf{r})\, \hat{\psi}_\alpha^\dagger(\mathbf{r})\, \hat{\psi}_\beta(\mathbf{r}) + \frac{1}{2}\iint d\mathbf{r} d\mathbf{r}' \Delta_{\alpha\beta}(\mathbf{r},\mathbf{r}')\, \hat{\psi}_\alpha^\dagger(\mathbf{r})\, \hat{\psi}_\beta^\dagger(\mathbf{r}') + \text{h.c.} \Big\}, \tag{12}$$

where the notation $K_{\alpha\beta}(\mathbf{r})$ is introduced in the first term as a shorthand for the quantity

$$\delta_{\alpha,\beta}\left(\frac{\hbar^2}{2m}\boldsymbol{\nabla}\, \hat{\psi}_\alpha^\dagger(\mathbf{r}) \cdot \nabla\, \hat{\psi}_\beta(\mathbf{r}') - \mu\, \hat{\psi}_\alpha^\dagger(\mathbf{r})\, \hat{\psi}_\beta(\mathbf{r})\right). \tag{13}$$

The operator $\hat{H}_{\text{mf}}$ does not conserve particle number, but does conserve the parity of the latter (i.e., it does not mix states with even and odd parity); note that the occurrence of terms proportional to (the operator) $\hat{\psi}^\dagger\hat{\psi}^\dagger$ leads rather naturally to eigenstates of the form (7).

In principle, we may try to use the mf Hamiltonian (12) for either of two purposes: (1) we could try to determine the form of the even-parity ground state [i.e., of the state vectors $\hat{\boldsymbol{\psi}}_N$ and coefficients C_N in Eq. (7)]. This is possible, but actually quite complicated (see e.g., Appendix A of Ref. 4), and in fact we need to proceed via the solution of task (2), namely to determine the **relation** between the even-parity ground state and the simplest ("single-fermion") odd-parity states. It is the latter problem which is central to the question of M.F., so I shall from now on specialize to that.

The "textbook" prescription for creating, from the even-parity ground state, the simplest odd-parity states goes as follows [for simplicity of notation I omit in the following the spin indices: in reality the quantities $\hat{\psi}\,(\mathbf{r})$, etc., are spinors in **real** spin space (not "Nambu space", see below)]: and the us and vs are matrices. We construct an operator of the form

$$\gamma_i^\dagger = \int d\mathbf{r}[u_i(\mathbf{r})\,\hat{\psi}^\dagger\,(\mathbf{r}) + v_i(\mathbf{r})\,\hat{\psi}\,(\mathbf{r})] \tag{14}$$

and determine the coefficients $u_i(\mathbf{r})$, $v_i(\mathbf{r})$ by requiring $\gamma_i^\dagger$ to satisfy the eigenvalue equation (with E_i the value of H_{mf} relative to the even-parity ground state).

$$[\hat{H}_{\mathrm{mf}}, \gamma_i^\dagger] = E_i\gamma_i^\dagger, \tag{15}$$

so that (with appropriate normalization of u_i and v_i, see below)

$$\hat{H}_{\mathrm{mf}} = \sum_i E_i\gamma_i^\dagger\gamma_i = \mathrm{const.} \tag{16}$$

The states $\gamma_i^\dagger|\ \hat{\psi}_{\mathrm{even}}\rangle$ are then the simplest ("single-fermion") odd-parity energy eigenstates of the system, and are said to contain a single "Bogoliubov quasiparticle".

When written out explicitly in terms of the amplitudes $u(\mathbf{r})$, $v(\mathbf{r})$ Eq. (14) becomes a set of linear equations which can be written **schematically** in the form

$$\begin{pmatrix} \hat{K} & \hat{\Delta} \\ \hat{\Delta} & -K \end{pmatrix}\begin{pmatrix} u \\ v \end{pmatrix} = E\begin{pmatrix} u \\ v \end{pmatrix}, \tag{17}$$

where the matrix elements $\hat{K}$ and $\hat{\Delta}$ are themselves matrices in the (real) spin space, and may also be integral operators in (real) coordinate space, cf. Eq. (12). $\hat{\Delta}$ must be calculated self-consistently according to Eq. (10). In the best-known case of simple BCS pairing, the equations take the possibly more familiar form

$$\begin{aligned} (\hat{H}_0 - \mu)u(\mathbf{r}) + \Lambda(\mathbf{r})u(\mathbf{r}) &= Eu(\mathbf{r}), \\ \Delta^*(\mathbf{r})u(\mathbf{r}) - (\hat{H}_0 - \mu)v(\mathbf{r}) &= Ev(\mathbf{r}), \end{aligned} \tag{18}$$

where $u(\mathbf{r})$, $v(\mathbf{r})$ and $\Delta(\mathbf{r})$ are all simple scalars. Equation (17) is called the BdG equations, and have been very widely used over the last 50 years in the theoretical literature on superconductivity, including cases more complicated than simple s-wave pairing. It is convenient to normalize the us and vs, so that

$$\int (|u|^2(\mathbf{r}) + |v|^2(\mathbf{r})) = 1, \tag{19}$$

whereupon the γ_is satisfy the standard fermionic ACRs,

$$\{\gamma_i, \gamma_j^\dagger\} = \delta_{ij}, \quad \{\gamma_i, \gamma_j\} = 0 \tag{20}$$

thereby justifying Eq. (16).

All the above is standard textbook stuff, but it is important to note a crucial point: in this mf treatment, the fermionic quasiparticles are quantum superpositions of a state corresponding to an extra real fermion [with probability amplitude $u(\mathbf{r})$] and one with an extra (real) hole [with amplitude $v(\mathbf{r})$]; this is usually justified by appealing to SBU(1)S. It is conventional to represent this situation by a spinor in the 2D "Nambu space" corresponding to the particle and the hole degrees of freedom:

$$\gamma_j^\dagger \to \begin{pmatrix} u(\mathbf{r}) \\ v(\mathbf{r}) \end{pmatrix} \tag{21}$$

and to regard it (to my mind misleadingly, cf. below) as analogous to the mixing of different bands in an insulator by spin–orbit coupling; then by analogy with the standard use of the term "topological insulator" for the latter case when the mixing is topologically nontrivial, in the corresponding case for a superconductor one talks about a "TS". In the recent literature, it has become common to refer to both of these cases, in the absence of more sophisticated terms than (12) in the Hamiltonian, as "noninteracting-electron" problems, although the occurrence of the self-consistency condition (10) makes this phrasing rather misleading.

Before proceeding, we need to note a couple of easily verified properties of the γ_is:

(1) $\gamma_i|\hat{\psi}_{\text{even}}\rangle = 0, \quad \forall i \ \text{ st.} E_i \neq 0.$
(2) If there exists a formal solution to the BdG equations with some positive energy E_i, then by interchanging $u(\mathbf{r})$ and $v(\mathbf{r})$ (and appropriate attention to signs and spin suffixes) we can always construct a solution with negative energy $-E_i$. These "negative-energy" solutions are often regarded (in my opinion misleadingly, cf. below) as analogous to the "positron" solutions in quantum field theory.

Let us now raise (for the moment for no obviously good reason!) the question: do there exist "Majorana-like" solutions to the BdG equations, that is, solutions such that

$$\gamma_i^\dagger = \gamma_i. \tag{22}$$

It is clear from the definition (14) that this requires $u_i^*(\mathbf{r}) = v_i(\mathbf{r})$, (with appropriate spin-dependences, see below) and it furthermore follows from (15) that the corresponding energy eigenvalue E_i is exactly zero. However, these two conditions are not sufficient for an "interesting" Majorana; they would, for example, be satisfied by an ordinary Bogoliubov quasiparticle created in a "polar" p-wave superfluid (which has no particularly interesting topological properties) exactly at a node of the gap $\Delta(\mathbf{k})$, but one would not normally call such a quasiparticle a "Majorana". What is missing in the above definition is that the solutions of the BdG equations are spatially **localized**, i.e., that both $u(\mathbf{r})$ and $v(\mathbf{r})$ tend to zero exponentially outside a small spatial region,[a] typically a volume $\sim \xi^d$ where ξ is the Cooper pair radius and d the spatial dimension. It turns out that in addition to the requirement that the pair spin structure is a triplet [necessary to guarantee the self-conjugacy condition (22)], such localized solutions only occur if the pair wavefunction (order parameter) has a sufficiently interesting structure, typically involving spontaneous breaking of time-reversal invariance.

At this point, rather than trying to review the general picture, let me specialize to what seems to be the most promising state for the real-life realization of M.F.s, namely the so-called "$p+ip$" state. This state is characterized by the fact that either there is only a single spin state involved, or the amplitudes for "$\uparrow\uparrow$" and "$\downarrow\downarrow$" pairs may be treated as independent, and that as the name implies, in homogeneous conditions the order parameter $\Lambda(\hat{p})$ written as a function of the direction of relative momentum $\mathbf{p}$ on the Fermi surface has the form.

$$\Delta(\hat{p}) \propto \hat{p}_x + i\hat{p}_y$$

thereby violating (*inter alia*) the time-reversal invariance of the M.F. Hamiltonian [or of the original one of Eqs. (3), which for now we assume does not include TRI-violating terms]. The "$p+ip$" state is believed to describe the A phase of superfluid ^{3}He and, probably, the superconducting state of the metal Sr_2RuO_4; in the future it may be possible to realize a $(p+ip)$ state of an ultracold Fermi gas which has only a single spin population paired.

In the case of ^{3}He-A and Sr_2RuO_4, which contain both types $|\uparrow\uparrow\rangle$ and $|\downarrow\downarrow\rangle$ of Cooper pairs, it is believed (and in the latter case consistent with experiment[5]) that the system may sustain so-called "half-quantum vortices" (HQVs), that is, configurations in which (say) the $\uparrow\uparrow$ pair wavefunction exhibits a simple vortex similar to those realized in the Bose system ^{4}He, while that of the $\downarrow\downarrow$ pairs remains perfectly uniform. If this is so, then it appears that Majorana-type solutions not only can but must exist in the vicinity of these vortices (and possibly on the boundary of the system, if the number of vortices is odd). The general proof of this statement follows from the celebrated Atiyah–Patodi–Singer theorem in algebraic topology, but for any specified form of the order parameter near the vortex the existence of these solutions can be established by directly solving the BdG equations (17) (see Refs. 6 and 7).

---

[a]Or in some cases away from the boundary.

The most exciting consequence of the existence of these M.F.s in a condensed-matter context is the prospect of possibly being able to use them for topological quantum computing (TQC), that is, a form of quantum computing in which the ubiquitous decoherence is frustrated by burying the relevant quantum information in strongly entangled many-body degrees of freedom in a way which is topologically protected. Why might M.F.s be a good prospect for TQC? There are essentially two major reasons (within the standard mf treatment, cf. below):

(1) Because a M.F. is an exactly equal superposition of particle and hole, it is undetectable by any local probe.
(2) M.F.s should behave under braiding as "Ising anyons": if two HQVs each carrying a M.F., are interchanged, the phase of the many-body wavefunction is changed (relative to the zero-M.F. state) by $\pi/2$: note this is different from the standard result of π which would obtain for ordinary fermions.

This then leads to the following scheme to create a TQC (for details see Ref. 4):

(1) Create a set of HQVs, with and without M.F.s on them.
(2) Braid the vortices adiabatically.
(3) Recombine them and "measure" the result.

Then provided that during the braiding processes, the HQVs are kept at mutual distance $\gg \xi$, all decoherence should be proportional to $\exp - L/\xi$ and thus at least a class[b,8] of quantum computing operations will be topologically protected.

The above is an extremely "potted" version of the standard wisdom concerning M.F. in condensed matter physics. Now for a few comments:

(1) What exactly is a M.F.?

Recall: its creation operator $\gamma_i^\dagger$ satisfies (since $E_i = 0$ exactly) the equation

$$[\hat{H}, \gamma_i^\dagger] \quad | \tilde{\psi}_{\text{even}}\rangle = 0.$$

But this equation actually has two possible interpretations: the more obvious one is that $\gamma_i^\dagger$ creates an odd-number-parity state (a Bogoliubov quasiparticle) with energy exactly equal to that of the even-parity one. However, an equal viable interpretation is that $\gamma_i^\dagger$ simply annihilates the ground state ("pure annihilator")! Now it is in fact easy to show (cf. Ref. 9) that neither of these possibilities can satisfy the self-conjugacy condition $\gamma_i^\dagger = \gamma_i$. To satisfy it, we must superpose the two possibilities: that is, formally,

> A M.F. is simply a quantum superposition of a "real" Bogoliubov quasipartcle with zero energy and a pure annihilator.

[b] As in other putative realizations of an Ising quantum computer, there is no complete topological protection: for that one needs to go to e.g., a "Fibonacci" QC, see e.g., Ref. 8, Sec. IV.B.

Consider now a situation in which we have just two HQVs, labeled 1 and 2, arbitrarily far apart. There exists on each vortex a Majorana solution, $\gamma_1^\dagger$ and $\gamma_2^\dagger$, respectively. As we have just seen, neither of these individually corresponds to creation of an extra zero-energy Bogoliubov quasiparticle. However, the combination

$$\alpha^+ \equiv \gamma_1^\dagger + i\gamma_2^\dagger \tag{23}$$

does so and satisfies the standard ACRs (4)! Thus, the two individual Majoranas, when suitably combined, indeed represent such a quasiparticle. The curious point is that this single extra fermion is spatially "split": it has support from two very different regions of space. In this light, the "braiding" of two HQVs each carrying an M.F. may be regarded as the rotation through π of this zero-energy fermion (note that the predicted phase change of $\pi/2$ is exactly the mean of what we would expect, for the "standard" double-well case, for the symmetric and antisymmetric combinations of the states localized on 1 and 2, respectively).

An intuitive appreciation of the nature of M.F.s may be obtained by considering the "Kitaev quantum wire", a set of equally spaced sites described by the BdG-type Hamiltonian

$$\hat{H} = \sum_i t_i(\alpha_i^+ a_{i+1} + \text{h.c.}) + \sum_i \Delta_i(a_i^+ a_{i+1} + \text{h.c.}) \tag{24}$$

with the special choice $t_i = \Delta_i \equiv X_i$ (but the X_i are not necessarily identical). Let us start with a ring geometry [Fig. 1(a)]. Then the Bogoliubov quasiparticles are created by the operators

$$a_i^+ \equiv (a_i^+ - a_i) + (a_{i-1}^+ + a_{i-1}) \tag{25}$$

and have excitation energies X_i: note that they are localized on the links rather than the sites.

Now imagine that we gradually turn down to zero the "strength" X_i of some link i, say O, [see Fig. 1(b)]. There is still a fermionic (Bogoliubov) excitation associated with this link, but the energy of the corresponding odd-parity many-body state is degenerate with that of the even-parity ground state. Finally, we break the ith link, and reconfigure the set of sites into Kitaev's original open "wire" [Fig. 1(c)]; since the energy associated with this link is zero, the structure of the fermionic states cannot change and we must still have the zero-energy state. But where is it localized? In view of the symmetry of the situation, it must be localized equally on the two open ends of the wire! It should be mentioned that while this argument seems rather straightforward, the actual wavefunctions of both the even- and odd-parity ground states individually are highly entangled and the derivation of the correct forms is not entirely trivial.

(2) The experimental situation (ultra-brief review).

In the metallic superconductor Sr_2RuO_4, there is some experimental evidence for HQVs but, at least so far, none for M.F.s.

In ^{3}He-A experiments have so far failed to detect HQVs, let alone M.F.s. In the B phase of superfluid ^{3}He, with its three coexisting pairing states

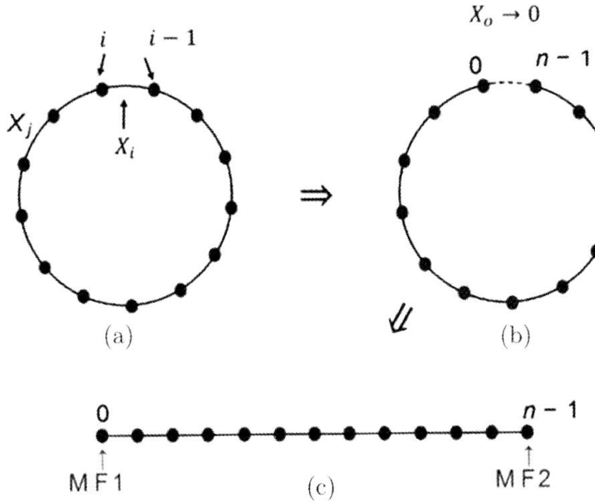

Fig. 1. Origin of two Majorana fermions in the Kitaev quantum wire.

$(\uparrow\uparrow, \downarrow\downarrow, 1/\sqrt{2}(\uparrow\downarrow + \downarrow\uparrow))$ theory has predicted that under certain conditions the **boundaries** of the sample may sustain a branch of M.F.s and experiments[10] on the transmission of ultrasound through the solid–liquid interface may provide some support for this.

An alternative proposal for the realization of M.F.s involves a semiconducting (e.g., InAs) nanowire with strong spin–orbit interaction in proximity to an ordinary s-wave superconductor: it turns out that this generates in effect an induced $(p+ip)$-like superconducting state in the wire. The experiments mostly look for signatures of Majoranas in the current–voltage (I–V) characteristics of the wire and in particular its dependence on magnetic field magnitude and direction, s-wave energy gap, temperature, etc. A number of experiments over the last four years have presented data which appear to agree with the theoretical predictions for M.F.s, but the trouble is that the argument is inevitably of the "what else could it be?" variety, and the answer given by various papers in the literature is "quite a lot"!

If one does not find spectroscopic experiments particularly conclusive, an alternative is to look for evidence of behavior which is predicted to be qualitatively different for M.F.s and for ordinary quasiparticles. One possibility which has been suggested concerns Josephson circuits whose geometry and topology permits M.F.s, for example, some generated using p-wave-like superconductivity: the current-phase relation $I(\varphi)$ of such circuits is predicted to have not just the straightforward "2π-periodicity" in the Cooper-pair phase drop φ, but an extra 4π–periodic component. Unfortunately, when translated into the language of flux Φ, this simply corresponds to the generic h/e periodicity required by the Byers–Yang theorem, so it is not clear that its observation would be particularly definitive.

At first sight, a unique signature of the presence of a pair of M.F.s might seem to be the phenomenon of "teleportation": since the corresponding single zero-energy

Bogoliubov quasiparticle is delocalized between two spatially very distant sites, one might imagine that a "cause" at HQV 1, say the addition of an extra electron, might have "effects" at HQV 2 within a very short time compared to the time $\Delta t \sim L/v_F$ for ordinary physical impulses to travel the distance L between the HQVs ($v_F =$ Fermi velocity). Unfortunately, there seem to be a number of reasons to doubt these conclusions, and the relevant theoretical community appears at the time of writing to be split/confused about it. In sum, while there is some circumstantial evidence for M.F.s in condensed matter physics, their existence cannot be said to have been established beyond reasonable doubt. No doubt that may change in the next few years.

The above discussion is wholly within the "standard" mf approach to superconductivity (or fermion superfluidity). In the last few minutes of my talk, I would like to explain briefly why I believe that this approach may itself be partly or wholly misleading. This is an ongoing research in collaboration with my student Yiruo Lin.

The problem lies in the very notion of spontaneously broken $U(1)$ symmetry (SBU(1)S). To recap, if we proceed as above using this idea, we relax particle number conservation and thus write the even-parity ground state in the form

$$\widehat{\psi}_{\text{even}} \sim \sum_{N\text{-even}} C_N \, \widehat{\psi}_N \tag{26}$$

this then allows us to seek the simplest odd-parity states in the form

$$\gamma_i^\dagger = \int d\mathbf{r}[u_i(\mathbf{r}) \, \widehat{\psi}^\dagger (\mathbf{r}) + v_i(\mathbf{r}) \, \widehat{\psi} \, (\mathbf{r})]. \tag{27}$$

But in real life,

SBU(1)S is a myth!

Indeed, if we consider a system contained in a well-defined volume, then either the boundaries of that volume are real physical walls, in which case a rigorous superselection rule forbids any nontrivial form of Eq. (26), or they are not, in which case, while the **reduced** density matrix of the system in the number representation may have different nonzero elements along the diagonal, ($\rho_{NN} \neq 0$ for several N), its **off-diagonal** elements ρ'_{NN}, $N \neq N'$ are rigorously zero and hence $\hat{\rho}$ cannot correspond to the ansatz (26). This state of affairs does not matter too much while we are considering only the even-parity situation, since we can use the "Anderson trick" to write

$$\widehat{\psi}_N \sim \int \widehat{\psi}_{\text{even}} \, (\varphi) \exp -iN\varphi \, d\varphi \tag{28}$$

[where φ is the overall phase of the pair wavefunction as defined under the assumption of SBU(1)S]. Similarly, it does not matter to the extent that we consider only the odd-parity situation (where a similar trick is possible). But if we want to discuss the **relation** between the even- and odd-parity states (which is the crux of what we have been doing above) it may be fatal! Two examples of situations where the use of (27) leads to contradictions are the theory of NMR of a surface M.F. in ^{3}He-B,[11] and a much more general problem, that of Galilean invariance (application of the "textbook" formula (14) for creation of a Bogoliubov quasiparticle successively in the rest frame of the condensate and one related to it by a Galilean boost violates this principle).

In both the above cases, the problem is easily fixed by properly taking into account the effect of the condensate, in particular by replacing formula (14) for $\gamma_i^\dagger$ by

$$\gamma_i^\dagger = \int d\mathbf{r}\{u(\mathbf{r})\,\widehat{\psi}^\dagger(\mathbf{r}) + v(\mathbf{r})\,\widehat{\psi}(\mathbf{r})C^\dagger\}, \tag{29}$$

where the operator $C^\dagger$ creates an extra Cooper pair so as to conserve total particle number. This makes essentially no difference in simple situations such as that of a BCS s-wave condensate at rest, in which the Cooper pairs have no "interesting" properties, but the moment they have a nontrivial degree of freedom (spin, center-of-mass momentum, angular momentum, etc.) it may change the picture qualitatively. But the very existence of M.F.s rests on the fact that the pairs indeed have "interesting" properties!

In this situation, we really need to go back to first principles and examine how the replacement of (14) by (29) (and also the question of whether $C^\dagger$ is simply the same quantity as appears in the even-parity ground state wavefunction) affects the "established" arguments about M.F.s and TQC. Our preliminary conclusion is that it probably does not affect the existence of M.F.s (though they are no longer strictly their own antiparticles), but it may well destroy much of the current orthodoxy concerning their local undertestability, behavior under braiding, etc. Wait for the next thrilling installment!

References

1. E. Majorana, *Nuovo Cimento* **14**, 171 (1937) (in Italian).
2. P. G. de Gennes, *Superconductivity of Metals and Alloys* (W. A. Benjamin, New York, 1966).
3. A. J. Leggett, *Quantum Liquids: Bose Condensation and Cooper Pairing in Condensed-Matter Systems* (Oxford University Press, Oxford, 2006).
4. M. Stone and S.-B. Chung, *Phys. Rev. B* **73**, 014505 (2006).
5. J. Jang *et al.*, *Science* **331**, 186 (2011).
6. N. B. Kopnin and M. M. Salomaa, *Phys. Rev. B* **44**, 9667 (1991).
7. G. E. Volovik, *JETP Lett.* **70**, 609 (1999).
8. C. Nayak *et al.*, *Rev. Mod. Phys.* **80**, 1083 (2008).
9. A. J. Leggett, Majorana fermions in fermi superfluid: A pedagogical note, in *Doing Physics: A Festscrift for Thomas Erber*, Chap. 16, ed. P. W. Johnson (IIT Press, 2010), pp. 173–184.
10. Y. Okuda and R. Nomura, *J. Phys. Condens. Matter* **24**, 343201 (2012).
11. E. Taylor *et al.*, *Phys. Rev.* **91**, 134505 (2015).

What Remains Invariant: Life Lessons from Abdus Salam

T. Z. Husain

tasneem.zehra@gmail.com
www.tasneemzehrahusain.com

Abdus Salam was a multi-dimensional man who straddled research and institution-building with enviable flair; he was both religious and iconoclastic, a true citizen of the world yet deeply nationalistic, a scientist and a lover of literature, a villager and a cosmopolitan. While the specific details of his life belong to the man alone, Salam's rich experiences exemplify certain values, attitudes and lessons that are universal.

In this talk, we attempt to draw out those truths, by looking through the lens of physics. Our analysis of Salam's personal journey mirrors the search for the invariants of a physical system in that we look beyond the particularities of his unique set of circumstances, to the essence that both categorizes, and transcends, explicit events.

Thus we move through Salam's life, collecting 'invariants' that apply as much to us today as they did to him several decades ago. Together, these constitute an enduring wisdom that can prove invaluable to young scientists — especially those from developing countries — as they navigate different cultures, manage diverse loyalties, and balance the lure of research with the demands of service.

1. Introduction

Glad as I am to have the opportunity to pay a tribute to Abdus Salam, I am keenly aware that I stand among a group of people who knew him extremely well. Several of the speakers here have been Salam's students or collaborators and, as such, have shared a deep, decades long association with him. I, on the other hand, cannot claim any familiarity with the intricacies of his research, and since I never set eyes on him, I obviously have no personal anecdotes to share either. In fact, my lack of an active relationship with Abdus Salam seemed quite an obstacle when I sat down to plan this talk.

I wanted to honor him by making a meaningful contribution to the conversation, yet I knew that any personal or intellectual episode I read up on, someone here would probably know in detail, from direct experience. I didn't want to merely recite a watered down second-hand history, and I struggled with what to say for quite a while, until I realized that the very fact that I feel such a deep intellectual and emotional bond with Salam — despite never having met him, or even having worked on his theories — is in itself worthy of exploration.

So, my talk today is about how a young scientist who grew up in Pakistan decades after Salam left the country, came to look upon him as a mentor, and why I feel he could play that role for so many other young people, particularly in the developing world.

2. 'Discovering' Abdus Salam

My initial awareness of Abdus Salam had much in common with his early impressions of the weak force. In Salam's own words:

> "When I was in school in about 1936, I remember the teacher giving us a lecture on the basic forces in Nature. He began with gravity. Of course we had all heard of gravity. Then he went on to say, 'Electricity. Now there is a force called electricity but it doesn't live in our town Jhang. It lives in the capital town of Lahore, a hundred miles to the east. He had just heard of the nuclear force, and he said 'That only exists in Europe'. We were told not to worry about it."

Growing up, I didn't hear all that much about Abdus Salam. The lack of public discourse was largely attributable to the political ramifications that came from his sect being declared non-Muslims, and the abstract nature of his research put it out of the realm of most social conversations. There was, however, a small group of people who either knew him personally, or knew enough physics to appreciate his stature, and among them, he was a larger than life figure. As my interest in theoretical physics grew, I began to come into contact with some of these people — college and university professors, mainly. Many of them had spent summers in Trieste, and all of them revered Salam. The stories they told were my first real introduction to him.

When I got to the ICTP, Salam was no longer physically there, but his presence was still tangible. Most daily affairs were conducted as if under his invisible eye. "Professor Salam would want it this way," or "Professor Salam would never stand for this," were phrases that could effectively end a debate. The office in the library lay untouched, as if waiting for him to walk in at any moment. Everyone had a story — or ten — to tell about him and, steeped in that atmosphere, studying the theories he had worked on, I began to piece together a image of Abdus Salam, the man.

But it was not until I returned to Pakistan as a founding faculty member in a new school of Science and Engineering, that I truly came to appreciate Salam. In those early days, when issues of education and institution building were very much on my mind, I discovered Ideals and Realities, a compendium of Abdus Salam's writings on many varied topics. Thus far, I had known Salam only second-hand, through stories and anecdotes related by others; this was my first direct interaction with his mind. Reading his words, thinking through his thoughts, I came to feel a distinct intellectual kinship with him.

I saw a multi-dimensional man who straddled research and institution-building with enviable flair; he was both religious and iconoclastic, a true citizen of the world yet deeply nationalistic, a scientist and a lover of literature, a villager and a cosmopolitan. I was struck by the breadth of his concerns and the depth of his analysis, and I couldn't help but admire the generosity of spirit that shone through the page.

I began to toy with the idea of making a short film on Salam, both to make him better known among the public, and to show young people from all backgrounds that it is possible to start from a "humble" beginning and rise all the way to the top. In order to tell his story compellingly and effectively to high-school students, I had to take a step back and view Salam's life and career as a whole. I had to articulate exactly what I thought a young aspiring scientist today could learn from Abdus Salam.

That project never got beyond the fund-raising stage, but it gave me two gifts. First, in the process of planning the film, I went through archival footage and videotaped interviews and 'saw' Salam for the first time; second, it planted the seed for this talk.

3. The Invariants

Our attempt to understand the laws of Nature is, in part, an effort to gain control over a world that sometimes seems completely random. A physical theory is said to explain certain phenomena if it models their behavior in such a way that it is able to predict the outcome of a situation; given a system in a certain initial condition, the theory should be able to tell us what comes next.

Some of our most brilliant and most successful theories — like the Standard Model — are gauge theories. Such theories are formulated in terms of invariants; quantities that remain fixed despite shifting appearances and changed circumstances. Identifying these invariants goes a long way towards dictating the form of the theory, and consequently, predicting the evolution of the system.

I sometimes wish we had such theories for life; that there were equations we could churn to calculate the outcomes of situations in which we find ourselves, that there was a degree of mathematical certainty to the evolution of personal circumstance. Such a formulation is probably not forthcoming, but in the interim, there is still some wisdom to be gleaned from gauge theories. They remind us to look beyond the vagaries of fate and fortune to the invariant essence of life, to the qualities and principles that can anchors us, as we navigate this wildly variable waters of this world.

And that's where Abdus Salam comes in. I know of few others who have covered as much ground, both intellectually and physically. Salam rose from what he called "humble beginnings" all the way to the top, and the breadth of his interests and concerns was almost inexhaustibly vast. While the specific details of his life belong to the man alone, his experiences exemplify certain values, attitudes and lessons that are universal.

And so, I sat down to analyze Salam's life in an effort to extricate the invariants; the principles that apply to us all.

Here's what I came up with:

4. Cultivate a Wide Intellectual Curiosity

That might seem a strange place to start, but his lively curiosity gave Abdus Salam a unique edge. Salam was so much more than a brilliant scientist. He was almost a one-man corporation! He was able to juggle so many things because he was deeply and widely curious, he was genuinely interested in how things work.

Most of the time, those of us who pursue the natural sciences, don't focus much on the humanities. This is particularly true in the developing world where you typically choose to study one at the expense of the other. In my personal opinion, that's a huge mistake. History, literature, and philosophy might not equip you to perform surgeries, help you make technical calculations or design tangible objects, but they enable you to reach out and make connections to people, which is an equally crucial ability.

Salam read up voraciously, on all sorts of subjects, from ballroom dancing to how to fly a plane, from the Marx brothers to Punjabi poetry. He had been a lover of English literature long before he set foot in England, but during his first year in Cambridge he apparently decided that his general knowledge was unsatisfactory, and so, he set out to remedy the situation. Salam read widely about the history of civilizations, and the different religions of the world.

His familiarity with the humanities and his wide range of interests served Salam well, whether he wanted to convince officials with decision making powers, or educate and empower the masses.

Salam's facility with literature and mythology shone through in what is perhaps the most engaging and accessible description of parity violation I have ever heard. Constructing an analogy between the neutrino and the cyclops, Salam said, it is as if we live in a world where all cyclops have eyes on the left side, and no right-eyed cyclops are allowed to exist. When a (necessarily left-eyed) cyclops looks into a mirror, it sees nothing — its reflection being ruled out by the laws of Nature!

Salam was always able to speak to his intended audience in their own language, weaving in little cultural nuggets that put people at ease. During an interview in Pakistan, Salam spoke reverently of a much loved folk tale; in an article for a British newspaper, he wove in the quintessentially English humor of P.G. Wodehouse; during a lecture he delivered in Turkey, he quoted examples from their illustrious history, and the list goes on. Whatever the cause, whatever the occasion, Salam always made people feel as if he was buying in; he was listened to because he spoke as an insider.

5. Honor Your Roots, but Grow Wings

The turban he wore to the Nobel award ceremony is an obvious symbol of the value Abdus Salam placed on his cultural heritage. Salam wore suits almost every day of his life; his decision to don traditional clothes was a deliberate statement about his pride in his roots. It was as loud a tribute as he could pay to the country of his birth.

Salam's love for Pakistan ran fathoms deep. Despite being treated extremely shabbily by the Government, Salam continued to go back 'home' at every opportunity. On one such trip, after he won the Nobel Prize, Salam gave an interview in Punjabi (the local language of his region) so that he could connect with people who might not necessarily be very well educated (and hence not be comfortable with the national language, Urdu). He never became too proud to bend, to make gracious gestures like that. In this interview, Salam answered basic questions, very patiently for hours. But while he honored the humblest part of his background, he never let it hold him back.

Leaving Pakistan was probably one of the most difficult decisions Salam ever had to make — for those of us who come from the developing world — it is an all too familiar dilemma. Each of us grapples with the consequences of leaving our countries. We struggle uneasily with the question: If I go abroad, am I abandoning my people? Salam managed to walk this tightrope to perfection, performing the ultimate balancing act.

Salam's solution to the dilemma was this: if there is a way to stay and contribute, by all means one should do so, but the crucial component that will drive change in our countries is our ability to make positive contributions, not our decision to stay rooted in the soil. Salam knew that if he stayed on at Government College, he wouldn't have access to the facilities that he needed to do research, so he made the choice to return to Cambridge. He realized that if he established himself and came from a position of strength, he could do far greater good to Pakistan — even from a distance — than if he stayed on and stagnated.

6. Be Open to Change. Respond to the World

View life as a conversation, not a prewritten script. Don't walk into situations with a prepared speech in mind. Let the people you meet and the circumstances you face, influence you.

At the beginning of his academic career, Salam's intention was to join the Indian Civil Service. He studied Mathematics because this was a subject in which it was possible to achieve a perfect score, and thus gain a competitive edge in the Civil Service Exams. But, even though he set off in pursuit of a definite goal in mind, Salam was not closed-minded. While in England, he had discussions with many people, Dirac among them. The greater Salam's exposure to physics, he more he responded to it, until eventually he came to the realization that this was what he wanted to do.

Had Salam not been open to new ideas, had he been unwilling to explore previously unconsidered opportunities, he could have spent his days as a bureaucrat in India — or Pakistan. Salam dodged that fate because he was alive to possibility.

He moved through the world deliberately, fully engaging with his surroundings. As his son Umer put it, "everything touched him. Everything reached him."

7. Acknowledge Your Debts

Salam always acknowledged his debts; a rare quality, and one for which I respect him immensely. To declare yourself indebted, you must be secure enough to not feel diminished by the power dynamic that is implicit in such a statement; you must be just enough to admit that your successes are not solely due to your own "superhuman" efforts; and you must be honorable enough to not attempt to hide your tracks. Abdus Salam had all these qualities.

Unlike a lot of grand people who gloss over the non-glamorous parts of their past, Salam made it a point to acknowledge everyone who had helped him along the way. He repeatedly expressed his gratitude to the teachers at whose hands he learnt in the modest Government-run school he attended in Jhang, and until the end of his life, he arranged for books and laboratory equipment to be sent to his old school.

Each time he went back to Lahore, Abdus Salam paid his respects to his professors at Government College and Punjab University. In giving voice to, even emphasizing, their contribution to his life, he gave his teachers the rare and generous gift of showing what they had meant to him, and by doing so, he raised the esteem in which society held them.

8. Be Loyal

Salam was deeply loyal, but what truly set him apart was his ability to be loyal to many different institutions. When his alma mater in Pakistan, Government College, celebrated its 150th Anniversary, Salam was at the peak of his career; he had won the Nobel Prize and was running the ICTP. Despite all the demands on his time, he wrote a piece for the commemorative version of the Ravi: the college magazine of which he had been an editor.

From Government College in Lahore, Abdus Salam went on to St John's College at Cambridge, and he was equally loyal to that too. Most of you know the anecdote about Salam being offered a fellowship at the more prestigious Trinity College, but electing to stay at St John's, saying "the roses here are so beautiful that I don't want to leave." I heard that version many times before I learnt the corollary, which for me, changed the entire tone of the story; after praising the roses at St. John's, Salam added, "and there is also something known as loyalty."

Salam believed in nurturing institutions. Not that St John's, or for that matter, any college in Cambridge, needed nurturing but that was the general principle according to which Salam lived his life: when an institution gives to you, you give back. That's how nations are built, that's how societies grow.

Salam did the same again with Imperial College. Imperial was far sighted enough to keep him on as faculty, despite the fact that he spent a large part of his time in Trieste, and Salam lived up to the trust that was placed in him. He took pride in Imperial College and in the group he had established there, till the end of his days. Salam arranged for many from his department to visit the new institute in Trieste, and as his son Ahmad phrases it, always "took great pride in flying the college colors." The distance in no way affected Salam's loyalty to his department in London, and actually, to British Science in general.

When, in the mid 1980s, the British Government was considering cutting particle physics funding for participation large international ventures — like CERN — Salam penned an impassioned protest. "Clearly, a country that has upheld fundamental science ... cannot lightly absolve itself and withdraw from supporting this most exciting adventure of ideas of our times," he said, in a newspaper article entitled "Particle Physics: Will Britain Kill its Own Creation."

One of Salam's greatest attributes was that he acknowledged and honored all aspects of his identity, subsuming them within himself without causing any conflict. Given his ability to compound loyalties, it should come as no surprise that when Abdus Salam moved to Trieste, he made the city his own. He did not merely live somewhere, he belonged. Italy was his country, as was England. Salam thought of the entire Muslim world as his, even after he was officially declared to lie outside its folds. The Government of Pakistan treated him in a deplorable fashion, but he never responded in kind. On the contrary, in an almost heartbreakingly touching gesture of loyalty, Abdus Salam held on to his Pakistani nationality until the end of his life.

9. Be Resilient

Resilience is not an easy quality to cultivate, particularly for those of us who come from developing countries, where speaking up and asking questions is not always encouraged.

I don't know if Salam was born with this gift, but — reading between the lines — it seems to me that this quality was fortified (if not created) as a result of the two-component neutrino saga. When Salam first came up with that idea, he excitedly wrote about it to Pauli — the de facto 'referee' of the theoretical physics community. Pauli was critical, even dismissive, of the theory so Salam refrained from writing it up.

I am sure that in later years he must have regretted following Pauli's advice, but Salam's reaction to the incident is what elevates this story from a thousand others about missed opportunities. Rather than getting stuck on this one — admittedly consequential — miss, Salam incorporated the experience into his worldview, and decided that in order to avoid such a situation in the future, he would voice all his ideas, and let the rules of natural selection dictate their survival.

Salam might have resolved to not let the opinion of authority keep him from publishing something ever again, but he wasn't irresponsible about letting his ideas loose on the world. One way he managed the outflow was to use a group of peers at Imperial as his sounding board. Every day at the faculty restaurant, he would regale John Moffatt, P.T. Matthews and John Taylor with his latest theories, which they would then pick apart, sometimes dismissing them entirely, and — with no affront to his ego — Salam would simply come back the next day with fresh ideas for his posse to dismantle. One of the secrets of Salam's success was his adherence to the adage, "the best way to have a good idea is to have a lot of ideas."

We all face setbacks. What separates the successful people from the might-have-beens is the ability to bounce back. Salam not only rebounded himself, he also encouraged a similar resilience in others. In the words of Peter West, after talking to Salam, "one came away feeling failure is a temporary phenomenon."

10. Be Resourceful

The ability to find a way around obstacles always stands one in good stead. On the surface, it appears that opportunity favored Salam, and it might well have done, but often with a little nudge from the man himself.

There is no argument about the fact that Cambridge was a turning point in Salam's life and career, and one could argue that — since the particular scholarship he was awarded lasted only one year, and he was its only recipient — it was sheer luck that landed Abdus Salam in Cambridge. But that version of the story ignores one crucial component.

The scholarship was set up for the sons of farmers. Abdus Salam's father was a school inspector not a large landowner. However, when the family found out about the scholarship, Salam's father bought a small tract of land that qualified his son to apply. I do want to point out here that Salam was playing by the rules. Without ripping anyone else off, he gave himself a fighting chance. The scholarship was awarded through a competitive process and had Salam not been the best candidate, he wouldn't have won it. On the other hand, he not exhibited some creative problem solving, Abdus Salam might have followed the career path he was then pursuing, and spent his life as an officer in the Indian Civil Service.

As another case in point, there was the time when Salam came across a problem while working on renormalizing meson theories in Cambridge. He called up Freeman Dyson, who was then in Birmingham, to help resolve the puzzle but Dyson said he was leaving for Princeton the next day. Salam immediately hopped on to the next train to Birmingham, only to join Dyson on his journey to London. The two discussed physics — uninterrupted — on their way down.

Or there was the time when Salam, upon his return to Pakistan, needed to secure a scholarship. With characteristic gusto, he went out to obtain it. Salam met with the Director Education and many other officials of the Punjab Government and was eventually advised to go directly to the Governor, Francis Moody. Like Salam, Moody was also a St. John's man, and Salam thought it would do his case no harm if the alumni connection were to become known. Not wanting to be gauche about it,

Salam merely ensured that he wore his college necktie to the meeting, and trusted Fate to take care of the rest.

Reading through his biography, one comes across many examples where destiny seems to favor Salam, but to merely dismiss it at that is to trivialize his own role in shaping his life. When faced with obstacles, Salam often came up with ingenious solutions, which he then enacted with aplomb.

11. Have a Sense of Fun

While it is of course very important to work hard and discharge our duties as best we can, it is also important to refrain from becoming impassive workhorses, particularly when we occupy a position of authority. The culture of every institution benefits from a degree of relaxation, informality, even straight out fun. These little breathers are what allow us to focus sharper and work harder.

Abdus Salam was by all accounts exceedingly hard working and remarkably focused. If you glance at a list of what he was able to accomplish in his life, you might be tempted to think that the man never relaxed; you would be wrong.

Abdus Salam had a lively sense of fun. When his daughter's school held a fundraiser, Salam — having just read a book on the subject — volunteered to be a palmist. He dressed up in his sherwani and turban, threw himself into the role, and was a huge hit.

One of Salam's favorite ways to unwind was to read P.G. Wodehouse, and apparently he laughed out loud while doing so. When asked why he enjoyed these books so much, Salam replied that he liked them because they were "inconsequential". Wodehouse's stories could be enjoyed in the moment, but they left no intellectual aftertaste. You didn't stay up nights worrying about the characters. As such, they provided the perfect break.

12. Do That Which, But for You, Might Not Be Done

When Abdus Salam was asked how it felt to straddle Physics and his administrative duties, he replied that he wished he had more time for Physics. What he left unsaid was that Physics, much as he loved it, was something he did for his own intellectual satisfaction and pleasure; in setting up the ICTP he was opening doors for others, and thereby discharging a responsibility placed upon him at birth. It is a testament to Salam's character that he never considered compromising the latter for the former.

The hours spent in administrative work obviously took away from research. Steven Weinberg, who shared the Nobel Prize with Salam, remarked at the amount of time he devoted to the ICTP "at the cost of physics." "I don't know that I would be able to do that," Weinberg said, but for Salam the matter went beyond choice.

Salam knew that he was uniquely positioned to do certain things; if he didn't do them, they wouldn't get done. Salam did not merely want to rest on his laurels as the first Muslim Nobel Laureate, he wanted to make sure that many came after him.

Growing up in a developing country, Salam was intimately acquainted with the perils and frustrations that come from a lack of access, and having crossed the oceans, he had also experienced the opportunities that the developed world enjoys. The sheer knowledge that such a huge chasm existed between his two worlds, was a call to action for Salam. He felt it incumbent upon him to bridge the gaping void, to try to give others born in circumstances similar to his own, the things he wished he had had — and the things he had not even known to wish for. Salam wanted to do his part to ensure that no one's potential go unfulfilled just because they were not born into plenty, and any time spent in this endeavor he did not consider wasted.

All of us occasionally come across tasks that we — with our particular knowledge and expertise — might be uniquely positioned to accomplish. This is especially true in developing countries, where there is so much progress still to be made. Such tasks often seem mundane, administrative, and not at all exciting on a personal level, but they have a value for the society into which we are born. The world would be a far better place if we were to accept these charges with good grace, and plunge into them as whole heartedly as Salam did with the ICTP.

13. Resist Bitterness

I am convinced that Salam was able to accomplish so much partly because he didn't waste energy being bitter, but instead channeled his resources into more positive endeavors.

As a case in point, Salam could have fretted endlessly — and unproductively — about missing out on publishing the theory of the two-component neutrino, but instead he learnt from the episode and diverted his thoughts to a far more constructive direction.

Similarly, when the Government of Pakistan declared Salam's (Ahmaddiya) sect to be non-Muslim, much as that hurt him personally, he did not let it sour his relationship with the country. Salam was able to mentally divorce the people and the Government, realizing that regardless of what the State said, the people loved him, and it was for their benefit that he worked.

Salam's faith in his compatriots was vindicated. When he died and his body arrived in Pakistan for the burial, it was greeted by throngs of people, lining the streets. Thousands upon thousands had turned out to pay their last respects to this illustrious son of the soil. The outpouring of love and pride was unfortunately not reflected in the attitude of the government, but Salam knew the sentiment existed and throughout his life he acted with the confidence and affection that came from this belief.

Perhaps the seed for his confidence was planted decades earlier, when Abdus Salam stood first in the Matriculation Exam. The result was declared on a hot summer afternoon, and as Salam biked back home after receiving the news, he saw the entire village out on the streets to greet him. This was particularly significant because most of the shopkeepers in the village were Hindus, and traditionally a boy from the (more educated) Hindu community stood first in this exam. But in Jhang that day, everyone, regardless of caste or creed, stood in the blazing sun to pay a tribute to this Muslim son of their village. I like to think that incident nurtured Salam's subsequent outlook, that he was strengthened by the memory of a time when there was unity across religious and ethnic divides, and he carried that feeling within him.

Regardless of what fueled it, Salam believed that his people loved him, and he reciprocated their emotion; he simply did not allow the Government any place in that relationship. While I am sure he would have appreciated some official acknowledgement and acceptance from his country, Salam did not let the lack thereof embitter him.

14. Develop Tolerance for a Multiplicity of Views

Science is remarkable not only as a body of knowledge, but also as a system of thought. The logical process, the emphasis on reproducibility, the international scrutiny, the exacting standard for proof: these are the reasons we trust scientific knowledge. Equally crucial is the iconoclastic nature of the discipline. In science, no authority is absolute, and no one's word is taken on reputation. There is no culture of bowing to precedence, or deferring to the Classics; every result is evaluated afresh and is subject to the same critical analysis.

All these attributes can be interpreted as virtues of science, but they are equally conditions that must be met in order for science to grow and thrive.

Salam knew just what a crucial role science plays in the growth of nations, and he considered — deeply, and with characteristic care — the problem of how best to strengthen science in developing countries. He wrote many, very well researched, policy papers on the subject, full of data, and from all his observations, he culled a few 'laws': basic ingredients that seemed to be common to almost all cases of social progress.

"The last hundred years have seen nation after nation start with something like our conditions and crush through the poverty barrier. The laws governing this type of transformation are now well understood," Salam said, at a talk he delivered in Pakistan.

He outlined many ways of engendering such a transformation. Here is one I find particularly poignant.

"Excellence in science is dependent on the freedom and openness within the scientific community, and not necessarily upon the society at large. In science, if there is a considerable body of persons who appreciate what one is engaged on, if there is a body of persons who are permitted to discuss freely and openly their doubts, express their reservations of each other's work as well as speak out their own visions, if there is a body of persons who can discard the beliefs they cherished, if empirical evidence goes against them, then science at the highest level will continue to be created," Salam said.

Scientists, in other words, could create their own mini-Utopia, regardless of any dissonance that existed around them. Within this little intellectual oasis, the rules of the discipline would reign rather than the rules of the country, thus ensuring scientific progress. Once a nation made solid advances in science, Salam felt, it was firmly on the road to development: technological advances would help solve many of the problems that beset society, and theoretical advances would raise the intellectual stature — and confidence — of the people.

It is important to point out that by suggesting that scientists focus first on their own little community, and on producing work of true merit, Salam was not advocating that academics seal themselves up in ivory towers, divorced from the problems around them; quite the contrary. He was, in fact, rousing the (often most highly) educated people to bring about — peaceful — social revolutions.

The idea, or at least the hope, was that the logical, problem solving, free-thinking, critical, thought process cultivated by scientists would eventually become strong enough to spread out into the wider community and take root in the culture at large.

Should this scenario play out, a society could undergo an incredible amount of progress, without incurring any of the casualties revolutions bring in their wake. It would be almost an ideal solution, and the seed for this incredible social change is planted by tolerance. The practice of science is an excellent way to cultivate the capacity to engage with multiple viewpoints, and develop tolerance for a vast range of ideas and opinions.

Salam had this ability in spades. He was able to see many different perspectives, and even hold what some might consider opposing views without being conflicted. Perhaps the most well known example is his religiosity. Salam was a devout Believer. He quoted verses from the Holy Quran in many of his speeches including, most notably, the Nobel acceptance speech. Salam never apologized for, or hid, his beliefs. He listened to recitations of the Quran while working technical calculations, balancing both the religious and scientific parts of himself in perfect harmony.

Scientists in general are not very religious, and Salam numbered among his dear friends many avowed, and sometimes quite vocal, atheists. Even in a matter like religion where arguments often become heated, Salam was able to maintain and honor his own beliefs, while being completely tolerant of the beliefs — or lack thereof — of others around him.

We have already spoken about Salam's simultaneous loyalty to many different institutions. That, too, would not have been possible had Salam not been able to hold multiple viewpoints.

In fact, Salam's prolific idea generation — which we discussed earlier — bears testimony to the fact that he was able to consider many different options. His resilience speaks volumes about his willingness to put his ideas out there, solicit the opinions of others, accept their criticism, and be ready to discard ideas that did not make the cut, without feeling personally affronted.

Salam's open-mindedness and intellectual tolerance played an essential role in his career, and he argued that it could play a parallel part in the growth of nations.

15. Carry Yourself with Dignity

Salam may have been "a humble man", but he was never servile. He competed, in every arena of his life, as an equal. He treated people with respect and consideration, but was not subservient or sycophantic. Maintaining his dignity was very important to Salam. He knew that in order to be taken seriously by the best, he had carry himself with confidence and grace, and of course excel at what he did.

The same logic holds for societies as well, Salam argued. It was a perpetual thorn in his side that developing countries kept importing ready-made technology, rather than investing in fundamental science. One can never hope to be respected if one is always seen with a begging bowl in hand. Salam wanted developing countries to take charge of their destiny through a cultivation of scientific knowledge.

Salam completely agreed with C.P. Snow that "there is no evidence that any country or race is better than any other in scientific teachability; there is a good deal of evidence that all are much alike. Tradition and technical background count for surprisingly little. There is no getting away from it; It is possible to carry out the scientific revolution [across the world] in fifty years. There is no excuse [...] not to know this."

Invest in science, Salam pleaded developing nations: take your place on the world stage as contributors, not just consumers. He advocated passionately for developing countries to engage "in the enterprise of creating scientific knowledge not just because Allah has endowed us with the urge to know, this is not just because in the conditions of today this knowledge is power and science in application the major instrument of material progress, it is also that as members of the international community, one feels that lash of contempt for us — unspoken, but still there — of those who create this knowledge. Why should others have an obligation to aid feed and keep alive nations who do not contribute to man's stock of knowledge?"

In order to enforce this crucial message, Salam came at it from another angle. Importing technology is a temporary tide-over, he said, not a proper solution; an understanding of the underlying science is crucial in order to manipulate and modify existing technology to meet your individual needs. The creation and mastery of science will fill you with a pride and joy that is not to be found in the utilization of technology, Salam reiterated over and over again.

Salam's belief in science as a tool for social betterment comes across loud and clear in the many speeches he delivered on the subject, and he makes excellent arguments to support his case. But there is at least one more reason behind his desire to see developing countries pursue lofty scientific goals.

Salam thought of science as "the joint heritage of all mankind." When we work for the advancement of our common pool of knowledge, differences of race, caste, creed, religion and politics become irrelevant. All the external labels we use to differentiate ourselves from others, simply fall away. We come to know each other on a level that surpasses these superficialities, simply as human beings, held together by the essential curiosity of our race, and our shared excitement in any progress that is made towards our common goal.

This might sound overly idealistic, but if you consider the demographics of at any substantial scientific endeavor today — be it CERN or LIGO or FermiLab — you will see a huge cross-section of nationalities, races and religions working together collegially, sparing no thought to the political or religious differences that supposedly divide them.

16. Salam's Other Masterpiece

Abdus Salam held that "the joint endeavour in science [could become] one of the unifying forces among the diverse people on the planet." And as every good scientist does, Salam constructed a toy model to demonstrate the features of his theory: he called it the International Centre for Theoretical Physics.

Rexhep Meidani, who later became the President of Albania, was first a physicist, who personally benefitted from his association with the ICTP. Meidani recognized exactly what Salam had set out to accomplish. Salam's "target was not only to create a scientific center," said Meidani, "but also to capture, through his human and spiritual imagination, the most advanced model of a community, and a new building of collaboration, independent from race, religion, nationality or political affiliation, and moreover to light the path of an integrated peaceful world of free people with an open heart and original mind."

For half a century now, people from all over the globe have flocked to Trieste, to meet there as equals. The ICTP truly is a little haven where people make it a point to help each other, where bureaucratic hurdles are minimized, where you meet people who face problems similar to your own, and have negotiated them in ways that perhaps did not occur to you. Scientists who feel isolated in their own countries, discover a network of peers. Connections made in Trieste often deepen into collaborations.

The ICTP is as much Salam's masterpiece as the electroweak theory, and there are some remarkable parallels between these vastly different endeavors. Electroweak theory uncovered a deep, hidden, symmetry between two apparently very different forces of nature; by looking back to a time when electromagnetism and the weak nuclear force had not yet adopted their current guises, Weinberg and Salam were able to show that these forces we now perceive as being fundamentally different, in fact stem from the same root. In its own way, the ICTP accomplishes the same. By looking beyond our apparent differences and focusing on what we have in common, the ICTP unifies people just as electroweak theory unified two forces.

There is an argument that comes up every now and again, about the scientist's role in discovering laws of Nature. Does the individual really have any 'creative freedom' in formulating a theory? If Einstein had not discovered relativity — so the canonical example goes — someone else would have eventually arrived at the same truths, but no one other than Beethoven could have written the 9th Symphony. While I personally believe that every great scientist leaves an 'artistic signature' on their work, I know that is a statement about style, not content. The strength of a scientific theory lies, after all, in its universality and reproducibility! And in fact it has often happened that scientists working completely independently — and in total ignorance — of each other, have stumbled upon the same scientific truth.

Salam was once asked what he thought of the electroweak theory. With characteristic humility, he replied "It is a good piece of work. Allah was very gracious in letting me get involved with something which, InshaAllah, will live."

One could perhaps concede Salam's point, and say that yes, there was an element of luck, or grace, in that he focused on just the right problem, at just the right time. Had he not done so, the 1979 Nobel Prize in Physics would probably still have been awarded, at the same time, for the same piece of work, but to Glashow and Weinberg only.

But there was something else Salam was involved in. Something great, that only he could do. Something in which there was much 'creative freedom' in the accomplishment of the task, and a deliberate choice in taking it on in the first place. If the electroweak theory came Salam's way, the ICTP is something he went out on a limb to create. The center would not have existed, had it not been for Salam. And this too, is a "good piece of work," and one that "Inshallah, will live."

Acknowledgments

I am deeply grateful to Lars Brink for inviting me to participate in this conference, and allowing me the opportunity to speak here. I am grateful, also, to Ahmad Ali, Elizabeth and Robert Delbourgo, Michael Duff, Chris Hull, Peter Jenni, Ahmad Salam, Kelley Stelle, Monika and Qaiser Shafi, Spenta Wadia and Peter West for many wonderful and enlightening discussions. Abdus Salam is even more alive to me now that I have seen him through their eyes.

References

1. Z. Hassan and C. H. Lai, *Ideals and Realities: Selected Essays of Abdus Salam* (World Scientific, 1983).
2. M. Duff, *Salam +50: Proceedings of The Conference* (Imperial College Press, 2008).
3. G. Fraser, *Cosmic Anger: Abdus Salam, the First Muslim Nobel Scientist* (Oxford University Press, 2012).
4. J. Singh, *Abdus Salam* (Penguin Books, 1992).
5. M. Kamran, *The Inspiring Life of Abdus Salam* (University of Punjab Press, 2013).

Banquet Speech at the Memorial Meeting for Abdus Salam's 90th Birthday

Ahmad Salam

In the Name of Allah Most Gracious, Ever Merciful.

It is indeed with great pleasure I stand before you to represent the entire Salam family.

Firstly a huge thank you to the organisers for their fantastic effort and hard work to arrange this amazing celebration in honour of my father, his life and his work. It is most humbling for me to see how many of his friends and admirers, from all around the world, have gathered here to commemorate his achievements.

I think everyone here has been touched, influenced or positively affected by Abdus Salam in some shape or form. It has been wonderful to see how these four days have been filled with such tributes and comment and discussion about his work.

It is of course so sad that his own country which he loved so much cannot find four minutes to celebrate the man and his achievements let alone four days.

I only have a few minutes to say a few words about my father and the older I get, the more amazed and inadequate I feel that he achieved so much in so little time, and yet also suffered so much for his cause.

My preference is not to talk about his physics to which I could never do justice to, but instead to talk briefly about what I sincerely believe to be his most valuable and enduring contribution to the world. His humanitarian work. It may be that if my father had not worked on his physics then someone else may have come along and accomplished the same academic work, but what very, very few could have done or indeed have done, was work so hard and diligently to bring education, science and technology to the developing world.

It may be a bit too simplistic to say but, perhaps, if more people of influence, those who father pleaded to, Presidents, Prime Ministers, Members of Royal Families and so on had listened to father 45 years ago with his plea to reduce the imbalance and gap between the developed and developing worlds, if more had been done to raise education budgets and reduce defence budgets, if the Gulf Arab Muslims had invested more in education as he suggested, and gave some of their huge wealth to fellow Muslims in the region, then maybe, just maybe, we would have reduced the inequality in the world and we may not have quite as many economic migrants as we have now, with people taking incredible risks and facing great danger to try and find a better life for themselves and their families. Maybe 45 years of investment in education would have produced economic growth and investment in local economies which would have provided oppurtunities for all. Just a thought.

Father's inspiration for his humanitarian work came when he had to leave a very basic and rudimentary Pakistan to try and find a better life for himself and his family. He left behind all he knew and loved and ventured into a very different world with no easy communication link to his adored family, save an occasional letter. Father's experience led him to a lifelong commitment to make a better more equitable world: to try and do something to reduce the imbalance and the extremes of inequality. It is sad to report now that the differences between the haves and have nots in this world is now greater than at any time.

As you all know father's great passion and, as the late Nigel Calder, so eloquently put it, his 'cosmic anger' drove him to believe and advocate that sustainable economic growth comes from investment in science and technology. He saw no earthly reason why just because you were born in Lahore, Lagos, or Lima, that you should not have access to the same intellectual ideas and stimulation as someone born in London, Lausanne or Los Angeles. He had the dream and vision of the ICTP and all the thousands of students who have passed through it, and greatly benefitted from it. But it was never to just be in one location. It was to be a global phenomenon. Indeed now 19 years after his death, the ongoing work to create satellite and joint venture ICTPs are all wonderful and lasting legacies to father. In addition to his physics I think that is what we should remember and commemorate.

The co-chairmen and the organisers, Mike Duff, Lars Brink and KK Phua and his team deserve our thanks and appreciation for their work to keep the memory and story of Abdus Salam alive.

I am often asked we know about the science and something of the man, but what was he like as a person. I don't think I can answer that question in the time I have left but let me give you a few random insights.

Father was a very simple man with simple tastes and simple habits. The Holy Quran was his greatest single inspiration. He was proud to be a Muslim and would always tell me of the stories of great Muslim teachers, scholars and scientists.

His favourite place to work at home either in Trieste or London, was a hot room, with a supply of hot tea, some digestive biscuits, cashew nuts and cheese next to him, an army surplus woollen cap on his head, and the Holy Quran being recited in the background on a record player or a cassette player. He would have incense burning giving a very heavy and scented atmosphere. He would be surrounded by books and papers, but he didn't really like to work at a desk, instead he preferred to sit in an armchair and bring his legs up under him, balancing his papers on his knees.

He had simple tastes in food and I don't think he ever actually tried to cook from scratch himself, not even a proverbial egg. He could warm up food in a microwave, he could fry fish fingers or sausages for breakfast, make tea and coffee, but not cook. His favourite fruit was Pakistani mango. He would say the only way to eat a mango was sitting in the bath!

He rarely watched television; indeed we did not get a TV at home until the mid-seventies as father regarded it as great waste of time. His favourite programmes were BBC comedies such as Dads Army and the films of the Marx brothers. He was a fan of Groucho Mark and also Charlie Chaplin.

Later in life he developed an appreciation of the Great Tenors and in his final years when he was bedridden with this most cruel disease, he would take pleasure from listening to the their works. Earlier in life he had been a fan of Gilbert and Sullivan. This was of course alongside his lifelong love of listening to the recitation of the Holy Quran.

But reading was his favourite pastime. His library in London included literature, fiction and non fiction. Some of the more interesting titles included "Teach Yourself Ballroom Dancing", and "Teach Yourself Air Navigation", "Russian Made Simple", (as he never learned Italian so I am not sure why he thought he would learn Russian), though to be fair, when he was asked about his Italian, he did promise to learn one word a year! He also had an interesting book "Physics Made Simple"! His other favourites included humour from PG Wodehouse, and the plays and works of George Bernard Shaw, biographies of Napoleon, Gandhi and Churchill to name but a few.

Father was a great book shopper and much preferred second hand book shops. He had some favourites in London and when he visited he would often buy a dozen or more books in one go. He was blessed as speed reader and so could devour books at an incredible pace

Literature was his escape. It was his first love, he had the mind of scientist but the heart of a poet.

His library would extend to the bathroom so he could read whilst he was there, so as to not waste time. I remember when I was 12 years old I made him a small book shelf to go in the toilet so he could store his books neatly in there.

Father was very very careful with time: he was always early to bed and early to rise. He would work between 4 am and 7.30 am as he believed this is the most productive time when he would not be disturbed by phone calls, etc. Of course in those days the telephone was the only interruption you could have. There were no emails, etc., to interrupt the flow of work and ideas. His father had taught him "If you lose an hour in the morning, you spend the rest of the day looking for it". Father taught us all that lesson too, and my children complain that I am forcing it on them too. On those rare occasions when father was in London, he would sleep on the ground floor and early in the morning, especially during the holidays and weekends, we would hear his heavy foot steps on the stairs and we would all panic to get up before he got to the top of the stairs, and try and look as if we had been up for some time. We pulled it off sometimes!

Father was not a good driver as he didn't really concentrate too much. I am sure he would often drive along thinking of some tricky problems. He told me when I was learning to drive not to worry about the rear view mirror: if someone hit you from behind it was always their fault. He would park using the time tested Italian way of only stopping when he touched the bumper of the car in front or behind.

I love one story in particular; of his driving test in 1957. I told this story recently at the Einstein Centenary of the Theory of Relativity at Imperial College, so apologies to those of you who may have heard it before: he left home at 9.30 am for an 11 am test. My mother expected him to be back at around 12.30 pm. 1 pm came, no sign of him. 1.30 pm, still no sign. Remember in those days there was no ability to check where someone was, so mother just had to sit and wait. She became more and more worried as 3 pm came, 4 pm and still no sign of father. Finally at about 5 pm, she heard the key in the lock and ran to the door to see father. He came in with a very dejected figure. Mother asked him where on earth he had been and what had happened. Father very simply replied he had failed his driving test.

At the age of 31 he had never failed anything in his life, so he had just wandered the streets of London trying to understand how he could cope with this new concept: failure. It was a shock for him.

There are many more stories I could tell of father but unfortunately time prevents me from doing so.

Let me just end by thanking you all from the bottom of my heart for showing my father such respect and affection by attending this event. He would have been thrilled and delighted, and equally humbled by the event itself, and by all of you giving up so much time and going to such great efforts to be here.

May Allah bless you.